Universitext

*Editorial Board
(North America):*

S. Axler
K.A. Ribet

T0202460

For other titles in this series, go to
www.springer.com/series/223

Walter G. Kelley • Allan C. Peterson

The Theory of
Differential Equations

Classical and Qualitative

Second Edition

 Springer

Walter G. Kelley
Department of Mathematics
University of Oklahoma
Norman, OK 73019
USA
wkelley@ou.edu

Allan C. Peterson
Department of Mathematics
University of Nebraska-Lincoln
Lincoln, NE 68588-0130
USA
apeterso@math.unl.edu

Editorial Board:
Sheldon Axler, San Francisco State University
Vincenzo Capasso, Università degli Studi di Milano
Carles Casacuberta, Universitat de Barcelona
Angus MacIntyre, Queen Mary, University of London
Kenneth Ribet, University of California, Berkeley
Claude Sabbah, CNRS, École Polytechnique
Endre Süli, University of Oxford
Wojbor Woyczyński, Case Western Reserve University

ISBN 978-1-4419-5782-5   e-ISBN 978-1-4419-5783-2
DOI 10.1007/978-1-4419-5783-2
Springer New York Dordrecht Heidelberg London

Library of Congress Control Number: 2010924820

Mathematics Subject Classification (2010): 34-XX, 34-01

© Springer Science+Business Media, LLC 2010
All rights reserved. This work may not be translated or copied in whole or in part without the written permission of the publisher (Springer Science+Business Media, LLC, 233 Spring Street, New York, NY 10013, USA), except for brief excerpts in connection with reviews or scholarly analysis. Use in connection with any form of information storage and retrieval, electronic adaptation, computer software, or by similar or dissimilar methodology now known or hereafter developed is forbidden.
The use in this publication of trade names, trademarks, service marks, and similar terms, even if they are not identified as such, is not to be taken as an expression of opinion as to whether or not they are subject to proprietary rights.

Printed on acid-free paper

Springer is part of Springer Science+Business Media (www.springer.com)

We dedicate this book to our families:
Marilyn and Joyce
and
Tina, Carla, David, and Carrie.

Contents

Preface   ix

Chapter 1   First-Order Differential Equations   1
1.1   Basic Results   1
1.2   First-Order Linear Equations   4
1.3   Autonomous Equations   5
1.4   Generalized Logistic Equation   10
1.5   Bifurcation   14
1.6   Exercises   16

Chapter 2   Linear Systems   23
2.1   Introduction   23
2.2   The Vector Equation $x' = A(t)x$ 27
2.3   The Matrix Exponential Function   42
2.4   Induced Matrix Norm   59
2.5   Floquet Theory   64
2.6   Exercises   76

Chapter 3   Autonomous Systems   87
3.1   Introduction   87
3.2   Phase Plane Diagrams   90
3.3   Phase Plane Diagrams for Linear Systems   96
3.4   Stability of Nonlinear Systems   107
3.5   Linearization of Nonlinear Systems   113
3.6   Existence and Nonexistence of Periodic
     Solutions   120
3.7   Three-Dimensional Systems   134
3.8   Differential Equations and *Mathematica*   145
3.9   Exercises   149

Chapter 4   Perturbation Methods   161
4.1   Introduction   161
4.2   Periodic Solutions   172
4.3   Singular Perturbations   178
4.4   Exercises   186

**Chapter 5   The Self-Adjoint Second-Order Differential
     Equation**   192
5.1   Basic Definitions   192

5.2	An Interesting Example	197
5.3	Cauchy Function and Variation of Constants Formula	199
5.4	Sturm-Liouville Problems	204
5.5	Zeros of Solutions and Disconjugacy	212
5.6	Factorizations and Recessive and Dominant Solutions	219
5.7	The Riccati Equation	229
5.8	Calculus of Variations	240
5.9	Green's Functions	251
5.10	Exercises	272

Chapter 6   Linear Differential Equations of Order n   281
6.1	Basic Results	281
6.2	Variation of Constants Formula	283
6.3	Green's Functions	287
6.4	Factorizations and Principal Solutions	297
6.5	Adjoint Equation	302
6.6	Exercises	307

Chapter 7   BVPs for Nonlinear Second-Order DEs   309
7.1	Contraction Mapping Theorem (CMT)	309
7.2	Application of the CMT to a Forced Equation	311
7.3	Applications of the CMT to BVPs	313
7.4	Lower and Upper Solutions	325
7.5	Nagumo Condition	334
7.6	Exercises	340

Chapter 8   Existence and Uniqueness Theorems   345
8.1	Basic Results	345
8.2	Lipschitz Condition and Picard-Lindelof Theorem	348
8.3	Equicontinuity and the Ascoli-Arzela Theorem	356
8.4	Cauchy-Peano Theorem	358
8.5	Extendability of Solutions	363
8.6	Basic Convergence Theorem	369
8.7	Continuity of Solutions with Respect to ICs	372
8.8	Kneser's Theorem	375
8.9	Differentiating Solutions with Respect to ICs	378
8.10	Maximum and Minimum Solutions	387
8.11	Exercises	396

Solutions to Selected Problems   403

Bibliography   415

Index   419

Preface

Differential equations first appeared in the late seventeenth century in the work of Isaac Newton, Gottfried Wilhelm Leibniz, and the Bernoulli brothers, Jakob and Johann. They occurred as a natural consequence of the efforts of these great scientists to apply the new ideas of the calculus to certain problems in mechanics, such as the paths of motion of celestial bodies and the brachistochrone problem, which asks along which path from point P to point Q a frictionless object would descend in the least time. For over 300 years, differential equations have served as an essential tool for describing and analyzing problems in many scientific disciplines. Their importance has motivated generations of mathematicians and other scientists to develop methods of studying properties of their solutions, ranging from the early techniques of finding exact solutions in terms of elementary functions to modern methods of analytic and numerical approximation. Moreover, they have played a central role in the development of mathematics itself since questions about differential equations have spawned new areas of mathematics and advances in analysis, topology, algebra, and geometry have often offered new perspectives for differential equations.

This book provides an introduction to many of the important topics associated with ordinary differential equations. The material in the first six chapters is accessible to readers who are familiar with the basics of calculus, while some undergraduate analysis is needed for the more theoretical subjects covered in the final two chapters. The needed concepts from linear algebra are introduced with examples, as needed. Previous experience with differential equations is helpful but not required. Consequently, this book can be used either for a second course in ordinary differential equations or as an introductory course for well-prepared students.

The first chapter contains some basic concepts and solution methods that will be used throughout the book. Since the discussion is limited to first-order equations, the ideas can be presented in a geometrically simple setting. For example, dynamics for a first-order equation can be described in a one-dimensional space. Many essential topics make an appearance here: existence, uniqueness, intervals of existence, variation of parameters, equilibria, stability, phase space, and bifurcations. Since proofs of existence-uniqueness theorems tend to be quite technical, they are reserved for the last chapter.

Systems of linear equations are the major topic of the second chapter. An unusual feature is the use of the Putzer algorithm to provide a constructive method for solving linear systems with constant coefficients. The study of stability for linear systems serves as a foundation for nonlinear systems in the next chapter. The important case of linear systems with periodic coefficients (Floquet theory) is included in this chapter.

Chapter 3, on autonomous systems, is really the heart of the subject and the foundation for studying differential equations from a dynamical viewpoint. The discussion of phase plane diagrams for two-dimensional systems contains many useful geometric ideas. Stability of equilibria is investigated by both Liapunov's direct method and the method of linearization. The most important methods for studying limit cycles, the Poincare-Bendixson theorem and the Hopf bifurcation theorem, are included here. The chapter also contains a brief look at complicated behavior in three dimensions and at the use of *Mathematica* for graphing solutions of differential equations. We give proofs of many of the results to illustrate why these methods work, but the more intricate verifications have been omitted in order to keep the chapter to a reasonable length and level of difficulty.

Perturbation methods, which are among the most powerful techniques for finding approximations of solutions of differential equations, are introduced in Chapter 4. The discussion includes singular perturbation problems, an important topic that is usually not covered in undergraduate texts.

The next two chapters return to linear equations and present a rich mix of classical subjects, such as self-adjointness, disconjugacy, Green's functions, Riccati equations, and the calculus of variations.

Since many applications involve the values of a solution at different input values, boundary value problems are studied in Chapter 7. The contraction mapping theorem and continuity methods are used to examine issues of existence, uniqueness, and approximation of solutions of nonlinear boundary value problems.

The final chapter contains a thorough discussion of the theoretical ideas that provide a foundation for the subject of differential equations. Here we state and prove the classical theorems that answer the following questions about solutions of initial value problems: Under what conditions does a solution exist, is it unique, what type of domain does a solution have, and what changes occur in a solution if we vary the initial condition or the value of a parameter? This chapter is at a higher level than the first six chapters of the book.

There are many examples and exercises throughout the book. A significant number of these involve differential equations that arise in applications to physics, biology, chemistry, engineering, and other areas. To avoid lengthy digressions, we have derived these equations from basic principles only in the simplest cases.

In this new edition we have added 81 new problems in the exercises. In Chapter 1 there is a new section on the generalized logistic equation, which has important applications in population dynamics. In Chapter 2 an additional theorem concerning fundamental matrices, a corresponding example and related exercises are now included. Also results on matrix norms in Section 2.4 are supplemented by the matrix norm induced by the Euclidean norm and Lozinski's measure with examples and exercises are included. In Chapter 3 an intuitive sketch of the proof that every cycle contains an equilibrium point in it's interior has been added. Also to supplement the results concerning periodic solutions, Liénard's Theorem is included with an application to van der pol's equation. Section 3.8 has been updated to be compatible with Mathematica, version 7.0. In Chapter 5, the integrated form of the Euler–Lagrange equation has been added with an application to minimizing the surface area obtain by rotating a curve about the x-axis. Liapunov's inequality is identified and an example and exercises are included.

We would like to thank Deborah Brandon, Chris Ahrendt, Ross Chiquet, Valerie Cormani, Lynn Erbe, Kirsten Messer, James Mosely, Mark Pinsky, Mohammad Rammaha, and Jacob Weiss for helping with the proof reading of this book. We would like to thank Lloyd Jackson for his influence on Chapters 7 and 8 in this book. We would also like to thank Ned Hummel and John Davis for their work on the figures that appear in this book. Allan Peterson would like to thank the National Science Foundation for the support of NSF Grant 0072505. We are very thankful for the great assistance that we got from the staff at Prentice Hall; in particular, we would like to thank our acquisitions editor, George Lobell; the production editor, Jeanne Audino; editorial assistant, Jennifer Brady; and copy editor, Patricia M. Daly, for the accomplished handling of this manuscript.

Walter Kelley
wkelley@math.ou.edu

Allan Peterson
apeterso@math.unl.edu

Chapter 1

First-Order Differential Equations

1.1 Basic Results

In the scientific investigation of any phenomenon, mathematical models are used to give quantitative descriptions and to derive numerical conclusions. These models can take many forms, and one of the most basic and useful is that of a differential equation, that is, an equation involving the rate of change of a quantity. For example, the rate of decrease of the mass of a radioactive substance, such as uranium, is known to be proportional to the present mass. If $m(t)$ represents the mass at time t, then we have that m satisfies the differential equation

$$m' = -km,$$

where k is a positive constant. This is an *ordinary differential equation* since it involves only the derivative of mass with respect to a single independent variable. Also, the equation is said to be of *first-order* because the highest order derivative appearing in the equation is first-order. An example of a second-order differential equation is given by Newton's second law of motion

$$mx'' = f(t, x, x'),$$

where m is the (constant) mass of an object moving along the x-axis and located at position $x(t)$ at time t, and $f(t, x(t), x'(t))$ is the force acting on the object at time t.

In this chapter, we will consider only first-order differential equations that can be written in the form

$$x' = f(t, x), \tag{1.1}$$

where $f : (a, b) \times (c, d) \rightarrow \mathbb{R}$ is continuous, $-\infty \leq a < b \leq \infty$, and $-\infty \leq c < d \leq \infty$.

Definition 1.1 We say that a function x is a *solution* of (1.1) on an interval $I \subset (a, b)$ provided $c < x(t) < d$ for $t \in I$, x is a continuously differentiable function on I, and

$$x'(t) = f(t, x(t)),$$

for $t \in I$.

W.G. Kelley and A.C. Peterson, *The Theory of Differential Equations: Classical and Qualitative*, Universitext 278, DOI 10.1007/978-1-4419-5783-2_1, © Springer Science+Business Media, LLC 2010

Definition 1.2 Let $(t_0, x_0) \in (a, b) \times (c, d)$ and assume f is continuous on $(a, b) \times (c, d)$. We say that the function x is a solution of the initial value problem (IVP)

$$x' = f(t, x), \quad x(t_0) = x_0, \tag{1.2}$$

on an interval $I \subset (a, b)$ provided $t_0 \in I$, x is a solution of (1.1) on I, and

$$x(t_0) = x_0.$$

The point t_0 is called the initial point for the IVP (1.2) and the number x_0 is called the initial value for the IVP (1.2).

Note, for example, that if $(a, b) = (c, d) = (-\infty, \infty)$, then the function m defined by $m(t) = 400e^{-kt}$, $t \in (-\infty, \infty)$ is a solution of the IVP

$$m' = -km, \quad m(0) = 400$$

on the interval $I = (-\infty, \infty)$.

Solving an IVP can be visualized (see Figure 1) as finding a solution of the differential equation whose graph passes through the given point (t_0, x_0).

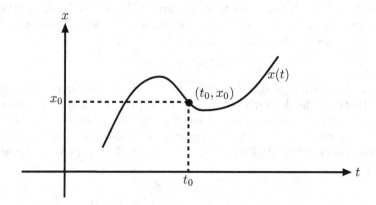

FIGURE 1. Graph of solution of IVP.

We state without proof the following important existence-uniqueness theorem for solutions of IVPs. Statements and proofs of some existence and uniqueness theorems will be given in Chapter 8.

Theorem 1.3 *Assume $f : (a, b) \times (c, d) \to \mathbb{R}$ is continuous, where $-\infty \leq a < b \leq \infty$ and $-\infty \leq c < d \leq \infty$. Let $(t_0, x_0) \in (a, b) \times (c, d)$, then the IVP (1.2) has a solution x with a maximal interval of existence $(\alpha, \omega) \subset (a, b)$, where $\alpha < t_0 < \omega$. If $a < \alpha$, then*

$$\lim_{t \to \alpha+} x(t) = c, \quad or \quad \lim_{t \to \alpha+} x(t) = d$$

and if $\omega < b$, then

$$\lim_{t \to \omega-} x(t) = c, \quad or \quad \lim_{t \to \omega-} x(t) = d.$$

If, in addition, the partial derivative of f with respect to x, f_x, is continuous on $(a, b) \times (c, d)$, then the preceding IVP has a unique solution.

We now give a couple of examples related to Theorem 1.3. The first example shows that if the hypothesis that the partial derivative f_x is continuous on $(a, b) \times (c, d)$ is not satisfied, then we might not have uniqueness of solutions of IVPs.

Example 1.4 (Nonuniqueness of Solutions to IVPs) If we drop an object from a bridge of height h at time $t = 0$ (assuming constant acceleration of gravity and negligible air resistance), then the height of the object after t units of time is $x(t) = -\frac{1}{2}gt^2 + h$. The velocity at time t is $x'(t) = -gt$, so by eliminating t, we are led to the IVP

$$x' = f(t, x) := -\sqrt{2g|h - x|}, \quad x(0) = h. \tag{1.3}$$

Note that this initial value problem has the constant solution $x(t) = h$, which corresponds to holding the object at bridge level without dropping it! We can find other solutions by separation of variables. If $h > x$, then

$$\int \frac{x'(t)\, dt}{\sqrt{2g(h - x(t))}} = -\int dt.$$

Computing the indefinite integrals and simplifying, we arrive at

$$x(t) = -\frac{g}{2}(t - C)^2 + h,$$

where C is an arbitrary constant. We can patch these solutions together with the constant solution to obtain for each $C > 0$

$$x(t) := \begin{cases} h, & \text{for} \quad t \le C \\ h - \frac{g}{2}(t - C)^2, & \text{for} \quad t > C. \end{cases}$$

Thus for each $C > 0$ we have a solution of the IVP (1.3) that corresponds to releasing the object at time C. Note that the function f defined by $f(t, x) = -\sqrt{2g|h - x|}$ is continuous on $(-\infty, \infty) \times (-\infty, \infty)$ so by Theorem 1.3 the IVP (1.3) has a solution, but f_x does not exist when $x = h$ so we cannot use Theorem 1.3 to get that the IVP (1.3) has a unique solution.   △

To see how bad nonuniqueness of solutions of initial value problems can be, we remark that in Hartman [18], pages 18–23, an example is given of a scalar equation $x' = f(t, x)$, where $f : \mathbb{R} \times \mathbb{R} \to \mathbb{R}$, is continuous, where for every IVP (1.2) there is more than one solution on $[t_0, t_0 + \epsilon]$ and $[t_0 - \epsilon, t_0]$ for arbitrary $\epsilon > 0$.

The next example shows even if the hypotheses of Theorem 1.3 hold the solution of the IVP might only exist on a proper subinterval of (a, b).

Example 1.5 Let k be any nonzero constant. The function $f : \mathbb{R}^2 \to \mathbb{R}$ defined by $f(t, x) = kx^2$ is continuous and has a continuous partial derivative with respect to x. By Theorem 1.3, the IVP

$$x' = kx^2, \quad x(0) = 1$$

has a unique solution with a maximal interval of existence (α, ω). Using separation of variables, as in the preceding example, we find

$$x(t) = \frac{1}{C - kt}.$$

When we apply the initial condition $x(0) = 1$, we have $C = 1$, so that the solution of the IVP is

$$x(t) = \frac{1}{1 - kt},$$

with maximal interval of existence $(-\infty, 1/k)$ if $k > 0$ and $(1/k, \infty)$ if $k < 0$. In either case, $x(t)$ goes to infinity as t approaches $1/k$ from the appropriate direction.

Observe the implications of this calculation in case $x(t)$ is the density of some population at time t. If $k > 0$, then the density of the population is growing, and we conclude that growth cannot be sustained at a rate proportional to the square of density because the density would have to become infinite in finite time! On the other hand, if $k < 0$, the density is declining, and it is theoretically possible for the decrease to occur at a rate proportional to the square of the density, since $x(t)$ is defined for all $t > 0$ in this case. Note that $\lim_{t \to \infty} x(t) = 0$ if $k < 0$. $\triangle$

1.2 First-Order Linear Equations

An important special case of a first-order differential equation is the first-order linear differential equation given by

$$x' = p(t)x + q(t), \tag{1.4}$$

where we assume that $p : (a, b) \to \mathbb{R}$ and $q : (a, b) \to \mathbb{R}$ are continuous functions, where $-\infty \le a < b \le \infty$. In Chapter 2, we will study systems of linear equations involving multiple unknown functions. The next theorem shows that a single linear equation can always be solved in terms of integrals.

Theorem 1.6 (Variation of Constants Formula) *If* $p : (a, b) \to \mathbb{R}$ *and* $q : (a, b) \to \mathbb{R}$ *are continuous functions, where* $-\infty \le a < b \le \infty$*, then the unique solution* x *of the IVP*

$$x' = p(t)x + q(t), \quad x(t_0) = x_0, \tag{1.5}$$

where $t_0 \in (a, b)$*,* $x_0 \in \mathbb{R}$*, is given by*

$$x(t) = e^{\int_{t_0}^{t} p(\tau)\, d\tau} x_0 + e^{\int_{t_0}^{t} p(\tau)\, d\tau} \int_{t_0}^{t} e^{-\int_{t_0}^{s} p(\tau)\, d\tau} q(s)\, ds,$$

$t \in (a, b)$*.*

Proof Here the function f defined by $f(t, x) = p(t)x + q(t)$ is continuous on $(a, b) \times (-\infty, \infty)$ and $f_x(t, x) = p(t)$ is continuous on $(a, b) \times (-\infty, \infty)$. Hence by Theorem 1.3 the IVP (1.5) has a unique solution with a maximal interval of existence $(\alpha, \omega) \subset (a, b)$ [the existence and uniqueness of the solution of the IVP (1.5) and the fact that this solution exists on the whole interval (a, b) follows from Theorem 8.65]. Let

$$x(t) := e^{\int_{t_0}^t p(\tau)\, d\tau} x_0 + e^{\int_{t_0}^t p(\tau)\, d\tau} \int_{t_0}^t e^{-\int_{t_0}^s p(\tau)\, d\tau} q(s)\, ds$$

for $t \in (a, b)$. We now show that x is the solution of the IVP (1.5) on the whole interval (a, b). First note that $x(t_0) = x_0$ as desired. Also,

$$
\begin{aligned}
x'(t) &= p(t)e^{\int_{t_0}^t p(\tau)\, d\tau} x_0 + p(t)e^{\int_{t_0}^t p(\tau)\, d\tau} \int_{t_0}^t e^{-\int_{t_0}^s p(\tau)d\tau} q(s)\, ds + q(t) \\
&= p(t)\left[e^{\int_{t_0}^t p(\tau)\, d\tau} x_0 + e^{\int_{t_0}^t p(\tau)\, d\tau} \int_{t_0}^t e^{-\int_{t_0}^s p(\tau)d\tau} q(s)\, ds \right] + q(t) \\
&= p(t)x(t) + q(t)
\end{aligned}
$$

for $t \in (a, b)$. $\qquad\square$

In Theorem 2.40, we generalize Theorem 1.6 to the vector case. We now give an application of Theorem 1.6.

Example 1.7 (Newton's Law of Cooling) Newton's law of cooling states that the rate of change of the temperature of an object is proportional to the difference between its temperature and the temperature of the surrounding medium. Suppose that the object has an initial temperature of 40 degrees. If the temperature of the surrounding medium is $70 + 20e^{-2t}$ degrees after t minutes and the constant of proportionality is $k = -2$, then the initial value problem for the temperature $x(t)$ of the object at time t is

$$x' = -2(x - 70 - 20e^{-2t}), \quad x(0) = 40.$$

By the variation of constants formula, the temperature of the object after t minutes is

$$
\begin{aligned}
x(t) &= 40e^{\int_0^t -2d\tau} + e^{\int_0^t -2d\tau} \int_0^t e^{\int_0^s 2d\tau}(140 + 40e^{-2s})\, ds \\
&= 40e^{-2t} + e^{-2t} \int_0^t (140e^{2s} + 40)\, ds \\
&= 40e^{-2t} + e^{-2t}[70(e^{2t} - 1) + 40t] \\
&= 10(4t - 3)e^{-2t} + 70.
\end{aligned}
$$

Sketch the graph of x. Does the temperature of the object exceed 70 degrees at any time t? $\qquad\triangle$

1.3 Autonomous Equations

If, in equation (1.1), f depends only on x, we get the autonomous differential equation

$$x' = f(x). \tag{1.6}$$

We always assume $f : \mathbb{R} \to \mathbb{R}$ is continuous and usually we assume its derivative is also continuous. The fundamental property of autonomous differential equations is that translating any solution of the autonomous differential equation along the t-axis produces another solution.

Theorem 1.8 *If x is a solution of the autonomous differential equation (1.6) on an interval (a, b), where $-\infty \le a < b \le \infty$, then for any constant c, the function y defined by $y(t) := x(t-c)$, for $t \in (a+c, b+c)$ is a solution of (1.6) on $(a + c, b + c)$.*

Proof Assume x is a solution of the autonomous differential equation (1.6) on (a, b); then x is continuously differentiable on (a, b) and

$$x'(t) = f(x(t)),$$

for $t \in (a, b)$. Replacing t by $t - c$ in this last equation, we get that

$$x'(t - c) = f(x(t - c)),$$

for $t \in (a + c, b + c)$. By the chain rule of differentiation we get that

$$\frac{d}{dt}[x(t - c)] = f(x(t - c)),$$

for $t \in (a + c, b + c)$. Hence if $y(t) := x(t - c)$ for $t \in (a + c, b + c)$, then y is continuously differentiable on $(a + c, b + c)$ and we get the desired result that

$$y'(t) = f(y(t)),$$

for $t \in (a + c, b + c)$. $\square$

Definition 1.9 *If $f(x_0) = 0$ we say that x_0 is an equilibrium point for the differential equation (1.6). If, in addition, there is a $\delta > 0$ such that $f(x) \ne 0$ for $|x - x_0| < \delta$, $x \ne x_0$, then we say x_0 is an isolated equilibrium point.*

Note that if x_0 is an equilibrium point for the differential equation (1.6), then the constant function $x(t) = x_0$ for $t \in \mathbb{R}$ is a solution of (1.6) on $\mathbb{R}$.

Example 1.10 (Newton's Law of Cooling) Consider again Newton's law of cooling as in Example 1.7, where in this case the temperature of the surrounding medium is a constant 70 degrees. Then we have that the temperature $x(t)$ of the object at time t satisfies the differential equation

$$x' = -2(x - 70).$$

Note that $x = 70$ is the only equilibrium point. All solutions can be written in the form

$$x(t) = De^{-2t} + 70,$$

where D is an arbitrary constant. If we translate a solution by a constant amount c along the t-axis, then

$$x(t - c) = De^{-2(t-c)} + 70 = De^{2c}e^{-2t} + 70$$

is also a solution, as predicted by Theorem 1.8. Notice that if the temperature of the object is initially greater than 70 degrees, then the temperature will decrease and approach the equilibrium temperature 70 degrees as t goes to infinity. Temperatures starting below 70 degrees will increase toward the limiting value of 70 degrees. A simple graphical representation of this behavior is a "phase line diagram," (see Figure 2) showing the equilibrium point and the direction of motion of the other solutions.

FIGURE 2. Phase line diagram of $x' = -2(x - 70)$.

$\triangle$

Definition 1.11 Let ϕ be a solution of (1.6) with maximal interval of existence (α, ω). Then the set

$$\{\phi(t) : t \in (\alpha, \omega)\}$$

is called an *orbit* for the differential equation (1.6).

Note that the orbits for

$$x' = -2(x - 70)$$

are the sets

$$(-\infty, 70), \quad \{70\}, \quad (70, \infty).$$

A convenient way of thinking about phase line diagrams is to consider $x(t)$ to be the position of a point mass moving along the x-axis and $x'(t) = f(x(t))$ to be its velocity. The phase line diagram then gives the direction of motion (as determined by the sign of the velocity). An orbit is just the set of all locations of a continuous motion.

Theorem 1.12 *Assume that $f : \mathbb{R} \to \mathbb{R}$ is continuously differentiable. Then two orbits of (1.6) are either disjoint sets or are the same set.*

Proof Let ϕ_1 and ϕ_2 be solutions of (1.6). We will show that if there are points t_1, t_2 such that

$$\phi_1(t_1) = \phi_2(t_2),$$

then the orbits corresponding to ϕ_1 and ϕ_2 are the same. Let

$$x(t) := \phi_1(t - t_2 + t_1);$$

then by Theorem 1.8 we have that x is a solution of (1.6). Since

$$x(t_2) = \phi_1(t_1) = \phi_2(t_2),$$

we have by the uniqueness theorem (Theorem 1.3) that x and ϕ_2 are the same solutions. Hence $\phi_1(t - t_2 + t_1)$ and $\phi_2(t)$ correspond to the same solution. It follows that the orbits corresponding to ϕ_1 and ϕ_2 are the same. $\square$

Example 1.13 (Logistic Growth) The logistic law of population growth
(Verhulst [52], 1838) is

$$N' = rN\left(1 - \frac{N}{K}\right), \tag{1.7}$$

where N is the number of individuals in the population, $r(1 - N/K)$ is the
per capita growth rate that declines with increasing population, and $K > 0$
is the *carrying capacity* of the environment. With $r > 0$, we get the phase
line diagram in Figure 3. What are the orbits of the differential equation
in this case?

FIGURE 3. Phase line diagram of $N' = rN(1 - N/K)$.

We can use the phase line diagram to sketch solutions of the logistic
equation. In order to make the graphs more accurate, let's first calculate
the second derivative of N by differentiating both sides of the differential
equation.

$$
\begin{aligned}
N'' &= rN'\left(1 - \frac{N}{K}\right) - r\frac{NN'}{K} \\
&= rN'\left(1 - \frac{2N}{K}\right) \\
&= r^2N\left(1 - \frac{N}{K}\right)\left(1 - \frac{2N}{K}\right).
\end{aligned}
$$

It follows that $N''(t) > 0$ if either $N(t) > K$ or $0 < N(t) < K/2$ and
$N''(t) < 0$ if either $N(t) < 0$ or $K/2 < N(t) < K$. With all this in mind,
we get the graph of some of the solutions of the logistic (Verhulst) equation
in Figure 4.

$\triangle$

Phase line diagrams are a simple geometric device for analyzing the
behavior of solutions of autonomous equations. In later chapters we will
study higher dimensional analogues of these diagrams, and it will be useful to have a number of basic geometric concepts for describing solution
behavior. The following definitions contain some of these concepts for the
one-dimensional case.

Definition 1.14 Assume $f : \mathbb{R} \to \mathbb{R}$ is continuously differentiable. Then
we let $\phi(\cdot, x_0)$ denote the solution of the IVP

$$x' = f(x), \quad x(0) = x_0.$$

Definition 1.15 We say that an equilibrium point x_0 of the differential
equation (1.6) is *stable* provided given any $\epsilon > 0$ there is a $\delta > 0$ such that

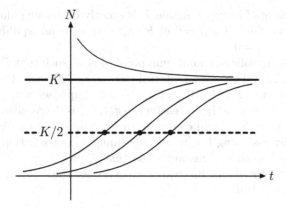

FIGURE 4. Graph of some solutions of $N' = rN(1 - N/K)$.

whenever $|x_1 - x_0| < \delta$ it follows that the solution $\phi(\cdot, x_1)$ exists on $[0, \infty)$ and
$$|\phi(t, x_1) - x_0| < \epsilon,$$
for $t \geq 0$. If, in addition, there is a $\delta_0 > 0$ such that $|x_1 - x_0| < \delta_0$ implies that
$$\lim_{t \to \infty} \phi(t, x_1) = x_0,$$
then we say that the equilibrium point x_0 is *asymptotically stable*. If an equilibrium point is not stable, then we say that it is *unstable*.

For the differential equation $N' = rN(1 - N/K)$ the equilibrium point $N_1 = 0$ is unstable and the equilibrium point $N_2 = K$ is asymptotically stable (see Figures 3 and 4).

Definition 1.16 We say that F is a *potential energy function* for the differential equation (1.6) provided
$$f(x) = -F'(x).$$

Theorem 1.17 *If F is a potential energy function for (1.6), then $F(x(t))$ is strictly decreasing along any nonconstant solution x. Also, x_0 is an equilibrium point of (1.6) iff $F'(x_0) = 0$. If x_0 is an isolated equilibrium point of (1.6) such that F has a local minimum at x_0, then x_0 is asymptotically stable.*

Proof Assume F is a potential energy function for (1.6), assume x is a nonconstant solution of (1.6), and consider
$$\begin{aligned} \frac{d}{dt}F(x(t)) &= F'(x(t))x'(t) \\ &= -f^2(x(t)) \\ &< 0. \end{aligned}$$

Hence the potential energy function F is strictly decreasing along noncon-stant solutions. Since $f(x_0) = 0$ iff $F'(x_0) = 0$, x_0 is an equilibrium point of (1.6) iff $F'(x_0) = 0$.

Let x_0 be an isolated equilibrium point of (1.6) such that F has a local minimum at x_0, and choose an interval $(x_0 - \delta, x_0 + \delta)$ such that $F'(x) > 0$ on $(x_0, x_0 + \delta)$ and $F'(x) < 0$ on $(x_0 - \delta, x_0)$. Suppose $x_1 \in (x_0, x_0 + \delta)$. Then $F(\phi(t, x_1))$ is strictly decreasing, so $\phi(t, x_1)$ is decreasing, remains in the interval $(x_0, x_0 + \delta)$, and converges to some limit $l \geq x_0$. We will show that $l = x_0$ by assuming $l > x_0$ and obtaining a contradiction. If $l > x_0$, then there is a positive constant C so that $F'(\phi(t, x_1)) \geq C$ for $t \geq 0$ (Why is the right maximal interval of existence for the solution $\phi(t, x_1)$ the interval $[0, \infty)$?). But

$$\phi(t, x_1) - x_1 = \int_0^t (-F'(\phi(s, x_1)))\ ds \leq -Ct,$$

for $t \geq 0$, which implies that $\phi(t, x_1) \to -\infty$ as $t \to \infty$, a contradiction. We conclude that $\phi(t, x_1) \to x_0$ as $t \to \infty$. Since the case $x_1 \in (x_0 - \delta, x_0)$ is similar, we have that x_0 is asymptotically stable. $\square$

Example 1.18 By finding a potential energy function for

$$x' = -2(x - 70),$$

draw the phase line diagram for this differential equation.

Here a potential energy function is given by

$$\begin{aligned} F(x) &= -\int_0^x f(u)\ du \\ &= -\int_0^x -2(u - 70)\ du \\ &= x^2 - 140x. \end{aligned}$$

In Figure 5 we graph $y = F(x)$ and using Theorem 1.17 we get the phase line diagram below the graph of the potential energy function. Notice that $x = 70$ is an isolated minimum for the potential energy function.

$\triangle$

1.4 Generalized Logistic Equation

We first do some calculations to derive what we will call the generalized logistic equation. Assume p and q are continuous functions on an interval I and let $x(t)$ be a solution of the first order linear differential equation

$$x' = -p(t)x + q(t) \tag{1.8}$$

with $x(t) \neq 0$ on I. Then set

$$y(t) = \frac{1}{x(t)}, \quad t \in I.$$

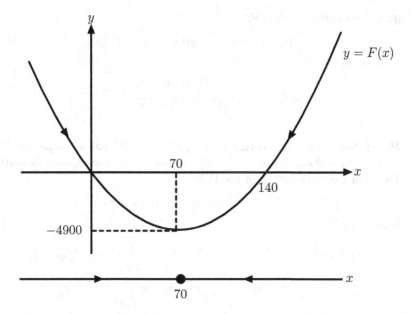

FIGURE 5. Potential energy function and phase line diagram for $x' = -2(x - 70)$.

It follows that

$$
\begin{aligned}
y'(t) &= -\frac{x'(t)}{x^2(t)} \\
&= [p(t)x(t) - q(t)]\, y^2(t) \\
&= [p(t) - q(t)y(t)]\, y(t), \quad t \in I.
\end{aligned}
$$

We call the differential equation

$$
y' = [p(t) - q(t)y]\, y \tag{1.9}
$$

the generalized logistic equation. Above we proved that if $x(t)$ is a nonzero solution of the linear equation (1.8) on I, then $y(t) = \frac{1}{x(t)}$ is a nonzero solution of the generalized logistic equation (1.9) on I. Conversely, if $y(t)$ is a nonzero solution of the generalized logistic equation (1.9) on I, then (Exercise 1.28) $x(t) = \frac{1}{y(t)}$ is a nonzero solution of the linear equation (1.8) on I.

We now state the following theorem.

Theorem 1.19 *If* $y_0 \neq 0$ *and*

$$
\frac{1}{y_0} + \int_{t_0}^{t} q(s)e^{\int_{t_0}^{s} p(\tau)\, d\tau}\, ds \neq 0, \quad t \in I,
$$

then the solution of the IVP

$$y' = [p(t) - q(t)y]\, y, \quad y(t_0) = y_0, \quad t_0 \in I \tag{1.10}$$

is given by

$$y(t) = \frac{e^{\int_{t_0}^t p(\tau)\, d\tau}}{\frac{1}{y_0} + \int_{t_0}^t q(s) e^{\int_{t_0}^s p(\tau)\, d\tau}\, ds}. \tag{1.11}$$

Proof Note that by Theorem 1.3 every IVP (1.10) has a unique solution. Assume $y_0 \neq 0$ and let $x_0 = \frac{1}{y_0}$. By the variation of constants formula in Theorem 1.6 the solution of the IVP

$$x' = -p(t)x + q(t), \quad x(t_0) = x_0$$

is given by

$$
\begin{aligned}
x(t) &= e^{-\int_{t_0}^t p(\tau)\, d\tau} x_0 + e^{-\int_{t_0}^t p(\tau)\, d\tau} \int_{t_0}^t e^{\int_{t_0}^s p(\tau)\, d\tau} q(s)\, ds \\
&= e^{-\int_{t_0}^t p(\tau)\, d\tau} \frac{1}{y_0} + e^{-\int_{t_0}^t p(\tau)\, d\tau} \int_{t_0}^t e^{\int_{t_0}^s p(\tau)\, d\tau} q(s)\, ds
\end{aligned}
$$

which is nonzero on I by assumption. It follows that the solution of the IVP (1.10) is given by

$$y(t) = \frac{1}{e^{-\int_{t_0}^t p(\tau)\, d\tau} \frac{1}{y_0} + e^{-\int_{t_0}^t p(\tau)\, d\tau} \int_{t_0}^t q(s) e^{\int_{t_0}^s p(\tau)\, d\tau}\, ds}.$$

Multiplying the numerator and denominator by $e^{\int_{t_0}^t p(\tau)\, d\tau}$ we get the desired result (1.11). □

In applications (e.g., population dynamics) one usually has that

$$p(t) = q(t)K$$

where $K > 0$ is a constant. In this case the generalized logistic equation becomes

$$y' = p(t)\left[1 - \frac{y}{K}\right] y. \tag{1.12}$$

The constant solutions $y(t) = 0$ and $y(t) = K$ are called equilibrium solutions of (1.12). The constant K is called the carrying capacity (saturation level).

We now state the following corollary of Theorem 1.19.

Corollary 1.20 *If $y_0 \neq 0$ and*

$$\frac{1}{y_0} - \frac{1}{K} + \frac{1}{K} e^{\int_{t_0}^t p(s)\, ds} \neq 0, \quad t \in I,$$

then the solution of the IVP

$$y' = p(t)\left[1 - \frac{y}{K}\right] y, \quad y(t_0) = y_0 \tag{1.13}$$

is given by

$$y(t) = \frac{e^{\int_{t_0}^t p(s)\,ds}}{\frac{1}{y_0} - \frac{1}{K} + \frac{1}{K}e^{\int_{t_0}^t p(s)\,ds}}. \tag{1.14}$$

Proof This follows from Theorem 1.19, where we use the fact that

$$\int_{t_0}^t e^{\int_{t_0}^s p(\tau)\,d\tau} q(s)\,ds = \frac{1}{K}\int_{t_0}^t e^{\int_{t_0}^s p(\tau)\,d\tau} p(s)\,ds$$

$$= \frac{1}{K}\left[e^{\int_{t_0}^t p(s)\,ds} - 1 \right].$$

$\square$

The following theorem gives conditions under which the solutions of the generalized logistic equation (1.12) with nonnegative initial conditions behave very similar to the corresponding solutions of the autonomous logistic equation (1.7).

Theorem 1.21 *Assume* $p : [t_0, \infty) \to [0, \infty)$ *is continuous and* $\int_{t_0}^\infty p(t)\,dt = \infty$. *Let* $y(t)$ *be the solution of the IVP* (1.13) *with* $y_0 > 0$, *then* $y(t)$ *exists on* $[t_0, \infty)$. *Also if* $0 < y_0 < K$, *then* $y(t)$ *is nondecreasing with* $\lim_{t\to\infty} y(t) = K$. *If* $y_0 > N$, *then* $y(t)$ *is nonincreasing with* $\lim_{t\to\infty} y(t) = K$.

Proof Let $y(t)$ be the solution of the IVP (1.13) with $y_0 > 0$. Then from (1.14)

$$y(t) = \frac{e^{\int_{t_0}^t p(s)\,ds}}{\frac{1}{y_0} - \frac{1}{K} + \frac{1}{K}e^{\int_{t_0}^t p(s)\,ds}}. \tag{1.15}$$

By the uniqueness of solutions of IVP's the solution $y(t)$ is bounded below by K and hence $y(t)$ remains positive to the right of t_0. But since

$$y'(t) = p(t)\left[1 - \frac{y(t)}{K} \right] y(t) \le 0,$$

$y(t)$ is decreasing. It follows from Theorem 1.3 that $y(t)$ exists on $[t_0, \infty)$ and from (1.15) we get that $\lim_{t\to\infty} y(t) = K$.

Next assume that $0 < y_0 < K$. Then by the uniqueness of solutions of IVP's we get that

$$0 < y(t) < K,$$

to the right of t_0. It follows that $y(t)$ is a solution on $[t_0, \infty)$ and by (1.15) we get that $\lim_{t\to\infty} y(t) = K$. Also $0 < y(t) < K$, implies that

$$y'(t) = p(t)\left[1 - \frac{y(t)}{K} \right] y(t) \ge 0,$$

so $y(t)$ is nondecreasing on $[t_0, \infty)$.

$\square$

1.5 Bifurcation

Any unspecified constant in a differential equation is called a *parameter*. One of the techniques that is used to study differential equations is to let a parameter vary and to observe the resulting changes in the behavior of the solutions. Any large scale change is called a *bifurcation* and the value of the parameter for which the change occurs is called a *bifurcation point*. We end this chapter with some simple examples of bifurcations.

Example 1.22 We consider the differential equation

$$x' = \lambda(x - 1),$$

where λ is a parameter. In Figure 6 the phase line diagrams for this differential equation when $\lambda < 0$ and $\lambda > 0$ are drawn. There is a drastic change in the phase line diagrams as λ passes through zero (the equilibrium point $x = 1$ loses its stability as λ increases through zero). Because of this we say bifurcation occurs when $\lambda = 0$.

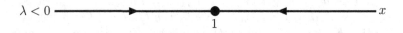

FIGURE 6. Phase line diagrams for $x' = \lambda(x - 1)$, $\lambda < 0$, $\lambda > 0$.

$\triangle$

Example 1.23 (Saddle-Node Bifurcation) Now consider the equation

$$x' = \lambda + x^2.$$

If $\lambda < 0$, then there is a pair of equilibrium points, one stable and one unstable. When $\lambda = 0$, the equilibrium points collide, and for $\lambda > 0$, there are no equilibrium points (see Figure 7). In this case, the bifurcation that occurs at $\lambda = 0$ is usually called a *saddle-node* bifurcation.

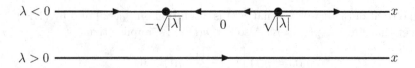

FIGURE 7. Phase line diagrams for $x' = \lambda + x^2$, $\lambda < 0$, $\lambda > 0$.

$\triangle$

Example 1.24 (Transcritical Bifurcation) Consider

$$x' = \lambda x - x^2.$$

For $\lambda < 0$, there is an unstable equilibrium point at $x = \lambda$ and a stable equilibrium point at $x = 0$. At $\lambda = 0$, the two equilibrium points coincide. For $\lambda > 0$, the equilibrium at $x = 0$ is unstable, while the one at $x = \lambda$ is stable, so the equilibrium points have switched stability! See Figure 8. This type of bifurcation is known as a *transcritical* bifurcation.

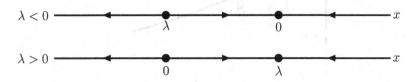

FIGURE 8. Phase line diagrams for $x' = \lambda x - x^2$, $\lambda < 0$, $\lambda > 0$.

$\triangle$

Example 1.25 (Pitchfork Bifurcation) We will draw the *bifurcation diagram* (see Figure 9) for the differential equation

$$x' = f(\lambda, x) := (x - 1)(\lambda - 1 - (x - 1)^2). \tag{1.16}$$

Note that the equations

$$x = 1, \quad \lambda = 1 + (x - 1)^2$$

give you the equilibrium points of the differential equation (1.16). We graph these two equilibrium curves in the λx-plane (see Figure 9). Note for each $\lambda \leq 1$ the differential equation (1.16) has exactly one equilibrium point and for each $\lambda > 1$ the differential equation (1.16) has exactly three equilibrium points. When part of an equilibrium curve is dashed it means the corresponding equilibrium points are unstable, and when part of an equilibrium curve is solid it means the corresponding equilibrium points are stable. To determine this stability of the equilibrium points note that at points on the pitchfork we have $f(\lambda, t) = 0$, at points above the pitchfork $f(\lambda, t) < 0$, at points below the pitchfork $f(\lambda, t) > 0$, at points between the top two forks of the pitchfork $f(\lambda, t) > 0$, and at points between the lower two forks of the pitchfork $f(\lambda, t) < 0$. Note that the equilibrium point $x = 1$ is asymptotically stable for $\lambda < 1$ and becomes unstable for $\lambda > 1$. We say we have *pitchfork bifurcation* at $\lambda = 1$.

$\triangle$

In the final example in this section we give an example of hysteresis.

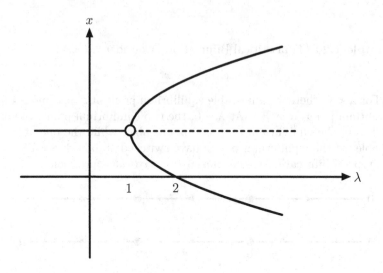

FIGURE 9. Bifurcation diagram for (1.16).

Example 1.26 (Hysteresis) The bifurcation diagram for the differential equation

$$x' = \lambda + x - x^3 \qquad (1.17)$$

is given in Figure 10. Note that if we start with $\lambda < \lambda_1 := -\frac{2}{3\sqrt{3}}$ and slowly increase λ, then for all practical purposes solutions stay close to the smallest equilibrium point until when λ passes through the value $\lambda_2 := \frac{2}{3\sqrt{3}}$, where the solution quickly approaches the largest equilibrium point for $\lambda > \lambda_2$. On the other hand, if we start with $\lambda > \lambda_2$ and start decreasing λ, solutions stay close to the largest equilibrium point until λ decreases through the value λ_1, where all of a sudden the solution approaches the smallest equilibrium point. $\triangle$

There are lots of interesting examples of hysteresis in nature. Murray [**37**], pages 4–8, discusses the possible existence of hysteresis in a population model for the spuce budworm. For an example of hysteresis concerning the temperature in a continuously stirred tank reactor see Logan [**33**], pages 430–434. Also for an interesting example concerning the buckling of a wire arc see Iooss and Joseph [**28**], pages 25–28. Hysteresis also occurs in the theory of elasticity.

1.6 Exercises

1.1 Find the maximal interval of existence for the solution of the IVP

$$x' = (\cos t)\,x^2, \quad x(0) = 2.$$

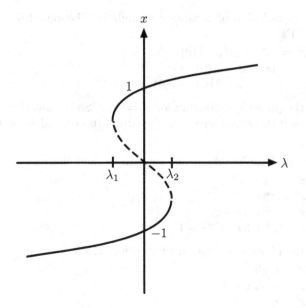

FIGURE 10. Hysteresis.

1.2 Find the maximal interval of existence for the solution of the IVP

$$x' = \frac{2tx^2}{1+t^2}, \quad x(0) = x_0$$

in terms of x_0.

1.3 Show that the IVP

$$x' = x^{\frac{1}{3}}, \quad x(0) = 0$$

has infinitely many solutions. Explain why Theorem 1.3 does not apply to give you uniqueness.

1.4 Assume that $f : \mathbb{R} \times (0, \infty) \to \mathbb{R}$ is defined by $f(t, x) = x^{\frac{1}{3}}$, for $(t, x) \in \mathbb{R} \times (0, \infty)$. Show that for any $(t_0, x_0) \in \mathbb{R} \times (0, \infty)$ the IVP

$$x' = f(t, x), \quad x(t_0) = x_0$$

has a unique solution. Find the maximum interval of existence for the solution of this IVP when $(t_0, x_0) = (1, 1)$.

1.5 Use the variation of constants formula in Theorem 1.6 to solve the following IVPs:

    (i) $x' = 2x + e^{2t}, \quad x(0) = 3$
    (ii) $x' = 3x + 2e^{3t}, \quad x(0) = 2$
    (iii) $x' = \tan(t)x + \sec(t), \quad x(0) = -1$
    (iv) $x' = \frac{2}{t}x + t, \quad x(1) = 2$

1.6   Use the variation of constants formula in Theorem 1.6 to solve the following IVPs:
  (i) $x' = -2x + e^{at}$, $x(0) = 2$
  (ii) $x' = 4x + 3t$, $x(0) = 1$
  (iii) $tx' - 3x = 1 - 3\ln t$, $x(0) = 1$

1.7 Draw the phase line diagram for $x' = -2x$. Show that there are infinitely many solutions that give you the orbits $(0, \infty)$ and $(-\infty, 0)$ respectively.

1.8 Draw the phase line diagrams for each of the following:
  (i) $x' = -x + x^3$
  (ii) $x' = x^4$
  (iii) $x' = x^2 + 4x + 2$
  (iv) $x' = x^3 - 3x^2 + 3x - 1$

1.9 Draw the phase line diagrams for each of the following:
  (i) $x' = \cosh x$
  (ii) $x' = \cosh x - 1$
  (iii) $x' = (x - a)^2$
  (iv) $x' = \sin x$
  (v) $x' = \sin(2x)$
  (vi) $x' = e^x$
  (vii) $x' = \sinh^2(x - b)$
  (viii) $x' = \cos x - 1$

1.10 Show that if x is a nonzero solution of the linear equation $x' = -rx + \frac{r}{K}$, where r and K are positive constants, then $N = \frac{1}{x}$ is a solution of the logistic equation (1.7). Use this and the variation of constants formula to solve the IVP $N' = rN\left(1 - \frac{N}{K}\right)$, $N(0) = \frac{K}{2}$.

1.11 Solve the logistic equation (1.7) by using the method of separation of variables. Note in general we can not solve the more general logistic equation (1.9) by the method of separation of variables.

1.12 (Bernoulli's Equation) The differential equation $y' = p(t)y + q(t)y^\alpha$, $\alpha \neq 0, 1$ is called *Bernoulli's equation*. Show that if y is a nonzero solution of Bernoulli's equation, then $x = y^{1-\alpha}$ is a solution of the linear equation $x' = (1 - \alpha)p(t)x + (1 - \alpha)q(t)$.

1.13 Use Exercise 1.12 to solve the following Bernoulli equations:
  (i) $x' = -\frac{1}{t}x + tx^2$
  (ii) $x' = x + e^t x^2$
  (iii) $x' = -\frac{1}{t}x + \frac{1}{tx^2}$

1.14 Assume that p and q are continuous functions on $\mathbb{R}$ which are periodic with period $T > 0$. Show that the linear differential equation (1.4) has a periodic solution x with positive period T iff the differential equation (1.4) has a solution x satisfying $x(0) = x(T)$.

1.15 Show that the equilibrium point $x_0 = 0$ for the differential equation $x' = 0$ is stable but not asymptotically stable.

1.16 Determine the stability (*stable, unstable, or asymptotically stable*) of the equilibrium points for each of the differential equations in Exercise 1.8.

1.17 A yam is put in a 200°C oven at time $t = 0$. Let $T(t)$ be the temperature of the yam in degrees Celsius at time t minutes later. According to Newton's law of cooling, $T(t)$ satisfies the differential equation

$$T' = -k(T - 200),$$

where k is a positive constant. Draw the phase diagram for this differential equation and then draw a possible graph of various solutions with various initial temperatures at $t = 0$. Determine the stability of all equilibrium points for this differential equation.

1.18 Given that the function F defined by $F(x) = x^3 + 3x^2 - x - 3$, for $x \in \mathbb{R}$ is a potential energy function for $x' = f(x)$, draw the phase line diagram for $x' = f(x)$.

1.19 Given that the function F defined by $F(x) = 4x^2 - x^4$, for $x \in \mathbb{R}$ is a potential energy function for $x' = f(x)$, draw the phase line diagram for $x' = f(x)$.

1.20 For each of the following differential equations find a potential energy function and use it to draw the phase line diagram:

   (i) $x' = x^2$
   (ii) $x' = 3x^2 - 10x + 6$
   (iii) $x' = 8x - 4x^3$
   (iv) $x' = \frac{1}{x^2+1}$

1.21 Find and graph a potential energy function for the equation

$$x' = -a(x - b), \quad a, b > 0$$

and use this to draw a phase line diagram for this equation.

1.22 Determine the stability (*stable, unstable, or asymptotically stable*) of the equilibrium points for each of the differential equations in Exercise 1.20.

1.23 Given that a certain population $x(t)$ at time t is known to satisfy the differential equation

$$x' = ax \ln\left(\frac{b}{x}\right),$$

when $x > 0$, where $a > 0$, $b > 0$ are constants, find the equilibrium population and determine its stability.

1.24 Use the phase line diagram and take into account the concavity of solutions to graph various solutions of each of the following differential equations

   (i) $x' = 6 - 5x + x^2$

(ii) $x' = 4x^2 + 3x^3 - x^4$

(iii) $x' = 2 + x - x^2$

1.25 A tank initially contains 1000 gallons of a solution of water and 5 pounds of some solute. Suppose that a solution with the same solute of concentration .1 pounds per gallon is flowing into the tank at the rate of 2 gallons per minute. Assume that the tank is constantly stirred so that the concentration of solute in the tank at each time t is essentially constant throughout the tank.

    (i) Suppose that the solution in the tank is being drawn off at the rate of 2 gallons per minute to maintain a constant volume of solution in the tank. Show that the number $x(t)$ of pounds of solute in the tank at time t satisfies the differential equation

$$x' = .2 - \frac{x}{500},$$

    and compute $x(t)$;

    (ii) Suppose now that the solution in the tank is being drawn off at the rate of 3 gallons per minute so that the tank is eventually drained. Show that the number $y(t)$ of pounds of solute in the tank at time t satisfies the equation

$$y' = .2 - \frac{3y}{1000 - t},$$

    for $0 < t < 1000$, and compute $y(t)$.

1.26 (Terminal Velocity) Let m be the mass of a large object falling rapidly toward the earth with velocity $v(t)$ at time t. (We take downward velocity to be positive in this problem.) If we take the force of gravity to be constant, the standard equation of motion is

$$mv' = mg - kv^2,$$

where g is the acceleration due to gravity and $-kv^2$ is the upward force due to air resistance:

    (i) Sketch the phase line diagram and determine which of the equilibrium points is asymptotically stable. Why is "terminal velocity" an appropriate name for this number?

    (ii) Assume the initial velocity is $v(0) = v_0$, and solve the IVP. Show that the solution v approaches the asymptotically stable equilibrium as $t \to \infty$.

1.27 Assume that $x_0 \in (a, b)$, $f(x) > 0$ for $a < x < x_0$, and $f(x) < 0$ for $x_0 < x < b$. Show that x_0 is an asymptotically stable equilibrium point for $x' = f(x)$.

1.28 Show that if $y(t)$ is a nonzero solution of the generalized logistic equation (1.9) on I, then $x(t) = \frac{1}{y(t)}$ is a nonzero solution of the linear equation (1.8) on I.

1.29 If in Theorem 1.21 we replace $\int_{t_0}^{\infty} p(t)\, dt = \infty$ by $\int_{t_0}^{\infty} p(t)\, dt = L$, where $0 \leq L < \infty$, what can we say about solutions of the IVP (1.13) with $y_0 > 0$

1.30 (Harvesting) Work each of the following:

(i) Explain why the following differential equation could serve as a model for logistic population growth with harvesting if λ is a positive parameter:

$$N' = rN\left(1 - \frac{N}{K}\right) - \lambda N.$$

(ii) If $\lambda < r$, compute the equilibrium points, sketch the phase line diagram, and determine the stability of the equilibria.

(iii) Show that a bifurcation occurs at $\lambda = r$. What type of bifurcation is this?

1.31 Solve the generalized logistic equation (1.13) by the method of separation of variables.

1.32 (Gene Activation) The following equation occurs in the study of gene activation:

$$x' = \lambda - x + \frac{4x^2}{1 + x^2}.$$

Here $x(t)$ is the concentration of gene product at time t.

(i) Sketch the phase line diagram for $\lambda = 1$.

(ii) There is a small value of λ, say λ_0, where a bifurcation occurs. Estimate λ_0, and sketch the phase line diagram for some $\lambda \in (0, \lambda_0)$.

(iii) Draw the bifurcation diagram for this differential equation.

1.33 For each of the following differential equations find values of λ where bifurcation occurs. Draw phase line diagrams for values of λ close to the value of λ where bifurcation occurs.

(i) $x' = \lambda - 4 - x^2$

(ii) $x' = x^3(\lambda - x)$

(iii) $x' = x^3 - x + \lambda$

1.34 Assume that the population $x(t)$ of rats on a farm at time t (in weeks) satisfies the differential equation $x' = -.1x(x - 25)$. Now assume that we decide to kill rats at a constant rate of λ rats per week. What would be the new differential equation that x satisfies? For what value of λ in this new differential equation does bifurcation occur? If the number of rats we kill per week is larger than this bifurcation value what happens to the population of rats?

1.35 Give an example of a differential equation with a parameter λ which has pitchfork bifurcation at $\lambda = 0$, where $x_0 = 0$ is an unstable equilibrium point for $\lambda < 0$ and is a stable equilibrium point for $\lambda > 0$.

1.36 Draw the bifurcation diagram for each of the following
- (i) $x' = \lambda - x^2$
- (ii) $x' = (\lambda - x)(\lambda - x^3)$

1.37 In each of the following draw the bifurcation diagram and give the value(s) of λ where bifurcation occurs:
- (i) $x' = (\lambda - x)(\lambda - x^2)$
- (ii) $x' = \lambda - 12 + 3x - x^3$
- (iii) $x' = \lambda - \frac{x^2}{1+x^2}$
- (iv) $x' = x \sin x + \lambda$

In which of these does hysteresis occur?

Chapter 2

Linear Systems

2.1 Introduction

In this chapter we will be concerned with linear systems of the form

$$
\begin{aligned}
x_1' &= a_{11}(t)x_1 + a_{12}(t)x_2 + \cdots + a_{1n}(t)x_n + b_1(t) \\
x_2' &= a_{21}(t)x_1 + a_{22}(t)x_2 + \cdots + a_{2n}(t)x_n + b_2(t) \\
&\cdots \\
x_n' &= a_{n1}(t)x_1 + a_{n2}(t)x_2 + \cdots + a_{nn}(t)x_n + b_n(t),
\end{aligned}
$$

where we assume that the functions $a_{ij}, 1 \leq i,j \leq n, b_i, 1 \leq i \leq n$, are continuous real-valued functions on an interval I. We say that the collection of n functions $x_1, x_2, \cdots, x_n$ is a solution on I of this linear system provided each of these n functions is continuously differentiable on I and

$$
\begin{aligned}
x_1'(t) &= a_{11}(t)x_1(t) + a_{12}(t)x_2(t) + \cdots + a_{1n}(t)x_n(t) + b_1(t) \\
x_2'(t) &= a_{21}(t)x_1(t) + a_{22}(t)x_2(t) + \cdots + a_{2n}(t)x_n(t) + b_2(t) \\
&\cdots \\
x_n'(t) &= a_{n1}(t)x_1(t) + a_{n2}(t)x_2(t) + \cdots + a_{nn}(t)x_n(t) + b_n(t),
\end{aligned}
$$

for $t \in I$.

This system can be written as an equivalent vector equation

$$
x' = A(t)x + b(t), \tag{2.1}
$$

where

$$
x := \begin{bmatrix} x_1 \\ x_2 \\ \vdots \\ x_n \end{bmatrix}, \quad x' := \begin{bmatrix} x_1' \\ x_2' \\ \vdots \\ x_n' \end{bmatrix},
$$

and

$$
A(t) := \begin{bmatrix} a_{11}(t) & \cdots & a_{1n}(t) \\ \vdots & \ddots & \vdots \\ a_{n1}(t) & \cdots & a_{nn}(t) \end{bmatrix}, \quad b(t) := \begin{bmatrix} b_1(t) \\ \vdots \\ b_n(t) \end{bmatrix},
$$

W.G. Kelley and A.C. Peterson, *The Theory of Differential Equations:*
Classical and Qualitative, Universitext 278, DOI 10.1007/978-1-4419-5783-2_2,
© Springer Science+Business Media, LLC 2010

for $t \in I$. Note that the matrix functions A and b are continuous on I (a matrix function is continuous on I if and only if (iff) all of its entries are continuous on I). We say that an $n \times 1$ vector function x is a solution of (2.1) on I provided x is a continuously differentiable vector function on I (iff each component of x is continuously differentiable on I) and

$$x'(t) = A(t)x(t) + b(t),$$

for all $t \in I$.

Example 2.1 It is easy to see that the pair of functions x_1, x_2 defined by $x_1(t) = 2 + \sin t$, $x_2(t) = -t + \cos t$, for $t \in \mathbb{R}$ is a solution on $\mathbb{R}$ of the linear system

$$\begin{aligned} x_1' &= x_2 + t, \\ x_2' &= -x_1 + 1 \end{aligned}$$

and the vector function

$$x := \left[\begin{array}{c} x_1 \\ x_2 \end{array} \right]$$

is a solution on $\mathbb{R}$ of the vector equation

$$x' = \left[\begin{array}{cc} 0 & 1 \\ -1 & 0 \end{array} \right] x + \left[\begin{array}{c} t \\ 1 \end{array} \right].$$

$\triangle$

The study of equation (2.1) includes the nth-order scalar differential equation

$$y^{(n)} + p_{n-1}(t)y^{(n-1)} + \cdots + p_0(t)y = r(t) \tag{2.2}$$

as a special case. To see this let y be a solution of (2.2) on I, that is, assume y has n continuous derivatives on I and

$$y^{(n)}(t) + p_{n-1}(t)y^{(n-1)}(t) + \cdots + p_0(t)y(t) = r(t), \quad t \in I.$$

Then let

$$x_i(t) := y^{(i-1)}(t),$$

for $t \in I$, $1 \le i \le n$. Then the $n \times 1$ vector function x with components x_i satisfies equation (2.1) on I if

$$A(t) = \left[\begin{array}{ccccc} 0 & 1 & 0 & \cdots & 0 \\ 0 & 0 & 1 & \cdots & 0 \\ \vdots & \vdots & \ddots & \ddots & \vdots \\ 0 & 0 & \cdots & 0 & 1 \\ -p_0(t) & -p_1(t) & -p_2(t) & \cdots & -p_{n-1}(t) \end{array} \right], \ b(t) = \left[\begin{array}{c} 0 \\ 0 \\ \vdots \\ 0 \\ r(t) \end{array} \right]$$

for $t \in I$. The matrix function A is called the *companion matrix* of the differential equation (2.2). Conversely, it can be shown that if x is a solution of the vector equation (2.1) on I, where A and b are as before, then it follows that the scalar function y defined by $y(t) := x_1(t)$, for $t \in I$ is a solution of (2.2) on I. We next give an interesting example that leads to a four-dimensional linear system.

Example 2.2 (Coupled Vibrations) Consider the system of two masses m_1 and m_2 in Figure 1 connected to each other by a spring with spring constant k_2 and to the walls by springs with spring constants k_1 and k_3 respectively. Let $u(t)$ be the displacement of m_1 from its equilibrium position at time t and $v(t)$ be the displacement of m_2 from its equilibrium at time t. (We are taking the positive direction to be to the right.) Let c be the coefficient of friction for the surface on which the masses slide. An application of Newton's second law yields

$$\begin{aligned} m_1 u'' &= -cu' - (k_1 + k_2)u + k_2 v, \\ m_2 v'' &= -cv' - (k_2 + k_3)v + k_2 u. \end{aligned}$$

Here we have a system of two second-order equations, and we define $x_1 := u$, $x_2 := u'$, $x_3 := v$, and $x_4 := v'$, obtaining the first-order system

$$\begin{bmatrix} x_1 \\ x_2 \\ x_3 \\ x_4 \end{bmatrix}' = \begin{bmatrix} 0 & 1 & 0 & 0 \\ -\frac{k_1+k_2}{m_1} & -\frac{c}{m_1} & \frac{k_2}{m_1} & 0 \\ 0 & 0 & 0 & 1 \\ \frac{k_2}{m_2} & 0 & -\frac{k_2+k_3}{m_2} & -\frac{c}{m_2} \end{bmatrix} \begin{bmatrix} x_1 \\ x_2 \\ x_3 \\ x_4 \end{bmatrix}.$$

$\triangle$

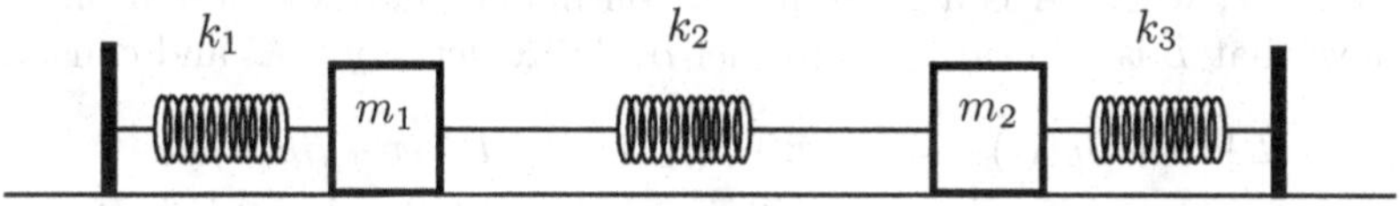

FIGURE 1. Coupled masses.

The following theorem is a special case of Theorem 8.65 (see also Corollary 8.18) .

Theorem 2.3 *Assume that the $n \times n$ matrix function A and the $n \times 1$ vector function b are continuous on an interval I. Then the IVP*

$$x' = A(t)x + b(t), \quad x(t_0) = x_0,$$

where $t_0 \in I$ and x_0 is a given constant $n \times 1$ vector, has a unique solution that exists on the whole interval I.

Note that it follows from Theorem 2.3 that in Example 2.2 if at any time t_0 the position and velocity of the two masses are known, then that uniquely determines the position and velocity of the masses at all other times.

We end this section by explaining why we call the differential equation (2.1) a *linear* vector differential equation.

Definition 2.4 A family of functions $\mathbb{A}$ defined on an interval I is said to be a *vector space* or *linear space* provided whenever $x, y \in \mathbb{A}$ it follows that for any constants $\alpha, \beta \in \mathbb{R}$

$$\alpha x + \beta y \in \mathbb{A}.$$

By the function $\alpha x + \beta y$ we mean the function defined by

$$(\alpha x + \beta y)(t) := \alpha x(t) + \beta y(t),$$

for $t \in I$. If $\mathbb{A}$ and $\mathbb{B}$ are vector spaces of functions defined on an interval I, then $L : \mathbb{A} \to \mathbb{B}$ is called a *linear operator* provided

$$L\left[\alpha x + \beta y\right] = \alpha L[x] + \beta L[y],$$

for all $\alpha, \beta \in \mathbb{R}$, $x, y \in \mathbb{A}$.

We now give an example of an important linear operator.

Example 2.5 Let $\mathbb{A}$ be the set of all $n \times 1$ continuously differentiable vector functions on an interval I and let $\mathbb{B}$ be the set of all $n \times 1$ continuous vector functions on an interval I and note that $\mathbb{A}$ and $\mathbb{B}$ are linear spaces. Define $L : \mathbb{A} \to \mathbb{B}$ by

$$Lx(t) = x'(t) - A(t)x(t),$$

for $t \in I$, where A is a given $n \times n$ continuous matrix function on I. To show that L is a linear operator, let $\alpha, \beta \in \mathbb{R}$, let $x, y \in \mathbb{A}$, and consider

$$
\begin{aligned}
L\left[\alpha x + \beta y\right](t) &= (\alpha x + \beta y)'(t) - A(t)(\alpha x + \beta y)(t) \\
&= \alpha x'(t) + \beta y'(t) - \alpha A(t)x(t) - \beta A(t)y(t) \\
&= \alpha\left[x'(t) - A(t)x(t)\right] + \beta\left[y'(t) - A(t)y(t)\right] \\
&= \alpha Lx(t) + \beta Ly(t) \\
&= (\alpha Lx + \beta Ly)(t),
\end{aligned}
$$

for $t \in I$. Hence

$$L\left[\alpha x + \beta y\right] = \alpha Lx + \beta Ly$$

and so $L : \mathbb{A} \to \mathbb{B}$ is a linear operator. $\triangle$

Since the differential equation (2.1) can be written in the form

$$Lx = b,$$

where L is the linear operator defined in Example 2.5, we call (2.1) a *linear vector differential equation*. If b is not the trivial vector function, then the equation $Lx = b$ is called a *nonhomogeneous* linear vector differential equation and $Lx = 0$ is called the corresponding *homogeneous* linear vector differential equation.

2.2 The Vector Equation $x' = A(t)x$

To solve the nonhomogeneous linear vector differential equation

$$x' = A(t)x + b(t)$$

we will see later that we first need to solve the corresponding homogeneous linear vector differential equation

$$x' = A(t)x. \tag{2.3}$$

Hence we will first study the homogeneous vector differential equation (2.3). Note that if the vector functions $\phi_1, \phi_2, \cdots, \phi_k$ are solutions of $x' = A(t)x$ (equivalently, of $Lx = 0$, where L is as in Example 2.5) on I, then

$$L\left[c_1\phi_1 + c_2\phi_2 + \cdots + c_k\phi_k\right]$$
$$= c_1 L\left[\phi_1\right] + c_2 L\left[\phi_2\right] + \cdots + c_n L\left[\phi_k\right]$$
$$= 0.$$

This proves that any linear combination of solutions of (2.3) on I is a solution of (2.3) on I. Consequently, the set of all such solutions is a vector space. To solve (2.3), we will see that we want to find n *linearly independent* solutions on I (see Definition 2.9).

Definition 2.6 We say that the constant $n \times 1$ vectors $\psi_1, \psi_2, \cdots, \psi_k$ are *linearly dependent* provided there are constants $c_1, c_2, \cdots, c_k$, not all zero, such that

$$c_1\psi_1 + c_2\psi_2 + \cdots + c_k\psi_k = 0,$$

where 0 denotes the $n \times 1$ zero vector. Otherwise we say that these k constant vectors are *linearly independent*.

Note that the constant $n \times 1$ vectors $\psi_1, \psi_2, \cdots, \psi_k$ are linearly independent provided that the only constants $c_1, c_2, \cdots, c_k$ that satisfy the equation

$$c_1\psi_1 + c_2\psi_2 + \cdots + c_k\psi_k = 0,$$

are $c_1 = c_2 = \cdots = c_k = 0$.

Theorem 2.7 *Assume we have exactly n constant $n \times 1$ vectors*

$$\psi_1, \psi_2, \cdots, \psi_n$$

and C is the column matrix $C = [\psi_1 \psi_2 \cdots \psi_n]$. Then $\psi_1, \psi_2, \cdots, \psi_n$ are linearly dependent iff $\det C = 0$.

Proof Let $\psi_1, \psi_2, \cdots, \psi_n$ and C be as in the statement of this theorem. Then

$$\det C = 0$$

if and only if there is a nontrivial vector

$$\begin{bmatrix} c_1 \\ c_2 \\ \vdots \\ c_n \end{bmatrix}$$

such that

$$C \begin{bmatrix} c_1 \\ c_2 \\ \vdots \\ c_n \end{bmatrix} = \begin{bmatrix} 0 \\ 0 \\ \vdots \\ 0 \end{bmatrix}$$

if and only if

$$c_1\psi_1 + c_2\psi_2 + \cdots + c_n\psi_n = 0,$$

where $c_1, c_2, \cdots, c_n$ are not all zero, if and only if

$$\psi_1, \psi_2, \cdots, \psi_n \quad \text{are linearly dependent.}$$

$\square$

Example 2.8 Since

$$\det \begin{bmatrix} 1 & 2 & -4 \\ 2 & 1 & 1 \\ -3 & -1 & -3 \end{bmatrix} = 0,$$

the vectors

$$\psi_1 = \begin{bmatrix} 1 \\ 2 \\ -3 \end{bmatrix}, \quad \psi_2 = \begin{bmatrix} 2 \\ 1 \\ -1 \end{bmatrix}, \quad \psi_3 = \begin{bmatrix} -4 \\ 1 \\ -3 \end{bmatrix}$$

are linearly dependent by Theorem 2.7. $\triangle$

Definition 2.9 Assume the $n \times 1$ vector functions $\phi_1, \phi_2, \cdots, \phi_k$ are defined on an interval I. We say that these k vector functions are *linearly dependent on* I provided there are constants $c_1, c_2, \cdots, c_k$, not all zero, such that

$$c_1\phi_1(t) + c_2\phi_2(t) + \cdots + c_k\phi_k(t) = 0,$$

for all $t \in I$. Otherwise we say that these k vector functions are *linearly independent on* I.

Note that the $n \times 1$ vector functions $\phi_1, \phi_2, \cdots, \phi_k$ are linearly independent on an interval I provided that the only constants $c_1, c_2, \cdots, c_k$ that satisfy the equation

$$c_1\phi_1(t) + c_2\phi_2(t) + \cdots + c_k\phi_k(t) = 0,$$

for all $t \in I$, are $c_1 = c_2 = \cdots = c_k = 0$.

Any three 2×1 constant vectors are linearly dependent, but in the following example we see that we can have three linearly independent 2×1 vector functions on an interval I.

Example 2.10 Show that the three vector functions ϕ_1, ϕ_2, ϕ_3 defined by

$$\phi_1(t) = \begin{bmatrix} t \\ t \end{bmatrix}, \quad \phi_2(t) = \begin{bmatrix} t^2 \\ t \end{bmatrix}, \quad \phi_3(t) = \begin{bmatrix} t^3 \\ t \end{bmatrix}$$

are linearly independent on any nondegenerate interval I (a nondegenerate interval is any interval containing at least two points).

To see this, assume c_1, c_2, c_3 are constants such that

$$c_1\phi_1(t) + c_2\phi_2(t) + c_3\phi_3(t) = 0,$$

for all $t \in I$. Then

$$c_1 \begin{bmatrix} t \\ t \end{bmatrix} + c_2 \begin{bmatrix} t^2 \\ t \end{bmatrix} + c_3 \begin{bmatrix} t^3 \\ t \end{bmatrix} = \begin{bmatrix} 0 \\ 0 \end{bmatrix},$$

for all $t \in I$. This implies that

$$c_1 t + c_2 t^2 + c_3 t^3 = 0, \tag{2.4}$$

for all $t \in I$. Taking three derivatives of both sides of equation (2.4), we have

$$6c_3 = 0.$$

Hence $c_3 = 0$. Letting $c_3 = 0$ in equation (2.4) and taking two derivatives of both sides of the resulting equation

$$c_1 t + c_2 t^2 = 0,$$

we get that

$$2c_2 = 0$$

and so $c_2 = 0$. It then follows that $c_1 = 0$. Hence the three vector functions ϕ_1, ϕ_2, ϕ_3 are linearly independent on I. $\triangle$

In the next theorem when we say (2.5) gives us a *general solution* of (2.3) we mean all functions in this form are solutions of (2.3) and all solutions of (2.3) can be written in this form.

Theorem 2.11 *The linear vector differential equation (2.3) has n linearly independent solutions on I, and if $\phi_1, \phi_2, \cdots, \phi_n$ are n linearly independent solutions on I, then*

$$x = c_1\phi_1 + c_2\phi_2 + \cdots + c_n\phi_n, \tag{2.5}$$

for $t \in I$, where $c_1, c_2, \cdots, c_n$ are constants, is a general solution of (2.3).

Proof Let $\psi_1, \psi_2, \cdots, \psi_n$ be n linearly independent constant $n \times 1$ vectors and let $t_0 \in I$. Then let ϕ_i be the solution of the IVP

$$x' = A(t)x, \quad x(t_0) = \psi_i,$$

for $1 \leq i \leq n$. Assume $c_1, c_2, \cdots, c_n$ are constants such that

$$c_1\phi_1(t) + c_2\phi_2(t) + \cdots + c_n\phi_n(t) = 0,$$

for all $t \in I$. Letting $t = t_0$ we have

$$c_1\phi_1(t_0) + c_2\phi_2(t_0) + \cdots + c_n\phi_n(t_0) = 0$$

or, equivalently,
$$c_1\psi_1 + c_2\psi_2 + \cdots + c_n\psi_n = 0.$$

Since $\psi_1, \psi_2, \cdots, \psi_n$ are n linearly independent constant vectors, we have that
$$c_1 = c_2 = \cdots = c_n = 0.$$

It follows that the vector functions $\phi_1, \phi_2, \cdots, \phi_n$ are n linearly independent solutions on I. Hence we have proved the existence of n linearly independent solutions.

Next assume the vector functions $\phi_1, \phi_2, \cdots, \phi_n$ are n linearly independent solutions of (2.3) on I. Since linear combinations of solutions of (2.3) are solutions of (2.3), any vector function x of the form
$$x = c_1\phi_1 + c_2\phi_2 + \cdots + c_n\phi_n$$

is a solution of (2.3). It remains to show that every solution of (2.3) is of this form. Let $t_0 \in I$ and let
$$\xi_i := \phi_i(t_0),$$

$1 \leq i \leq n$. Assume $b_1, b_2, \cdots, b_n$ are constants such that
$$b_1\xi_1 + b_2\xi_2 + \cdots + b_n\xi_n = 0.$$

Then let
$$v(t) := b_1\phi_1(t) + b_2\phi_2(t) + \cdots + b_n\phi_n(t),$$

for $t \in I$. Then v is a solution of (2.3) on I with $v(t_0) = 0$. It follows from the uniqueness theorem (Theorem 2.3) that v is the trivial solution and hence
$$b_1\phi_1(t) + b_2\phi_2(t) + \cdots + b_n\phi_n(t) = 0,$$

for $t \in I$. But $\phi_1, \phi_2, \cdots, \phi_n$ are linearly independent on I, so
$$b_1 = b_2 = \cdots = b_n = 0.$$

But this implies that the constant vectors $\xi_1 := \phi_1(t_0)$, $\xi_2 := \phi_2(t_0)$, $\cdots$, $\xi_n := \phi_n(t_0)$ are linearly independent.

Let z be an arbitrary but fixed solution of (2.3). Let $t_0 \in I$, since $z(t_0)$ is an $n \times 1$ constant vector and $\phi_1(t_0), \phi_2(t_0), \cdots, \phi_n(t_0)$ are linearly independent $n \times 1$ constant vectors, there are constants $a_1, a_2, \cdots, a_n$ such that
$$a_1\phi_1(t_0) + a_2\phi_2(t_0) + \cdots + a_n\phi_n(t_0) = z(t_0).$$

By the uniqueness theorem (Theorem 2.3) we have that
$$z(t) = a_1\phi_1(t) + a_2\phi_2(t) + \cdots + a_n\phi_n(t), \quad \text{for} \quad t \in I.$$

Hence
$$x = c_1\phi_1 + c_2\phi_2 + \cdots + c_n\phi_n$$

is a general solution of (2.3). $\square$

First we will see how to solve the vector differential equation

$$x' = Ax,$$

where A is a constant $n \times n$ matrix. We recall the definitions of eigenvalues and eigenvectors for a square matrix A.

Definition 2.12 Let A be a given $n \times n$ constant matrix and let x be a column unknown n-vector. For any number λ the vector equation

$$Ax = \lambda x \qquad (2.6)$$

has the solution $x = 0$ called the *trivial solution* of the vector equation. If λ_0 is a number such that the vector equation (2.6) with λ replaced by λ_0 has a nontrivial solution x_0, then λ_0 is called an *eigenvalue* of A and x_0 is called a corresponding *eigenvector*. We say λ_0, x_0 is an *eigenpair* of A.

Assume λ is an eigenvalue of A, then equation (2.6) has a nontrivial solution. Therefore,

$$(A - \lambda I) x = 0$$

has a nontrivial solution. From linear algebra we get that the *characteristic equation*

$$\det (A - \lambda I) = 0$$

is satisfied. If λ_0 is an eigenvalue, then to find a corresponding eigenvector we want to find a nonzero vector x so that

$$Ax = \lambda_0 x$$

or, equivalently,

$$(A - \lambda_0 I) x = 0.$$

Example 2.13 Find eigenpairs for

$$A = \begin{bmatrix} 0 & 1 \\ -2 & -3 \end{bmatrix}.$$

The characteristic equation of A is

$$\det (A - \lambda I) = \begin{vmatrix} -\lambda & 1 \\ -2 & -3 - \lambda \end{vmatrix} = 0.$$

Simplifying, we have

$$\lambda^2 + 3\lambda + 2 = (\lambda + 2)(\lambda + 1) = 0.$$

Hence the eigenvalues are $\lambda_1 = -2$, $\lambda_2 = -1$. To find an eigenvector corresponding to $\lambda_1 = -2$, we solve

$$(A - \lambda_1 I)x = (A + 2I)x = 0$$

or

$$\begin{bmatrix} 2 & 1 \\ -2 & -1 \end{bmatrix} \begin{bmatrix} x_1 \\ x_2 \end{bmatrix} = \begin{bmatrix} 0 \\ 0 \end{bmatrix}.$$

It follows that

$$-2, \begin{bmatrix} 1 \\ -2 \end{bmatrix}$$

is an eigenpair for A. Similarly, we get that

$$-1, \begin{bmatrix} 1 \\ -1 \end{bmatrix}$$

is an eigenpair for A.   △

Theorem 2.14 *If λ_0, x_0 is an eigenpair for the constant $n \times n$ matrix A, then*

$$x(t) = e^{\lambda_0 t} x_0, \quad t \in \mathbb{R},$$

defines a solution x of

$$x' = Ax \tag{2.7}$$

on $\mathbb{R}$.

Proof Let

$$x(t) = e^{\lambda_0 t} x_0,$$

then

$$\begin{aligned} x'(t) &= \lambda_0 e^{\lambda_0 t} x_0 \\ &= e^{\lambda_0 t} \lambda_0 x_0 \\ &= e^{\lambda_0 t} A x_0 \\ &= A e^{\lambda_0 t} x_0 \\ &= A x(t), \end{aligned}$$

for $t \in \mathbb{R}$.   □

Example 2.15 Solve the differential equation

$$x' = \begin{bmatrix} 0 & 1 \\ -2 & -3 \end{bmatrix} x.$$

From Example 2.13 we get that the eigenpairs for

$$A := \begin{bmatrix} 0 & 1 \\ -2 & -3 \end{bmatrix}$$

are

$$-2, \begin{bmatrix} 1 \\ -2 \end{bmatrix} \quad \text{and} \quad -1, \begin{bmatrix} 1 \\ -1 \end{bmatrix}.$$

Hence by Theorem 2.14 the vector functions ϕ_1, ϕ_2 defined by

$$\phi_1(t) = e^{-2t} \begin{bmatrix} 1 \\ -2 \end{bmatrix}, \quad \phi_2(t) = e^{-t} \begin{bmatrix} 1 \\ -1 \end{bmatrix},$$

are solutions on $\mathbb{R}$. Since the vector functions ϕ_1, ϕ_2 are linearly independent (show this) on $\mathbb{R}$, a general solution x is given by

$$x(t) = c_1 e^{-2t} \begin{bmatrix} 1 \\ -2 \end{bmatrix} + c_2 e^{-t} \begin{bmatrix} 1 \\ -1 \end{bmatrix},$$

$t \in \mathbb{R}$.   △

Theorem 2.16 *If $x = u + iv$ is a complex vector-valued solution of (2.3), where u, v are real vector-valued functions, then u, v are real vector-valued solutions of (2.3).*

Proof Assume x is as in the statement of the theorem. Then

$$x'(t) = u'(t) + iv'(t) = A(t)\left[u(t) + iv(t)\right], \quad \text{for} \quad t \in I,$$

or

$$u'(t) + iv'(t) = A(t)u(t) + iA(t)v(t), \quad \text{for} \quad t \in I.$$

Equating real and imaginary parts, we have the desired results

$$u'(t) = A(t)u(t), \quad v'(t) = A(t)v(t), \quad \text{for} \quad t \in I. \quad \square$$

Example 2.17 Solve the differential equation

$$x' = \begin{bmatrix} 3 & 1 \\ -13 & -3 \end{bmatrix} x. \tag{2.8}$$

The characteristic equation of the coefficient matrix is

$$\lambda^2 + 4 = 0$$

and the eigenvalues are

$$\lambda_1 = 2i, \quad \lambda_2 = -2i.$$

To find an eigenvector corresponding to $\lambda_1 = 2i$, we solve

$$(A - 2iI)x = 0$$

or

$$\begin{bmatrix} 3 - 2i & 1 \\ -13 & -3 - 2i \end{bmatrix} \begin{bmatrix} x_1 \\ x_2 \end{bmatrix} = \begin{bmatrix} 0 \\ 0 \end{bmatrix}.$$

Hence we want

$$(3 - 2i)x_1 + x_2 = 0.$$

It follows that

$$2i, \quad \begin{bmatrix} 1 \\ -3 + 2i \end{bmatrix}$$

is an eigenpair for A. Hence by Theorem 2.14 the vector function ϕ defined by

$$
\begin{aligned}
\phi(t) &= e^{2it} \begin{bmatrix} 1 \\ -3 + 2i \end{bmatrix} \\
&= [\cos(2t) + i\sin(2t)] \begin{bmatrix} 1 \\ -3 + 2i \end{bmatrix} \\
&= \begin{bmatrix} \cos(2t) \\ -3\cos(2t) - 2\sin(2t) \end{bmatrix} + i \begin{bmatrix} \sin(2t) \\ 2\cos(2t) - 3\sin(2t) \end{bmatrix}
\end{aligned}
$$

is a solution. Using Theorem 2.16, we get that the vector functions ϕ_1, ϕ_2 defined by

$$\phi_1(t) = \begin{bmatrix} \cos(2t) \\ -3\cos(2t) - 2\sin(2t) \end{bmatrix}, \quad \phi_2(t) = \begin{bmatrix} \sin(2t) \\ 2\cos(2t) - 3\sin(2t) \end{bmatrix}$$

are real vector-valued solutions of (2.8). Since ϕ_1, ϕ_2 are linearly independent (show this) on $\mathbb{R}$, we have by Theorem 2.11 that a general solution x of (2.8) is given by

$$x(t) = c_1 \begin{bmatrix} \cos(2t) \\ -3\cos(2t) - 2\sin(2t) \end{bmatrix} + c_2 \begin{bmatrix} \sin(2t) \\ 2\cos(2t) - 3\sin(2t) \end{bmatrix},$$

for $t \in \mathbb{R}$. △

Example 2.18 Let's solve the system in Example 2.2 involving two masses attached to springs for the special case that all the parameters are equal to one.

In this case we have

$$A = \begin{bmatrix} 0 & 1 & 0 & 0 \\ -2 & -1 & 1 & 0 \\ 0 & 0 & 0 & 1 \\ 1 & 0 & -2 & -1 \end{bmatrix}.$$

By expanding $\det(A - \lambda I)$ along the first row, we get the characteristic equation

$$\begin{aligned} 0 = \det(A - \lambda I) &= \lambda(\lambda+1)(\lambda^2 + \lambda + 2) + 2(\lambda^2 + \lambda + 2) - 1 \\ &= (\lambda^2 + \lambda + 2)^2 - 1 \\ &= (\lambda^2 + \lambda + 1)(\lambda^2 + \lambda + 3). \end{aligned}$$

Hence the eigenvalues of A are

$$\lambda = -\frac{1}{2} \pm \frac{\sqrt{3}}{2}i, \quad -\frac{1}{2} \pm \frac{\sqrt{11}}{2}i.$$

As in the Example 2.17, the eigenpairs are computed to be

$$-\frac{1}{2} \pm \frac{\sqrt{3}}{2}i, \quad \begin{bmatrix} 1 \\ -\frac{1}{2} \pm \frac{\sqrt{3}}{2}i \\ 1 \\ -\frac{1}{2} \pm \frac{\sqrt{3}}{2}i \end{bmatrix} \quad \text{and} \quad -\frac{1}{2} \pm \frac{\sqrt{11}}{2}i, \quad \begin{bmatrix} 1 \\ -\frac{1}{2} \pm \frac{\sqrt{11}}{2}i \\ -1 \\ \frac{1}{2} \mp \frac{\sqrt{11}}{2}i \end{bmatrix}$$

(use your calculator to check this). Complex vector solutions corresponding to the first two eigenpairs are

$$e^{(-\frac{1}{2} \pm \frac{\sqrt{3}}{2}i)t} \begin{bmatrix} 1 \\ -\frac{1}{2} \pm \frac{\sqrt{3}}{2}i \\ 1 \\ -\frac{1}{2} \pm \frac{\sqrt{3}}{2}i \end{bmatrix},$$

and complex vector solutions for the remaining eigenpairs are obtained in a similar way. Finally, by multiplying out the complex solutions and taking real and imaginary parts, we obtain four real solutions. The first two are

$$
e^{-\frac{t}{2}}
\begin{bmatrix}
\cos\frac{\sqrt{3}}{2}t \\
-\frac{1}{2}\cos\frac{\sqrt{3}}{2}t - \frac{\sqrt{3}}{2}\sin\frac{\sqrt{3}}{2}t \\
\cos\frac{\sqrt{3}}{2}t \\
-\frac{1}{2}\cos\frac{\sqrt{3}}{2}t - \frac{\sqrt{3}}{2}\sin\frac{\sqrt{3}}{2}t
\end{bmatrix}
, \quad
e^{-\frac{t}{2}}
\begin{bmatrix}
\sin\frac{\sqrt{3}}{2}t \\
\frac{\sqrt{3}}{2}\cos\frac{\sqrt{3}}{2}t - \frac{1}{2}\sin\frac{\sqrt{3}}{2}t \\
\sin\frac{\sqrt{3}}{2}t \\
\frac{\sqrt{3}}{2}\cos\frac{\sqrt{3}}{2}t - \frac{1}{2}\sin\frac{\sqrt{3}}{2}t
\end{bmatrix}
,
$$

and the second two are

$$
e^{-\frac{t}{2}}
\begin{bmatrix}
\cos(at) \\
-\frac{1}{2}\cos(at) - a\sin(at) \\
-\cos(at) \\
\frac{1}{2}\cos(at) + a\sin(at)
\end{bmatrix}
, \quad
e^{-\frac{t}{2}}
\begin{bmatrix}
\sin(at) \\
a\cos(at) - \frac{1}{2}\sin(at) \\
-\sin(at) \\
-a\cos(at) + \frac{1}{2}\sin(at)
\end{bmatrix}
,
$$

where $a = \frac{\sqrt{11}}{2}$. We will show later (Example 2.26) that these four solutions are linearly independent on $\mathbb{R}$, so a general linear combination of them gives a general solution of the mass-spring problem. Note that as $t \to \infty$, the masses must experience exponentially decreasing oscillations about their equilibrium positions. $\triangle$

If the matrix A has n linearly independent eigenvectors, then Theorem 2.14 can be used to generate a general solution of $x' = Ax$ (see Exercise 2.15). The following example shows that an $n \times n$ constant matrix may have fewer than n linearly independent eigenvectors.

Example 2.19 Consider the vector differential equation $x' = Ax$, where

$$
A := \begin{bmatrix} 1 & 1 \\ -1 & 3 \end{bmatrix}.
$$

The characteristic equation for A is

$$
\lambda^2 - 4\lambda + 4 = 0,
$$

so $\lambda_1 = \lambda_2 = 2$ are the eigenvalues. Corresponding to the eigenvalue 2 there is only one linearly independent eigenvector, and so we cannot use Theorem 2.14 to solve this differential equation. Later (see Example 2.36) we will use Putzer's algorithm (Theorem 2.35) to solve this differential equation. $\triangle$

We now get some results for the linear vector differential equation (2.3). We define the matrix differential equation

$$
X' = A(t)X, \tag{2.9}
$$

where

$$
X := \begin{bmatrix}
x_{11} & x_{12} & \cdots & x_{1n} \\
x_{21} & x_{22} & \cdots & x_{2n} \\
\cdots & & \cdots & \\
x_{n1} & x_{n2} & \cdots & x_{nn}
\end{bmatrix}
$$

and

$$
X' := \begin{bmatrix}
x'_{11} & x'_{12} & \cdots & x'_{1n} \\
x'_{21} & x'_{22} & \cdots & x'_{2n} \\
\cdots & & \cdots & \\
x'_{n1} & x'_{n2} & \cdots & x'_{nn}
\end{bmatrix}
$$

are $n \times n$ matrix variables and A is a given $n \times n$ continuous matrix function on an interval I, to be the matrix differential equation corresponding to the vector differential equation (2.3). We say that a matrix function Φ is a solution of (2.9) on I provided Φ is a continuously differentiable $n \times n$ matrix function on I and

$$
\Phi'(t) = A(t)\Phi(t),
$$

for $t \in I$. The following theorem gives a relationship between the vector differential equation (2.3) and the matrix differential equation (2.9).

Theorem 2.20 *Assume A is a continuous $n \times n$ matrix function on an interval I and assume that Φ defined by*

$$
\Phi(t) = [\phi_1(t), \phi_2(t), \cdots, \phi_n(t)], \quad t \in I,
$$

is the $n \times n$ matrix function with columns $\phi_1(t), \phi_2(t), \cdots, \phi_n(t)$. Then Φ is a solution of the matrix differential equation (2.9) on I iff each column ϕ_i is a solution of the vector differential equation (2.3) on I for $1 \leq i \leq n$. Furthermore, if Φ is a solution of the matrix differential equation (2.9), then

$$
x(t) = \Phi(t)c
$$

is a solution of the vector differential equation (2.3) for any constant $n \times 1$ vector c.

Proof Assume $\phi_1, \phi_2, \cdots, \phi_n$ are solutions of (2.3) on I and define the $n \times n$ matrix function Φ by

$$
\Phi(t) = [\phi_1(t), \phi_2(t), \cdots, \phi_n(t)], \quad t \in I.
$$

Then Φ is a continuously differentiable matrix function on I and

$$
\begin{aligned}
\Phi'(t) &= [\phi'_1(t), \phi'_2(t), \cdots, \phi'_n(t)] \\
&= [A(t)\phi_1(t), A(t)\phi_2(t), \cdots, A(t)\phi_n(t)] \\
&= A(t)[\phi_1(t), \phi_2(t), \cdots, \phi_n(t)] \\
&= A(t)\Phi(t),
\end{aligned}
$$

for $t \in I$. Hence Φ is a solution of the matrix differential equation (2.9) on I. We leave it to the reader to show if Φ is a solution of the matrix differential equation (2.9) on I, then its columns are solutions of the vector differential equation (2.3) on I.

Next assume that the $n \times n$ matrix function Φ is a solution of the matrix differential equation (2.9) on I and let

$$
x(t) := \Phi(t)c, \quad \text{for} \quad t \in I,
$$

where c is a constant $n \times 1$ vector. Then

$$
\begin{aligned}
x'(t) &= \Phi'(t)c \\
&= A(t)\Phi(t)c \\
&= A(t)x(t),
\end{aligned}
$$

for $t \in I$. This proves the last statement in this theorem. $\qquad\square$

Theorem 2.21 (Existence-Uniqueness Theorem) *Assume A is a continuous matrix function on an interval I. Then the IVP*

$$
X' = A(t)X, \quad X(t_0) = X_0,
$$

where $t_0 \in I$ and X_0 is an $n \times n$ constant matrix, has a unique solution X that is a solution on the whole interval I.

Proof This theorem follows from Theorem 2.3 and the fact that X is a solution of the matrix equation (2.9) iff each of its columns is a solution of the vector equation (2.3). $\qquad\square$

We will use the following definition in the next theorem.

Definition 2.22 Let

$$
A(t) = \begin{bmatrix} a_{11}(t) & a_{12}(t) & \cdots & a_{1n}(t) \\ a_{21}(t) & a_{22}(t) & \cdots & a_{2n}(t) \\ \cdots & & \cdots & \\ a_{n1}(t) & a_{n2}(t) & \cdots & a_{nn}(t) \end{bmatrix}.
$$

Then we define the *trace* of $A(t)$ by

$$
\operatorname{tr}[A(t)] = a_{11}(t) + a_{22}(t) + \cdots + a_{nn}(t).
$$

Theorem 2.23 (Liouville's Theorem) *Assume $\phi_1, \phi_2, \cdots, \phi_n$ are n solutions of the vector differential equation (2.3) on I and Φ is the matrix function with columns $\phi_1, \phi_2, \cdots, \phi_n$. Then if $t_0 \in I$,*

$$
\det \Phi(t) = e^{\int_{t_0}^{t} \operatorname{tr}[A(s)] \, ds} \det \Phi(t_0),
$$

for $t \in I$.

Proof We will just prove this theorem in the case when $n = 2$. Assume ϕ_1, ϕ_2 are $n = 2$ solutions of the vector equation (2.1) on I and $\Phi(t)$ is the matrix with columns

$$
\phi_1(t) = \begin{bmatrix} \phi_{11}(t) \\ \phi_{21}(t) \end{bmatrix}, \ \phi_2(t) = \begin{bmatrix} \phi_{12}(t) \\ \phi_{22}(t) \end{bmatrix}.
$$

Let

$$
u(t) = \det \Phi(t) = \begin{vmatrix} \phi_{11}(t) & \phi_{12}(t) \\ \phi_{21}(t) & \phi_{22}(t) \end{vmatrix},
$$

for $t \in I$. Taking derivatives we get

$$
\begin{aligned}
u'(t) &= \begin{vmatrix} \phi'_{11}(t) & \phi'_{12}(t) \\ \phi_{21}(t) & \phi_{22}(t) \end{vmatrix} + \begin{vmatrix} \phi_{11}(t) & \phi_{12}(t) \\ \phi'_{21}(t) & \phi'_{22}(t) \end{vmatrix} \\
&= \begin{vmatrix} a_{11}(t)\phi_{11}(t) + a_{12}(t)\phi_{21}(t) & a_{11}(t)\phi_{12}(t) + a_{12}(t)\phi_{22}(t) \\ \phi_{21}(t) & \phi_{22}(t) \end{vmatrix} \\
&\quad + \begin{vmatrix} \phi_{11}(t) & \phi_{12}(t) \\ a_{21}(t)\phi_{11}(t) + a_{22}(t)\phi_{21}(t) & a_{21}(t)\phi_{12}(t) + a_{22}(t)\phi_{22}(t) \end{vmatrix} \\
&= \begin{vmatrix} a_{11}(t)\phi_{11}(t) & a_{11}(t)\phi_{12}(t) \\ \phi_{21}(t) & \phi_{22}(t) \end{vmatrix} + \begin{vmatrix} \phi_{11}(t) & \phi_{12}(t) \\ a_{22}(t)\phi_{21}(t) & a_{22}(t)\phi_{22}(t) \end{vmatrix} \\
&= a_{11}(t)\begin{vmatrix} \phi_{11}(t) & \phi_{12}(t) \\ \phi_{21}(t) & \phi_{22}(t) \end{vmatrix} + a_{22}(t)\begin{vmatrix} \phi_{11}(t) & \phi_{12}(t) \\ \phi_{21}(t) & \phi_{22}(t) \end{vmatrix} \\
&= [a_{11}(t) + a_{22}(t)]\begin{vmatrix} \phi_{11}(t) & \phi_{12}(t) \\ \phi_{21}(t) & \phi_{22}(t) \end{vmatrix} \\
&= \operatorname{tr}[A(t)] \det \Phi(t) \\
&= \operatorname{tr}[A(t)]u(t).
\end{aligned}
$$

Solving the differential equation $u' = \operatorname{tr}[A(t)]u$, we get

$$
u(t) = u(t_0)e^{\int_{t_0}^{t} \operatorname{tr}[A(s)]\, ds}, \quad \text{for} \quad t \in I,
$$

or, equivalently,

$$
\det \Phi(t) = e^{\int_{t_0}^{t} \operatorname{tr}[A(s)]\, ds} \det \Phi(t_0), \quad \text{for} \quad t \in I.
$$

$\square$

Corollary 2.24 *Assume* $\phi_1, \phi_2, \cdots, \phi_n$ *are* n *solutions of the vector equation* (2.3) *on* I *and* Φ *is the matrix function with columns* $\phi_1, \phi_2, \cdots, \phi_n$. *Then either*
(a) $\det \Phi(t) = 0$, *for all* $t \in I$,
  or
(b) $\det \Phi(t) \neq 0$, *for all* $t \in I$.
Case (a) *holds iff the solutions* $\phi_1, \phi_2, \cdots, \phi_n$ *are linearly dependent on* I, *while case* (b) *holds iff the solutions* $\phi_1, \phi_2, \cdots, \phi_n$ *are linearly independent on* I.

Proof The first statement of this theorem follows immediately from Liouville's formula in Theorem 2.23. The proof of the statements concerning linear independence and linear dependence is left as an exercise (see Exercise 2.24). $\square$

It follows from Corollary 2.24 that if $\phi_1, \phi_2, \cdots, \phi_n$ are n solutions of the vector equation (2.3) on I and Φ is the matrix function with columns $\phi_1, \phi_2, \cdots, \phi_n$ and $t_0 \in I$, then $\phi_1, \phi_2, \cdots, \phi_n$ are linearly independent on I iff

$$
\det \Phi(t_0) \neq 0, \quad \text{for any } t_0 \in I.
$$

We show how to use this in the next example.

Example 2.25 Show that the vector functions ϕ_1, ϕ_2 defined by

$$\phi_1(t) = \begin{bmatrix} \cos(2t) \\ -3\cos(2t) - 2\sin(2t) \end{bmatrix}, \quad \phi_2(t) = \begin{bmatrix} \sin(2t) \\ 2\cos(2t) - 3\sin(2t) \end{bmatrix},$$

for $t \in \mathbb{R}$ are linearly independent on $\mathbb{R}$.

In Example 2.17 we saw that ϕ_1, ϕ_2 are solutions of the vector differential equation (2.8) on $\mathbb{R}$. Let Φ be the matrix function with columns ϕ_1 and ϕ_2, respectively. Then

$$\det \Phi(0) = \begin{vmatrix} 1 & 0 \\ -3 & 2 \end{vmatrix} = 2 \neq 0.$$

Hence ϕ_1, ϕ_2 are linearly independent on R. △

Example 2.26 Let's show that the four real solutions computed in Example 2.18 involving the oscillations of two masses are linearly independent on $\mathbb{R}$.

If we evaluate each solution at $t = 0$, then we obtain the following determinant:

$$\begin{vmatrix} 1 & 0 & 1 & 0 \\ -\frac{1}{2} & \frac{\sqrt{3}}{2} & -\frac{1}{2} & \frac{\sqrt{11}}{2} \\ 1 & 0 & -1 & 0 \\ -\frac{1}{2} & \frac{\sqrt{3}}{2} & \frac{1}{2} & -\frac{\sqrt{11}}{2} \end{vmatrix},$$

which has value $\sqrt{33}$, so linear independence is established. △

In the preceding examples, it was essential that the vector functions were solutions of a vector equation of the form (2.3). In the following example, two linearly independent vector functions on an interval I are shown to constitute a matrix with zero determinant at a point $t_0 \in I$.

Example 2.27 Show that the vector functions ϕ_1, ϕ_2 defined by

$$\phi_1(t) = \begin{bmatrix} t^2 \\ 1 \end{bmatrix}, \quad \phi_2(t) = \begin{bmatrix} t \cdot |t| \\ 1 \end{bmatrix},$$

for $t \in \mathbb{R}$ are linearly independent on $\mathbb{R}$.

Assume c_1, c_2 are constants such that

$$c_1\phi_1(t) + c_2\phi_2(t) = 0,$$

for $t \in \mathbb{R}$. Then

$$c_1 \begin{bmatrix} t^2 \\ 1 \end{bmatrix} + c_2 \begin{bmatrix} t \cdot |t| \\ 1 \end{bmatrix} = \begin{bmatrix} 0 \\ 0 \end{bmatrix},$$

for $t \in \mathbb{R}$. This implies that

$$c_1 t^2 + c_2 t \cdot |t| = 0,$$

for $t \in \mathbb{R}$. Letting $t = 1$ and $t = -1$, we get the two equations

$$c_1 + c_2 = 0$$
$$c_1 - c_2 = 0,$$

respectively. This implies that $c_1 = c_2 = 0$, which gives us that ϕ_1, ϕ_2 are linearly independent on $\mathbb{R}$. Notice that if $\Phi(t) = [\phi_1(t), \phi_2(t)]$, then

$$\det \Phi(0) = \begin{vmatrix} 0 & 0 \\ 1 & 1 \end{vmatrix} = 0.$$

It must follow that

$$\phi_1(t) = \begin{bmatrix} t^2 \\ 1 \end{bmatrix}, \quad \phi_2(t) = \begin{bmatrix} t \cdot |t| \\ 1 \end{bmatrix}$$

are not solutions of a two-dimensional vector differential equation of the form (2.3) on the interval $I = \mathbb{R}$.   △

Definition 2.28 An $n \times n$ matrix function Φ is said to be a *fundamental matrix* for the vector differential equation (2.3) provided Φ is a solution of the matrix equation (2.9) on I and $\det \Phi(t) \neq 0$ on I.

Theorem 2.29 *An $n \times n$ matrix function Φ is a fundamental matrix for the vector differential equation (2.3) iff the columns of Φ are n linearly independent solutions of (2.3) on I. If Φ is a fundamental matrix for the vector differential equation (2.3), then a general solution x of (2.3) is given by*

$$x(t) = \Phi(t)c, \quad t \in I,$$

where c is an arbitrary $n \times 1$ constant vector. There are infinitely many fundamental matrices for the differential equation (2.3).

Proof Assume Φ is an $n \times n$ matrix function whose columns are linearly independent solutions of (2.3) on I. Since the columns of Φ are solutions of (2.3), we have by Theorem 2.20 that Φ is a solution of the matrix equation (2.9). Since the columns of Φ are linearly independent solutions of (2.9), we have by Corollary 2.24 that $\det \Phi(t) \neq 0$ on I. Hence Φ is a fundamental matrix for the vector equation (2.3). We leave it to the reader to show that if Φ is a fundamental matrix for the vector differential equation (2.3), then the columns of Φ are linearly independent solutions of (2.3) on I. There are infinitely many fundamental matrices for (2.3) since for any nonsingular $n \times n$ matrix X_0 the solution of the IVP

$$X' = A(t)X, \quad X(t_0) = X_0,$$

is a fundamental matrix for (2.3) (a nonsingular matrix is a matrix whose determinant is different than zero).

Next assume that Φ is a fundamental matrix for (2.3). Then by Theorem 2.20

$$x(t) = \Phi(t)c,$$

for any $n \times 1$ constant vector c, is a solution of (2.3). Now let z be an arbitrary but fixed solution of (2.3). Let $t_0 \in I$ and define

$$c_0 = \Phi^{-1}(t_0)z(t_0).$$

Then z and Φc_0 are solutions of (2.3) with the same vector value at t_0. Hence by the uniqueness of solutions to IVPs (Theorem 2.3),

$$z(t) = \Phi(t)c_0.$$

Therefore,

$$x(t) = \Phi(t)c, \quad \text{for } t \in I,$$

where c is an arbitrary $n \times 1$ constant vector defines a general solution of (2.3). □

Example 2.30 Find a fundamental matrix Φ for

$$x' = \begin{bmatrix} -2 & 3 \\ 2 & 3 \end{bmatrix} x. \tag{2.10}$$

Verify that Φ is a fundamental matrix and then write down a general solution of this vector differential equation in terms of this fundamental matrix.

The characteristic equation is

$$\lambda^2 - \lambda - 12 = 0$$

and so the eigenvalues are $\lambda_1 = -3$, $\lambda_2 = 4$. Corresponding eigenvectors are

$$\begin{bmatrix} 3 \\ -1 \end{bmatrix} \quad \text{and} \quad \begin{bmatrix} 1 \\ 2 \end{bmatrix}.$$

Hence the vector functions ϕ_1, ϕ_2 defined by

$$\phi_1(t) = e^{-3t} \begin{bmatrix} 3 \\ -1 \end{bmatrix} \quad \text{and} \quad \phi_2(t) = e^{4t} \begin{bmatrix} 1 \\ 2 \end{bmatrix},$$

for $t \in \mathbb{R}$ are solutions of (2.10). It follows from Theorem 2.20 that the matrix function Φ defined by

$$\Phi(t) = [\phi_1(t), \phi_2(t)] = \begin{bmatrix} 3e^{-3t} & e^{4t} \\ -e^{-3t} & 2e^{4t} \end{bmatrix},$$

for $t \in \mathbb{R}$ is a matrix solution of the matrix equation corresponding to (2.9). Since

$$\det \Phi(t) = \begin{vmatrix} 3e^{-3t} & e^{4t} \\ -e^{-3t} & 2e^{4t} \end{vmatrix} = 7e^t \neq 0,$$

for all $t \in \mathbb{R}$, Φ is a fundamental matrix of (2.10) on $\mathbb{R}$. It follows from Theorem 2.29 that a general solution x of (2.10) is given by

$$x(t) = \Phi(t)c = \begin{bmatrix} 3e^{-3t} & e^{4t} \\ -e^{-3t} & 2e^{4t} \end{bmatrix} c,$$

for $t \in \mathbb{R}$, where c is an arbitrary 2×1 constant vector. △

Theorem 2.31 *If Φ is a fundamental matrix for (2.3), then $\Psi = \Phi C$ where C is an arbitrary $n \times n$ nonsingular constant matrix is a general fundamental matrix of (2.3).*

Proof Assume Φ is a fundamental matrix for (2.3) and set

$$\Psi = \Phi C,$$

where C is an $n \times n$ constant matrix. Then Ψ is continuously differentiable on I and

$$
\begin{aligned}
\Psi'(t) &= \Phi'(t)C \\
&= A(t)\Phi(t)C \\
&= A(t)\Psi(t).
\end{aligned}
$$

Hence $\Psi = \Phi C$ is a solution of the matrix equation (2.9). Now assume that C is also nonsingular. Since

$$
\begin{aligned}
\det[\Psi(t)] &= \det[\Phi(t)C] \\
&= \det[\Phi(t)]\det[C] \\
&\neq 0
\end{aligned}
$$

for $t \in I$, $\Psi = \Phi C$ is a fundamental matrix of (2.9). It remains to show any fundamental matrix is of the correct form. Assume Ψ is an arbitrary but fixed fundamental matrix of (2.3). Let $t_0 \in I$ and let

$$C_0 := \Phi^{-1}(t_0)\Psi(t_0).$$

Then C_0 is a nonsingular constant matrix and

$$\Psi(t_0) = \Phi(t_0)C_0$$

and so by the uniqueness theorem (Theorem 2.21)

$$\Psi(t) = \Phi(t)C_0, \quad \text{for } t \in I.$$

$\square$

2.3 The Matrix Exponential Function

In this section, we show how to compute a fundamental matrix for the linear system with constant coefficients

$$x' = Ax.$$

Specifically, we will compute the special fundamental matrix whose initial value is the identity matrix. This matrix function turns out to be an extension of the familiar exponential function from calculus. Here is the definition:

Definition 2.32 Let A be an $n \times n$ constant matrix. Then we define the matrix exponential function by e^{At} is the solution of the IVP

$$X' = AX, \quad X(0) = I,$$

where I is the $n \times n$ identity matrix.

Before we give a formula for e^{At} (see Theorem 2.35) we recall without proof the following very important result from linear algebra. In Exercise 2.29 the reader is asked to prove Theorem 2.33 for 2×2 matrices.

Theorem 2.33 (Cayley-Hamilton Theorem) *Every $n \times n$ constant matrix satisfies its characteristic equation.*

Example 2.34 Verify the Cayley-Hamilton Theorem (Theorem 2.33) directly for the matrix

$$A = \begin{bmatrix} 2 & 3 \\ 4 & 1 \end{bmatrix}.$$

The characteristic equation for A is

$$\begin{vmatrix} 2 - \lambda & 3 \\ 4 & 1 - \lambda \end{vmatrix} = \lambda^2 - 3\lambda - 10 = 0.$$

Now

$$
\begin{aligned}
A^2 - 3A - 10I &= \begin{bmatrix} 16 & 9 \\ 12 & 13 \end{bmatrix} - \begin{bmatrix} 6 & 9 \\ 12 & 3 \end{bmatrix} - \begin{bmatrix} 10 & 0 \\ 0 & 10 \end{bmatrix} \\
&= \begin{bmatrix} 0 & 0 \\ 0 & 0 \end{bmatrix},
\end{aligned}
$$

which is what we mean by A satisfies its characteristic equation.   △

Theorem 2.35 (Putzer Algorithm for Finding e^{At}) *Let $\lambda_1, \lambda_2, \cdots, \lambda_n$ be the (not necessarily distinct) eigenvalues of the matrix A. Then*

$$e^{At} = \sum_{k=0}^{n-1} p_{k+1}(t) M_k,$$

where $M_0 := I$,

$$M_k := \prod_{i=1}^{k} (A - \lambda_i I),$$

for $1 \le k \le n$ and the vector function p defined by

$$p(t) = \begin{bmatrix} p_1(t) \\ p_2(t) \\ \cdots \\ p_n(t) \end{bmatrix},$$

for $t \in \mathbb{R}$, is the solution of the IVP

$$p' = \begin{bmatrix} \lambda_1 & 0 & 0 & \cdots & 0 \\ 1 & \lambda_2 & 0 & \cdots & 0 \\ 0 & 1 & \lambda_3 & \cdots & 0 \\ \vdots & \ddots & \ddots & \ddots & \vdots \\ 0 & \cdots & 0 & 1 & \lambda_n \end{bmatrix} p, \quad p(0) = \begin{bmatrix} 1 \\ 0 \\ 0 \\ \vdots \\ 0 \end{bmatrix}.$$

Proof Let

$$\Phi(t) := \sum_{k=0}^{n-1} p_{k+1}(t) M_k, \quad \text{for } t \in \mathbb{R},$$

where p_k, $1 \leq k \leq n$ and M_k, $0 \leq k \leq n$ are as in the statement of this theorem. Then by the uniqueness theorem (Theorem 2.21) it suffices to show that Φ satisfies the IVP

$$X' = AX, \quad X(0) = I.$$

First note that

$$
\begin{aligned}
\Phi(0) &= \sum_{k=0}^{n-1} p_{k+1}(0) M_k \\
&= p_1(0) I \\
&= I.
\end{aligned}
$$

Hence Φ satisfies the correct initial condition. Note that since the vector function p defined by

$$
p(t) := \begin{bmatrix} p_1(t) \\ p_2(t) \\ \cdots \\ p_n(t) \end{bmatrix}
$$

is the solution of the IVP

$$
p' = \begin{bmatrix} \lambda_1 & 0 & 0 & \cdots & 0 \\ 1 & \lambda_2 & 0 & \cdots & 0 \\ 0 & 1 & \lambda_3 & \cdots & 0 \\ \vdots & \ddots & \ddots & \ddots & \vdots \\ 0 & \cdots & 0 & 1 & \lambda_n \end{bmatrix} p, \quad p(0) = \begin{bmatrix} 1 \\ 0 \\ 0 \\ \vdots \\ 0 \end{bmatrix},
$$

we get that

$$
\begin{aligned}
p_1'(t) &= \lambda_1 p_1(t), \\
p_i'(t) &= p_{i-1}(t) + \lambda_i p_i(t),
\end{aligned}
$$

for $t \in \mathbb{R}$, $2 \leq i \leq n$.

Now consider, for $t \in \mathbb{R}$,

$$\Phi'(t) - A\Phi(t)$$

$$= \sum_{k=0}^{n-1} p'_{k+1}(t) M_k - A \sum_{k=0}^{n-1} p_{k+1}(t) M_k$$

$$= \lambda_1 p_1(t) M_0 + \sum_{k=1}^{n-1} [\lambda_{k+1} p_{k+1}(t) + p_k(t)] M_k - \sum_{k=0}^{n-1} p_{k+1}(t) A M_k$$

$$= \lambda_1 p_1(t) M_0 + \sum_{k=1}^{n-1} [\lambda_{k+1} p_{k+1}(t) + p_k(t)] M_k$$

$$\quad - \sum_{k=0}^{n-1} p_{k+1}(t) [M_{k+1} + \lambda_{k+1} I M_k]$$

$$= \sum_{k=1}^{n-1} p_k(t) M_k - \sum_{k=0}^{n-1} p_{k+1}(t) M_{k+1}$$

$$= -p_n(t) M_n$$

$$= 0,$$

since $M_n = 0$, by the Cayley-Hamilton theorem (Theorem 2.33). $\square$

Example 2.36 Use the Putzer algorithm (Theorem 2.35) to find e^{At} when

$$A := \begin{bmatrix} 1 & 1 \\ -1 & 3 \end{bmatrix}.$$

The characteristic equation for A is

$$\lambda^2 - 4\lambda + 4 = 0,$$

so $\lambda_1 = \lambda_2 = 2$ are the eigenvalues. By the Putzer algorithm (Theorem 2.35),

$$e^{At} = \sum_{k=0}^{1} p_{k+1}(t) M_k = p_1(t) M_0 + p_2(t) M_1.$$

Now

$$M_0 = I = \begin{bmatrix} 1 & 0 \\ 0 & 1 \end{bmatrix}$$

and

$$M_1 = A - \lambda_1 I = A - 2I = \begin{bmatrix} -1 & 1 \\ -1 & 1 \end{bmatrix}.$$

Now the vector function p given by

$$p(t) := \begin{bmatrix} p_1(t) \\ p_2(t) \end{bmatrix},$$

for $t \in \mathbb{R}$ is the solution of the IVP

$$p' = \begin{bmatrix} 2 & 0 \\ 1 & 2 \end{bmatrix} p, \quad p(0) = \begin{bmatrix} 1 \\ 0 \end{bmatrix}.$$

Hence the first component p_1 of p solves the IVP

$$p_1' = 2p_1, \quad p_1(0) = 1$$

and so we get that

$$p_1(t) = e^{2t}.$$

Next the second component p_2 of p is a solution of the IVP

$$p_2' = e^{2t} + 2p_2, \quad p_2(0) = 0.$$

It follows that

$$p_2(t) = te^{2t}.$$

Hence

$$
\begin{aligned}
e^{At} &= p_1(t)M_0 + p_2(t)M_1 \\
&= e^{2t}\begin{bmatrix} 1 & 0 \\ 0 & 1 \end{bmatrix} + te^{2t}\begin{bmatrix} -1 & 1 \\ -1 & 1 \end{bmatrix} \\
&= e^{2t}\begin{bmatrix} 1-t & t \\ -t & 1+t \end{bmatrix}.
\end{aligned}
$$

It follows from Defintion 2.28 and Theorem 2.29 that a general solution of the vector differential equation

$$x' = \begin{bmatrix} 1 & 1 \\ -1 & 3 \end{bmatrix} x \tag{2.11}$$

is given by

$$
\begin{aligned}
x(t) &= e^{At}c \\
&= c_1 e^{2t}\begin{bmatrix} 1-t \\ -t \end{bmatrix} + c_2 e^{2t}\begin{bmatrix} t \\ 1+t \end{bmatrix},
\end{aligned}
$$

for $t \in \mathbb{R}$. Note that in Example 2.19 we pointed out that we can not solve the vector differential equation (2.11) using Theorem 2.14. △

Example 2.37 (Complex Eigenvalues) Use the Putzer algorithm (Theorem 2.35) to find e^{At} when

$$A := \begin{bmatrix} 1 & -1 \\ 5 & -1 \end{bmatrix}.$$

The characteristic equation for A is

$$\lambda^2 + 4 = 0,$$

so $\lambda_1 = 2i, \lambda_2 = -2i$ are the eigenvalues. By the Putzer algorithm (Theorem 2.35),

$$e^{At} = \sum_{k=0}^{1} p_{k+1}(t)M_k = p_1(t)M_0 + p_2(t)M_1.$$

Now

$$M_0 = I = \begin{bmatrix} 1 & 0 \\ 0 & 1 \end{bmatrix}$$

and

$$M_1 = A - \lambda_1 I = \begin{bmatrix} 1 - 2i & -1 \\ 5 & -1 - 2i \end{bmatrix}.$$

Now the vector function p given by

$$p(t) := \begin{bmatrix} p_1(t) \\ p_2(t) \end{bmatrix}$$

for $t \in \mathbb{R}$ must be a solution of the IVP

$$p' = \begin{bmatrix} 2i & 0 \\ 1 & -2i \end{bmatrix} p, \quad p(0) = \begin{bmatrix} 1 \\ 0 \end{bmatrix}.$$

Hence p_1 is the solution of the IVP

$$p_1' = (2i)p_1, \quad p_1(0) = 1,$$

and we get that

$$p_1(t) = e^{2it}.$$

Next p_2 is the solution of the IVP

$$p_2' = e^{2it} - (2i)p_2, \quad p_2(0) = 0.$$

It follows that

$$
\begin{aligned}
p_2(t) &= \frac{1}{4i} e^{2it} - \frac{1}{4i} e^{-2it} \\
&= \frac{1}{2} \sin(2t),
\end{aligned}
$$

for $t \in \mathbb{R}$. Hence

$$
\begin{aligned}
e^{At} &= p_1(t)M_0 + p_2(t)M_1 \\
&= e^{2it} \begin{bmatrix} 1 & 0 \\ 0 & 1 \end{bmatrix} + \frac{1}{2} \sin(2t) \begin{bmatrix} 1 - 2i & -1 \\ 5 & -1 - 2i \end{bmatrix} \\
&= [\cos(2t) + i\sin(2t)] \begin{bmatrix} 1 & 0 \\ 0 & 1 \end{bmatrix} + \frac{1}{2} \sin(2t) \begin{bmatrix} 1 - 2i & -1 \\ 5 & -1 - 2i \end{bmatrix} \\
&= \begin{bmatrix} \cos(2t) + \frac{1}{2}\sin(2t) & -\frac{1}{2}\sin(2t) \\ \frac{5}{2}\sin(2t) & \cos(2t) - \frac{1}{2}\sin(2t) \end{bmatrix},
\end{aligned}
$$

for $t \in \mathbb{R}$. △

Example 2.38 Use the Putzer algorithm (Theorem 2.35) to help you solve the vector differential equation

$$x' = \begin{bmatrix} 2 & 0 & 0 \\ 1 & 2 & 0 \\ 1 & 0 & 3 \end{bmatrix} x. \tag{2.12}$$

Let A be the coefficient matrix in (2.12). The characteristic equation for A is

$$(\lambda - 2)^2 (\lambda - 3) = 0,$$

so $\lambda_1 = \lambda_2 = 2, \lambda_3 = 3$ are the eigenvalues of A. By the Putzer algorithm (Theorem 2.35),

$$e^{At} = \sum_{k=0}^{2} p_{k+1}(t) M_k = p_1(t) M_0 + p_2(t) M_1 + p_3(t) M_2.$$

Now

$$M_0 = I = \begin{bmatrix} 1 & 0 & 0 \\ 0 & 1 & 0 \\ 0 & 0 & 1 \end{bmatrix},$$

$$M_1 = (A - \lambda_1 I) = \begin{bmatrix} 0 & 0 & 0 \\ 1 & 0 & 0 \\ 1 & 0 & 1 \end{bmatrix},$$

and

$$\begin{aligned} M_2 &= (A - \lambda_2 I)(A - \lambda_1 I) \\ &= \begin{bmatrix} 0 & 0 & 0 \\ 1 & 0 & 0 \\ 1 & 0 & 1 \end{bmatrix} \begin{bmatrix} 0 & 0 & 0 \\ 1 & 0 & 0 \\ 1 & 0 & 1 \end{bmatrix} \\ &= \begin{bmatrix} 0 & 0 & 0 \\ 0 & 0 & 0 \\ 1 & 0 & 1 \end{bmatrix}. \end{aligned}$$

Now the vector function p given by

$$p(t) := \begin{bmatrix} p_1(t) \\ p_2(t) \\ p_3(t) \end{bmatrix},$$

for $t \in \mathbb{R}$ solves the IVP

$$p' = \begin{bmatrix} 2 & 0 & 0 \\ 1 & 2 & 0 \\ 0 & 1 & 3 \end{bmatrix} p, \quad p(0) = \begin{bmatrix} 1 \\ 0 \\ 0 \end{bmatrix}.$$

Since p_1 is the solution of the IVP

$$p_1' = 2p_1, \quad p_1(0) = 1,$$

we get that

$$p_1(t) = e^{2t}.$$

Next p_2 is the solution of the IVP

$$p_2' = e^{2t} + 2p_2, \quad p_2(0) = 0.$$

It follows that

$$p_2(t) = te^{2t}.$$

Finally, p_3 is the solution of the IVP

$$p_3' = te^{2t} + 3p_3, \quad p_3(0) = 0.$$

Solving this IVP, we obtain

$$p_3(t) = -te^{2t} - e^{2t} + e^{3t}.$$

Hence

$$
\begin{aligned}
e^{At} &= p_1(t)M_0 + p_2(t)M_1 + p_3(t)M_2 \\
&= e^{2t}\begin{bmatrix} 1 & 0 & 0 \\ 0 & 1 & 0 \\ 0 & 0 & 1 \end{bmatrix} + te^{2t}\begin{bmatrix} 0 & 0 & 0 \\ 1 & 0 & 0 \\ 1 & 0 & 1 \end{bmatrix} \\
&\quad + \left(-te^{2t} - e^{2t} + e^{3t}\right)\begin{bmatrix} 0 & 0 & 0 \\ 0 & 0 & 0 \\ 1 & 0 & 1 \end{bmatrix} \\
&= \begin{bmatrix} e^{2t} & 0 & 0 \\ te^{2t} & e^{2t} & 0 \\ e^{3t} - e^{2t} & 0 & e^{3t} \end{bmatrix}.
\end{aligned}
$$

Hence a general solution x of (2.12) is given by

$$
\begin{aligned}
x(t) &= e^{At}c \\
&= \begin{bmatrix} e^{2t} & 0 & 0 \\ te^{2t} & e^{2t} & 0 \\ e^{3t} - e^{2t} & 0 & e^{3t} \end{bmatrix} c \\
&= c_1\begin{bmatrix} e^{2t} \\ te^{2t} \\ e^{3t} - e^{2t} \end{bmatrix} + c_2\begin{bmatrix} 0 \\ e^{2t} \\ 0 \end{bmatrix} + c_3\begin{bmatrix} 0 \\ 0 \\ e^{3t} \end{bmatrix},
\end{aligned}
$$

for $t \in \mathbb{R}$. △

In the following theorem we give some properties of the matrix exponential.

Theorem 2.39 *Assume A and B are $n \times n$ constant matrices. Then*

(i) $\frac{d}{dt}e^{At} = Ae^{At}$, *for $t \in \mathbb{R}$,*

(ii) $\det\left[e^{At}\right] \neq 0$, *for $t \in \mathbb{R}$ and e^{At} is a fundamental matrix for* (2.7),

(iii) $e^{At}e^{As} = e^{A(t+s)}$, *for $t, s \in \mathbb{R}$,*

(iv) $\{e^{At}\}^{-1} = e^{-At}$, *for $t \in \mathbb{R}$ and, in particular,*

$$\{e^A\}^{-1} = e^{-A},$$

(v) *if $AB = BA$, then $e^{At}B = Be^{At}$, for $t \in \mathbb{R}$ and, in particular,*

$$e^A B = Be^A,$$

(vi) *if $AB = BA$, then $e^{At}e^{Bt} = e^{(A+B)t}$, for $t \in \mathbb{R}$ and, in particular,*

$$e^A e^B = e^{(A+B)},$$

(vii) $e^{At} = I + A\frac{t}{1!} + A^2\frac{t^2}{2!} + \cdots + A^k\frac{t^k}{k!} + \cdots$, *for $t \in \mathbb{R}$,*

(viii) *if P is a nonsingular matrix, then $e^{PBP^{-1}} = Pe^B P^{-1}$.*

Proof The result (i) follows immediately from the definition of e^{At}.

Since e^{At} is the identity matrix at $t = 0$ and $\det(I) = 1 \neq 0$, we get from Corollary 2.24 that $\det(e^{At}) \neq 0$, for all $t \in \mathbb{R}$ and so (ii) holds.

We now prove that (iii) holds. Fix $s \in \mathbb{R}$, let t be a real variable and let

$$\Phi(t) := e^{At}e^{As} - e^{A(t+s)}.$$

Then

$$
\begin{aligned}
\Phi'(t) &= Ae^{At}e^{As} - Ae^{A(t+s)} \\
&= A\left[e^{At}e^{As} - e^{A(t+s)}\right] \\
&= A\Phi(t),
\end{aligned}
$$

for $t \in \mathbb{R}$. So Φ is a solution of the matrix equation $X' = AX$. Also, $\Phi(0) = e^{As} - e^{As} = 0$, so by the uniqueness theorem, Theorem 2.21, $\Phi(t) = 0$ for $t \in \mathbb{R}$. Hence

$$e^{At}e^{As} = e^{A(t+s)},$$

for $t \in \mathbb{R}$. Since $s \in \mathbb{R}$ is arbitrary, (iii) holds.

To show that (iv) holds, we get by using (iii)

$$
\begin{aligned}
e^{At}e^{-At} &= e^{At}e^{A(-t)} \\
&= e^{A(t+(-t))} \\
&= I,
\end{aligned}
$$

for $t \in \mathbb{R}$. This implies that

$$\{e^{At}\}^{-1} = e^{-At},$$

for $t \in \mathbb{R}$. Letting $t = 1$, we get that

$$\{e^{A}\}^{-1} = e^{-A},$$

and so (iv) holds.

The proof of (v) is similar to the proof of (iii) and is left as an exercise (see Exercise 2.39).

To prove (vi), assume $AB = BA$ and let

$$\Phi(t) := e^{At}e^{Bt} - e^{(A+B)t}.$$

Then using the product rule and using (iv),

$$
\begin{aligned}
\Phi'(t) &= Ae^{At}e^{Bt} + e^{At}Be^{Bt} - (A+B)e^{(A+B)t} \\
&= Ae^{At}e^{Bt} + Be^{At}e^{Bt} - (A+B)e^{(A+B)t} \\
&= (A+B)\left[e^{At}e^{Bt} - e^{(A+B)t}\right] \\
&= (A+B)\Phi(t),
\end{aligned}
$$

for $t \in \mathbb{R}$. Also, $\Phi(0) = I - I = 0$, so by the uniqueness theorem, Theorem 2.21, $\Phi(t) = 0$ for $t \in \mathbb{R}$. Hence

$$e^{(A+B)t} = e^{At}e^{Bt},$$

for $t \in \mathbb{R}$. Letting $t = 1$, we get that

$$e^{(A+B)} = e^A e^B$$

and hence (vi) holds.

We now prove (vii). It can be shown that the infinite series of matrices

$$I + A\frac{t}{1!} + A^2\frac{t^2}{2!} + \cdots + A^n\frac{t^n}{n!} + \cdots$$

converges for $t \in \mathbb{R}$ and that this infinite series of matrices can be differentiated term by term. Let

$$\Phi(t) := I + A\frac{t}{1!} + A^2\frac{t^2}{2!} + \cdots + A^n\frac{t^n}{n!} + \cdots ,$$

for $t \in \mathbb{R}$. Then

$$
\begin{aligned}
\Phi'(t) &= A + A^2\frac{t}{1!} + A^3\frac{t^2}{2!} + \cdots \\
&= A\left[I + A\frac{t}{1!} + A^2\frac{t^2}{2!} + \cdots\right] \\
&= A\Phi(t),
\end{aligned}
$$

for $t \in \mathbb{R}$. Since $\Phi(0) = I$, we have by the uniqueness theorem, Theorem 2.21, $\Phi(t) = e^{At}$, for $t \in \mathbb{R}$. Hence

$$e^{At} = I + A\frac{t}{1!} + A^2\frac{t^2}{2!} + \cdots ,$$

for $t \in \mathbb{R}$ and so (vii) holds.

Finally, (viii) follows from (vii). $\square$

Theorem 2.40 (Variation of Constants Formula) *Assume that A is an $n \times n$ continuous matrix function on an interval I, b is a continuous $n \times 1$ vector function on I, and Φ is a fundamental matrix for (2.3). Then the solution of the IVP*

$$x' = A(t)x + b(t), \quad x(t_0) = x_0,$$

where $t_0 \in I$ and $x_0 \in \mathbb{R}^n$, is given by

$$x(t) = \Phi(t)\Phi^{-1}(t_0)x_0 + \Phi(t)\int_{t_0}^{t} \Phi^{-1}(s)b(s)\, ds,$$

for $t \in I$.

Proof The uniqueness of the solution of the given IVP follows from Theorem 2.3. Let Φ be a fundamental matrix for (2.3) and set

$$x(t) = \Phi(t)\Phi^{-1}(t_0)x_0 + \Phi(t)\int_{t_0}^{t} \Phi^{-1}(s)b(s)\, ds,$$

for $t \in I$. Then

$$
\begin{aligned}
x'(t) &= \Phi'(t)\Phi^{-1}(t_0)x_0 + \Phi'(t)\int_{t_0}^{t} \Phi^{-1}(s)b(s)\, ds + \Phi(t)\Phi^{-1}(t)b(t) \\
&= A(t)\Phi(t)\Phi^{-1}(t_0)x_0 + A(t)\Phi(t)\int_{t_0}^{t} \Phi^{-1}(s)b(s)\, ds + b(t) \\
&= A(t)\left[\Phi(t)\Phi^{-1}(t_0)x_0 + \Phi(t)\int_{t_0}^{t} \Phi^{-1}(s)b(s)\, ds\right] + b(t) \\
&= A(t)x(t) + b(t),
\end{aligned}
$$

for $t \in I$. Also,

$$
\begin{aligned}
x(t_0) &= \Phi(t_0)\Phi^{-1}(t_0)x_0 \\
&= x_0.
\end{aligned}
$$

$\square$

Corollary 2.41 *Assume A is an $n \times n$ constant matrix and b is a continuous $n \times 1$ vector function on an interval I. Then the solution x of the IVP*

$$
x' = Ax + b(t), \quad x(t_0) = x_0,
$$

where $t_0 \in I, x_0 \in \mathbb{R}^n$ is given by (the reader should compare this to the variation of constants formula in Theorem 1.6)

$$
x(t) = e^{A(t-t_0)}x_0 + \int_{t_0}^{t} e^{A(t-s)}b(s)\, ds,
$$

for $t \in I$.

Proof Letting $\Phi(t) = e^{At}$ in the general variation of constants formula in Theorem 2.40, we get, using the fact that $\{e^{At}\}^{-1} = e^{-At}$,

$$
\begin{aligned}
x(t) &= \Phi(t)\Phi^{-1}(t_0)x_0 + \Phi(t)\int_{t_0}^{t} \Phi^{-1}(s)b(s)\, ds \\
&= e^{At}e^{-At_0}x_0 + e^{At}\int_{t_0}^{t} e^{-As}b(s)\, ds \\
&= e^{A(t-t_0)}x_0 + \int_{t_0}^{t} e^{A(t-s)}b(s)\, ds,
\end{aligned}
$$

for $t \in I$. $\square$

In the next theorem we see under a strong assumption (2.13) we can find a fundamental matrix for the nonautonomous case $x' = A(t)x$.

Theorem 2.42 *Assume $A(t)$ is a continuous $n \times n$ matrix function on an interval I. If*

$$
A(t)A(s) = A(s)A(t) \tag{2.13}
$$

for all $t, s \in I$, *then*

$$\Phi(t) := e^{\int_{t_0}^t A(s) \, ds}$$

(defined by it's power series (2.14)) is a fundamental matrix for $x' = A(t)x$ *on* I.

Proof Let

$$\Phi(t) := e^{\int_{t_0}^t A(s) \, ds} := \sum_{k=0}^{\infty} \frac{1}{k!} \left[\int_{t_0}^t A(s) ds \right]^k. \tag{2.14}$$

We leave it to the reader to show that the infinite series of matrices in (2.14) converges uniformly on each closed subinterval of I. Differentiating term by term and using the fact that (2.13) implies that $A(t) \int_{t_0}^t A(s) \, ds = \int_{t_0}^t A(s) \, ds \, A(t)$ we get

$$
\begin{aligned}
\Phi'(t) &= A(t) \sum_{k=1}^{\infty} \frac{1}{(k-1)!} \left[\int_{t_0}^t A(s) ds \right]^{k-1} \\
&= A(t) \sum_{k=0}^{\infty} \frac{1}{k!} \left[\int_{t_0}^t A(s) ds \right]^k \\
&= A(t)\Phi(t).
\end{aligned}
$$

Since

$$\det \Phi(t_0) = \det I = 1 \neq 0,$$

we have that $\Phi(t)$ is a fundamental matrix for the vector differential equation $x' = A(t)x$. $\square$

Note that condition (2.13) holds when either $A(t)$ is a diagonal matrix or $A(t) \equiv A$, a constant matrix.

Example 2.43 Use the variation of constants formula to solve the IVP

$$x' = \begin{bmatrix} 1 & 1 \\ -1 & 3 \end{bmatrix} x + \begin{bmatrix} e^{2t} \\ 2e^{2t} \end{bmatrix}, \quad x(0) = \begin{bmatrix} 2 \\ 1 \end{bmatrix}.$$

Since

$$A := \begin{bmatrix} 1 & 1 \\ -1 & 3 \end{bmatrix},$$

we have from Example 2.36 that

$$e^{At} = e^{2t} \begin{bmatrix} 1-t & t \\ -t & 1+t \end{bmatrix}.$$

From the variation of constants formula given in Corollary 2.41,

$$
\begin{aligned}
x(t) &= e^{2t}\begin{bmatrix} 1-t & t \\ -t & 1+t \end{bmatrix}\begin{bmatrix} 2 \\ 1 \end{bmatrix} \\
&+ \int_0^t e^{2(t-s)}\begin{bmatrix} 1-t+s & t-s \\ -t+s & 1+t-s \end{bmatrix}\begin{bmatrix} e^{2s} \\ 2e^{2s} \end{bmatrix} ds \\
&= e^{2t}\begin{bmatrix} 2-2t+t \\ -2t+1+t \end{bmatrix} + e^{2t}\int_0^t \begin{bmatrix} 1-t+s+2t-2s \\ -t+s+2+2t-2s \end{bmatrix} ds \\
&= e^{2t}\begin{bmatrix} 2-t \\ 1-t \end{bmatrix} + e^{2t}\int_0^t \begin{bmatrix} 1+t-s \\ 2+t-s \end{bmatrix} ds \\
&= e^{2t}\begin{bmatrix} 2-t \\ 1-t \end{bmatrix} + e^{2t}\begin{bmatrix} t+\frac{t^2}{2} \\ 2t+\frac{t^2}{2} \end{bmatrix} \\
&= e^{2t}\begin{bmatrix} 2+\frac{t^2}{2} \\ 1+t+\frac{t^2}{2} \end{bmatrix},
\end{aligned}
$$

for $t \in \mathbb{R}$.   △

Example 2.44 Use the variation of constants formula to solve the IVP

$$
x' = \begin{bmatrix} 1 & 2 & 0 \\ 0 & 1 & 2 \\ 0 & 0 & 1 \end{bmatrix} x + \begin{bmatrix} 2te^t \\ 0 \\ 0 \end{bmatrix}, \quad x(0) = \begin{bmatrix} 2 \\ 1 \\ 0 \end{bmatrix}.
$$

Let

$$
A := \begin{bmatrix} 1 & 2 & 0 \\ 0 & 1 & 2 \\ 0 & 0 & 1 \end{bmatrix}.
$$

We will find e^{At} by an alternate method. Note that

$$
e^{At} = e^{(B+C)t},
$$

where

$$
B = \begin{bmatrix} 1 & 0 & 0 \\ 0 & 1 & 0 \\ 0 & 0 & 1 \end{bmatrix}, \quad C = \begin{bmatrix} 0 & 2 & 0 \\ 0 & 0 & 2 \\ 0 & 0 & 0 \end{bmatrix}.
$$

Since $BC = CB$,

$$
e^{At} = e^{(B+C)t} = e^{Bt}e^{Ct}.
$$

It follows from Exercise 2.41 that

$$
e^{Bt} = \begin{bmatrix} e^t & 0 & 0 \\ 0 & e^t & 0 \\ 0 & 0 & e^t \end{bmatrix}.
$$

Also,

$$
\begin{aligned}
e^{Ct} &= I + C\frac{t}{1!} + C^2\frac{t^2}{2!} + C^3\frac{t^3}{3!} + \cdots \\
&= \begin{bmatrix} 1 & 0 & 0 \\ 0 & 1 & 0 \\ 0 & 0 & 1 \end{bmatrix} + \begin{bmatrix} 0 & 2 & 0 \\ 0 & 0 & 2 \\ 0 & 0 & 0 \end{bmatrix} t + \begin{bmatrix} 0 & 0 & 4 \\ 0 & 0 & 0 \\ 0 & 0 & 0 \end{bmatrix} \frac{t^2}{2!} \\
&= \begin{bmatrix} 1 & 2t & 2t^2 \\ 0 & 1 & 2t \\ 0 & 0 & 1 \end{bmatrix}.
\end{aligned}
$$

Hence

$$
\begin{aligned}
e^{At} &= e^{Bt}e^{Ct} \\
&= \begin{bmatrix} e^t & 0 & 0 \\ 0 & e^t & 0 \\ 0 & 0 & e^t \end{bmatrix} \begin{bmatrix} 1 & 2t & 2t^2 \\ 0 & 1 & 2t \\ 0 & 0 & 1 \end{bmatrix} \\
&= \begin{bmatrix} e^t & 2te^t & 2t^2e^t \\ 0 & e^t & 2te^t \\ 0 & 0 & e^t \end{bmatrix}.
\end{aligned}
$$

From the variation of constants formula given in Corollary 2.41

$$
\begin{aligned}
x(t) &= \begin{bmatrix} e^t & 2te^t & 2t^2e^t \\ 0 & e^t & 2te^t \\ 0 & 0 & e^t \end{bmatrix} \begin{bmatrix} 2 \\ 1 \\ 0 \end{bmatrix} \\
&\quad + \int_0^t \begin{bmatrix} e^{t-s} & 2(t-s)e^{t-s} & 2(t-s)^2e^{t-s} \\ 0 & e^{t-s} & 2(t-s)e^{t-s} \\ 0 & 0 & e^{t-s} \end{bmatrix} \begin{bmatrix} 2se^s \\ 0 \\ 0 \end{bmatrix} ds \\
&= \begin{bmatrix} 2e^t + 2te^t \\ e^t \\ 0 \end{bmatrix} + e^t \int_0^t \begin{bmatrix} 2s \\ 0 \\ 0 \end{bmatrix} ds \\
&= \begin{bmatrix} 2e^t + 2te^t \\ e^t \\ 0 \end{bmatrix} + \begin{bmatrix} t^2e^t \\ 0 \\ 0 \end{bmatrix} \\
&= \begin{bmatrix} 2e^t + 2te^t + t^2e^t \\ e^t \\ 0 \end{bmatrix},
\end{aligned}
$$

for $t \in \mathbb{R}$. △

Example 2.45 Use Theorem 2.40 and Theorem 2.42 to help you solve the IVP

$$
x' = \begin{bmatrix} \frac{1}{t} & 0 \\ 0 & \frac{2}{t} \end{bmatrix} x + \begin{bmatrix} t^2 \\ t \end{bmatrix}, \quad x(1) = \begin{bmatrix} 1 \\ -2 \end{bmatrix}.
$$

on the interval $I = (0, \infty)$. By Theorem 2.42,

$$\Phi(t) = e^{\int_1^t A(s)\,ds} = e^{\int_1^t \begin{bmatrix} \frac{1}{s} & 0 \\ 0 & \frac{2}{s} \end{bmatrix} ds}$$

$$= e^{\begin{bmatrix} \ln t & 0 \\ 0 & 2\ln t \end{bmatrix}}$$

$$= \begin{bmatrix} t & 0 \\ 0 & t^2 \end{bmatrix}$$

is a fundamental matrix for

$$x' = \begin{bmatrix} \frac{1}{t} & 0 \\ 0 & \frac{2}{t} \end{bmatrix} x.$$

Using the variation of constants formula in Theorem 2.40 we get the solution $x(t)$ of the given IVP is given by

$$
\begin{aligned}
x(t) &= \Phi(t)\Phi^{-1}(1)x_0 + \Phi(t)\int_1^t \Phi^{-1}(s)b(s)ds \\
&= \begin{bmatrix} t & 0 \\ 0 & t^2 \end{bmatrix} I \begin{bmatrix} 1 \\ -2 \end{bmatrix} + \begin{bmatrix} t & 0 \\ 0 & t^2 \end{bmatrix} \int_1^t \begin{bmatrix} \frac{1}{s} & 0 \\ 0 & \frac{1}{s^2} \end{bmatrix} \begin{bmatrix} s^2 \\ s \end{bmatrix} ds \\
&= \begin{bmatrix} t \\ -2t^2 \end{bmatrix} + \begin{bmatrix} t & 0 \\ 0 & t^2 \end{bmatrix} \int_1^t \begin{bmatrix} s \\ \frac{1}{s} \end{bmatrix} ds \\
&= \begin{bmatrix} t \\ -2t^2 \end{bmatrix} + \begin{bmatrix} \frac{1}{2}t^3 - \frac{1}{2}t \\ t^2 \ln t \end{bmatrix} \\
&= \begin{bmatrix} \frac{1}{2}t^3 + \frac{1}{2}t \\ -2t^2 + t^2 \ln t \end{bmatrix}.
\end{aligned}
$$

$\triangle$

Definition 2.46 Let $\mathbb{R}^n$ denote the set of all $n \times 1$ constant vectors. Then a *norm* on $\mathbb{R}^n$ is a function $\|\cdot\| : \mathbb{R}^n \to \mathbb{R}$ having the following properties:

(i) $\|x\| \geq 0$, for all $x \in \mathbb{R}^n$,

(ii) $\|x\| = 0$ iff $x = 0$,

(iii) $\|cx\| = |c| \cdot \|x\|$ for all $c \in \mathbb{R}$, $x \in \mathbb{R}^n$,

(iv) (triangle inequality) $\|x + y\| \leq \|x\| + \|y\|$ for all $x, y \in \mathbb{R}^n$.

Example 2.47 Three important examples of norms on $\mathbb{R}^n$ are

(i) the *Euclidean norm* (l_2 norm) defined by

$$\|x\|_2 := \sqrt{x_1^2 + x_2^2 + \cdots + x_n^2},$$

(ii) the *maximum norm* (l_∞ norm) defined by

$$\|x\|_\infty := \max\{|x_i| : 1 \leq i \leq n\},$$

(iii) the *traffic norm* (l_1 norm) defined by

$$\|x\|_1 := |x_1| + |x_2| + \cdots + |x_n|.$$

We leave it to the reader to check that these examples are actually norms.

$$\triangle$$

A sequence $\{x^k\}_{n=1}^{\infty}$ in $\mathbb{R}^n$ is said to converge with respect to a norm $\|\cdot\|$ on $\mathbb{R}^n$ if there is a $x_0 \in \mathbb{R}^n$ such that

$$\lim_{k\to\infty} \|x^k - x_0\| = 0.$$

It can be shown that a sequence in $\mathbb{R}^n$ converges with respect to one norm on $\mathbb{R}^n$ iff it converges with respect to any norm on $\mathbb{R}^n$ (think about this for the three norms just listed). Unless otherwise stated we will let $\|\cdot\|$ represent any norm on $\mathbb{R}^n$.

Definition 2.48 Let $\phi(t, x_0) = e^{At}x_0$ denote the solution of the IVP

$$x' = Ax, \quad x(0) = x_0.$$

(i) We say that the trivial solution of (2.7) is *stable on* $[0, \infty)$ provided given any $\epsilon > 0$ there is a $\delta > 0$ such that if $\|y_0\| < \delta$, then

$$\|\phi(t, y_0)\| < \epsilon,$$

for $t \geq 0$.

(ii) We say that the trivial solution of (2.7) is *unstable on* $[0, \infty)$ provided it is not stable on $[0, \infty)$.

(iii) We say that the trivial solution of (2.7) is *globally asymptotically stable on* $[0, \infty)$ provided it is stable on $[0, \infty)$ and for any $y_0 \in \mathbb{R}^n$,

$$\lim_{t\to\infty} \phi(t, y_0) = 0.$$

The next theorem shows how the eigenvalues of A determine the stability of the trivial solution of $x' = Ax$.

Theorem 2.49 (Stability Theorem) *Assume A is an $n \times n$ constant matrix.*

(i) *If A has an eigenvalue with positive real part, then the trivial solution is unstable on $[0, \infty)$.*

(ii) *If all the eigenvalues of A with zero real parts are simple (multiplicity one) and all other eigenvalues of A have negative real parts, then the trivial solution is stable on $[0, \infty)$.*

(iii) *If all the eigenvalues of A have negative real parts, then the trivial solution of $x' = Ax$ is globally asymptotically stable on $[0, \infty)$.*

Proof We will prove only part *(iii)* of this theorem. The proof of part *(i)* is Exercise 2.46. A proof of part *(ii)* can be based on the Putzer algorithm (Theorem 2.35) and is similar to the proof of part *(iii)* of this theorem.

We now prove part *(iii)* of this theorem. By part *(ii)* the trivial solution is stable on $[0, \infty)$, so it remains to show that every solution approaches the zero vector as $t \to \infty$. Let $\lambda_1, \cdots, \lambda_n$ be the eigenvalues of A, and choose

$\delta > 0$ so that $\Re(\lambda_k) \leq -\delta < 0$ for $k = 1, 2, \cdots, n$. Let $x_0 \in \mathbb{R}^n$, then by Putzer's algorithm (Theorem 2.35)

$$\phi(t, x_0) = e^{At}x_0 = \sum_{k=0}^{n-1} p_{k+1}(t) M_k x_0.$$

Since p_1 solves the IVP

$$p_1' = \lambda_1 p_1, \quad p_1(0) = 1,$$

we get

$$|p_1(t)| = |e^{\lambda_1 t}| \leq e^{-\delta t},$$

for all $t \geq 0$. Next, p_2 satisfies

$$p_2' = \lambda_2 p_2 + p_1, \quad p_2(0) = 0,$$

so by the variation of constants formula (Theorem 1.6 or Corollary 2.41),

$$p_2(t) = \int_0^t e^{\lambda_2(t-s)} e^{\lambda_1 s} \, ds.$$

Since $|e^{\lambda_1 s}| \leq e^{-\delta s}$ and $|e^{\lambda_2(t-s)}| \leq e^{-\delta(t-s)}$ for $t \geq s$,

$$|p_2(t)| \leq \int_0^t e^{-\delta(t-s)} e^{-\delta s} \, ds = t e^{-\delta t}.$$

We can continue in this way (by induction) to show

$$|p_k(t)| \leq \frac{t^{k-1}}{(k-1)!} e^{-\delta t},$$

for $k = 1, 2, \cdots, n$. It follows that each $p_k(t) \to 0$, as $t \to \infty$, and consequently that

$$\phi(t, x_0) = \sum_{k=0}^{n-1} p_{k+1}(t) M_k x_0 \to 0,$$

as $t \to \infty$. $\qquad\qquad\qquad\qquad\qquad\qquad\qquad\qquad\qquad\qquad\square$

Example 2.50 In the example involving the vibration of two coupled masses (see Example 2.18), we showed that the eigenvalues were $\lambda = -\frac{1}{2} \pm \frac{\sqrt{3}}{2}i$ and $\lambda = -\frac{1}{2} \pm \frac{\sqrt{11}}{2}i$. Since all four eigenvalues have negative real parts, the origin is globally asymptotically stable. $\qquad\qquad\qquad\qquad\triangle$

Example 2.51 Determine the stability of the trivial solution of

$$x' = \begin{bmatrix} 0 & -1 \\ 1 & 0 \end{bmatrix} x,$$

on $[0, \infty)$.

The characteristic equation is

$$\lambda^2 + 1 = 0,$$

and hence the eigenvalues are $\lambda_1 = i$, $\lambda_2 = -i$. Since both eigenvalues have zero real parts and both eigenvalues are simple, the trivial solution is stable on $[0, \infty)$. $\qquad\qquad\qquad\qquad\triangle$

Example 2.52 Determine the stability of the trivial solution of

$$x' = \begin{bmatrix} -2 & 1 & 0 \\ -1 & -2 & 0 \\ 0 & 0 & 1 \end{bmatrix} x,$$

on $[0, \infty)$.

The characteristic equation is

$$(\lambda^2 + 4\lambda + 5)(\lambda - 1) = 0$$

and hence the eigenvalues are $\lambda_1 = -2 + i$, $\lambda_2 = -2 - i$, and $\lambda_3 = 1$. Since the eigenvalue λ_3 has a positive real part, the trivial solution is unstable on $[0, \infty)$.   △

2.4 Induced Matrix Norm

Definition 2.53 Assume that $\| \cdot \|$ is a norm on $\mathbb{R}^n$. Let M_n denote the set of all $n \times n$ real matrices. We define the matrix norm on M_n induced by the vector norm by

$$\|A\| := \sup_{\|x\|=1} \|Ax\|.$$

Note that we use the same notation for the vector norm and the corresponding matrix norm, since from context it should be clear which norm we mean. To see that $\|A\|$ is well defined, assume there is a sequence of points $\{x_k\}$ in $\mathbb{R}^n$ with $\|x_k\| = 1$ such that

$$\lim_{k \to \infty} \|Ax_k\| = \infty.$$

Since $\|x_k\| = 1$, $k = 1, 2, 3, \cdots$, there is a convergent subsequence $\{x_{k_j}\}$. Let

$$x_0 := \lim_{j \to \infty} x_{k_j}.$$

But then

$$\lim_{j \to \infty} \|Ax_{k_j}\| = \|Ax_0\|,$$

which gives us a contradiction. Using a similar argument (see Exercise 2.51), it is easy to prove that

$$\|A\| := \max_{\|x\|=1} \|Ax\|.$$

This induced matrix norm is a norm on M_n (see Exercise 2.50).

Theorem 2.54 *The matrix norm induced by the vector norm* $\| \cdot \|$ *is given by*

$$\|A\| = \max_{x \neq 0} \frac{\|Ax\|}{\|x\|}.$$

In particular,

$$\|Ax\| \leq \|A\| \cdot \|x\|,$$

for all $x \in \mathbb{R}^n$.

Proof This result follows from the following statement. If $x \neq 0$, then

$$\frac{\|Ax\|}{\|x\|} = \left\|A\left(\frac{x}{\|x\|}\right)\right\| = \|Ay\|,$$

where $y = \frac{x}{\|x\|}$ is a unit vector. □

Theorem 2.55 *The matrix norm induced by the traffic norm (l_1 norm) is given by*

$$\|A\|_1 = \max_{1 \leq j \leq n} \sum_{i=1}^{n} |a_{ij}|.$$

Proof Let $\| \cdot \|_1$ be the traffic norm on $\mathbb{R}^n$, let $A \in M_n$, $x \in \mathbb{R}^n$, and consider

$$
\begin{aligned}
\|Ax\|_1 &= \left\| \begin{pmatrix} \sum_{j=1}^{n} a_{1j}x_j \\ \cdots \\ \sum_{j=1}^{n} a_{nj}x_j \end{pmatrix} \right\|_1 \\
&= \left| \sum_{j=1}^{n} a_{1j}x_j \right| + \cdots + \left| \sum_{j=1}^{n} a_{nj}x_j \right| \\
&\leq \sum_{j=1}^{n} |a_{1j}||x_j| + \cdots + \sum_{j=1}^{n} |a_{nj}||x_j| \\
&= \sum_{i=1}^{n} |a_{i1}||x_1| + \cdots + \sum_{i=1}^{n} |a_{in}||x_n| \\
&\leq \max_{1 \leq j \leq n} \sum_{i=1}^{n} |a_{ij}| \sum_{j=1}^{n} |x_j| \\
&= \max_{1 \leq j \leq n} \sum_{i=1}^{n} |a_{ij}| \|x\|_1.
\end{aligned}
$$

It follows that

$$\|A\|_1 \leq \max_{1 \leq j \leq n} \sum_{i=1}^{n} |a_{ij}|.$$

We next prove the reverse inequality. To see this, pick j_0, $1 \leq j_0 \leq n$, so that

$$\sum_{i=1}^{n} |a_{ij_0}| = \max_{1 \leq j \leq n} \sum_{i=1}^{n} |a_{ij}|.$$

Let e_{j_0} be the unit vector in $\mathbb{R}^n$ in the j_0 direction. Then

$$
\begin{aligned}
\|A\|_1 &= \max_{\|x\|=1} \|Ax\|_1 \\
&\geq \|Ae_{j_0}\|_1 \\
&= \left\| \begin{pmatrix} a_{1j_0} \\ \cdots \\ a_{nj_0} \end{pmatrix} \right\|_1 \\
&= \sum_{i=1}^{n} |a_{ij_0}| \\
&= \max_{1 \leq j \leq n} \sum_{i=1}^{n} |a_{ij}|.
\end{aligned}
$$

It follows that

$$
\|A\|_1 = \max_{1 \leq j \leq n} \sum_{i=1}^{n} |a_{ij}|.
$$

$\square$

The proof of the next theorem is Exercise 2.53.

Theorem 2.56 *The matrix norm induced by the maximum vector norm is given by*

$$
\|A\|_\infty = \max_{1 \leq i \leq n} \sum_{j=1}^{n} |a_{ij}|.
$$

One could also prove the following result.

Theorem 2.57 *The matrix norm induced by the Euclidean vector norm is given by*

$$
\|A\|_2 = \sqrt{\lambda_0},
$$

where λ_0 is the largest eigenvaue of $A^T A$.

Note that if I is the identity matrix, then

$$
\|I\| = \max_{\|x\|=1} \|Ix\| = \max_{\|x\|=1} \|x\| = 1.
$$

This fact will be used frequently throughout the remainder of this section.

We now define the Lozinski measure of a matrix A and we will see that this measure can sometimes be used in determining the global asymptotic stability of the vector equation $x' = Ax$.

Definition 2.58 Assume $\|\cdot\|$ is a matrix norm on M_n induced by a vector norm $\|\cdot\|$ on $\mathbb{R}^n$. Then we define the Lozinski measure $\mu : M_n \to \mathbb{R}$ by

$$
\mu(A) = \lim_{h \to 0+} \frac{\|I + hA\| - 1}{h}.
$$

The limit in the preceding definition exists by Exercise 2.49.

The Lozinski measure is not a norm, but satisfies the properties given in the following theorem.

Theorem 2.59 (Properties of μ) *The function μ satisfies the following properties:*

(i) $\mu(\alpha A) = \alpha \mu(A)$, *for* $A \in M_n$, $\alpha \geq 0$,
(ii) $|\mu(A)| \leq \|A\|$ *for* $A \in M_n$,
(iii) $\mu(A + B) \leq \mu(A) + \mu(B)$ *for* $A, B \in M_n$,
(iv) $|\mu(A) - \mu(B)| \leq \|A - B\|$ *for* $A, B \in M_n$,
(v) $Re(\lambda) \leq \mu(A)$ *for all eigenvalues λ of A.*

Proof Since

$$\mu(0A) = \mu(0) = \lim_{h \to 0+} \frac{\|I + h0\| - 1}{h} = 0 = \|0\|,$$

part (i) is true for $\alpha = 0$. Now assume that $\alpha > 0$ and consider

$$\begin{aligned}
\mu(\alpha A) &= \lim_{h \to 0+} \frac{\|I + h\alpha A\| - 1}{h} \\
&= \lim_{k \to 0+} \frac{\|I + kA\| - 1}{\frac{k}{\alpha}} \\
&= \alpha \lim_{k \to 0+} \frac{\|I + kA\| - 1}{k} \\
&= \alpha \mu(A)
\end{aligned}$$

and hence (i) is true. Part (ii) follows from the inequalities

$$|(\|I + hA\| - 1)| \leq \|I\| + h\|A\| - 1 = h\|A\|$$

and the definition of $\mu(A)$. To see that (iii) is true, consider

$$\begin{aligned}
\mu(A + B) &= \lim_{h \to 0+} \frac{\|I + h(A + B)\| - 1}{h} \\
&= \lim_{k \to 0+} \frac{\|I + \frac{k}{2}(A + B)\| - 1}{\frac{k}{2}} \\
&= \lim_{k \to 0+} \frac{\|2I + k(A + B)\| - 2}{k} \\
&\leq \lim_{k \to 0+} \frac{\|I + kA\| + \|I + kB\| - 2}{k} \\
&= \lim_{k \to 0+} \frac{\|I + kA\| - 1}{k} + \lim_{k \to 0+} \frac{\|I + kB\| - 1}{k} \\
&= \mu(A) + \mu(B).
\end{aligned}$$

To prove (iv), note that

$$\begin{aligned}
\mu(A) &= \mu(A - B + B) \\
&\leq \mu(A - B) + \mu(B).
\end{aligned}$$

Hence
$$\mu(A) - \mu(B) \le \|A - B\|.$$
Interchanging A and B, we have that
$$\mu(B) - \mu(A) \le \|B - A\| = \|A - B\|.$$
Altogether we get the desired result
$$|\mu(A) - \mu(B)| \le \|A - B\|.$$
Finally, to prove (v), let λ_0 be an eigenvalue of A. Let x_0 be a corresponding eigenvector with $\|x_0\| = 1$. Consider

$$
\begin{aligned}
\lim_{h \to 0+} \frac{\|(I + hA)x_0\| - 1}{h} &= \lim_{h \to 0+} \frac{\|(1 + h\lambda_0)x_0\| - 1}{h} \\
&= \lim_{h \to 0+} \frac{|1 + h\lambda_0| \cdot \|x_0\| - 1}{h} \\
&= \lim_{h \to 0+} \frac{|1 + h\lambda_0| - 1}{h} \\
&= \lim_{h \to 0+} \left(\frac{|1 + h\lambda_0| - 1}{h} \cdot \frac{|1 + h\lambda_0| + 1}{|1 + h\lambda_0| + 1} \right) \\
&= \lim_{h \to 0+} \frac{|1 + h\lambda_0|^2 - 1}{h[|1 + h\lambda_0| + 1]} \\
&= \lim_{h \to 0+} \frac{(1 + h\lambda_0)(1 + h\bar{\lambda}_0) - 1}{h[|1 + h\lambda_0| + 1]} \\
&= \lim_{h \to 0+} \frac{(\lambda_0 + \bar{\lambda}_0)h + h^2 \lambda_0 \bar{\lambda}_0}{h[|1 + h\lambda_0| + 1]} \\
&= \frac{\lambda_0 + \bar{\lambda}_0}{2} \\
&= Re(\lambda_0).
\end{aligned}
$$

On the other hand,
$$
\begin{aligned}
\lim_{h \to 0+} \frac{\|(I + hA)x_0\| - 1}{h} &\le \lim_{h \to 0+} \frac{\|(I + hA)\| \cdot \|x_0\| - 1}{h} \\
&= \lim_{h \to 0+} \frac{\|I + hA\| - 1}{h} \\
&= \mu(A).
\end{aligned}
$$

So we have that (v) holds. $\square$

Corollary 2.60 *If $\mu(A) < 0$, then the trivial solution of the vector equation $x' = Ax$ is globally asymptotically stable.*

Proof This follows from part (v) of Theorem 2.59 and Theorem 2.49. $\square$

Theorem 2.61 *The following hold:*

(i) *if μ_1 corresponds to the traffic vector norm, then, for $A \in M_n$*

$$\mu_1(A) = \max_{1 \le j \le n} \left\{ a_{jj} + \sum_{i=1, i \ne j}^{n} |a_{ij}| \right\};$$

(ii) *if μ_∞ corresponds to the maximum vector norm, then*

$$\mu_\infty(A) = \max_{1 \le i \le n} \left\{ a_{ii} + \sum_{j=1, j \ne i}^{n} |a_{ij}| \right\};$$

(iii) *if μ_2 corresponds to the Euclidean norm, then*

$$\mu_2(A) = \max\{\lambda : \lambda \text{ is an eigenvalue of } \tfrac{1}{2}(A + A^T)\}.$$

Proof We will prove part (i) here. The proof of part (ii) is Exercise 2.59 and the proof of part (iii) is nontrivial, but is left to the reader. Using Theorem 2.55

$$
\begin{aligned}
\mu_1(A) &= \lim_{h \to 0+} \frac{\|I + hA\|_1 - 1}{h} \\
&= \lim_{h \to 0+} \max_{1 \le j \le n} \frac{\sum_{i=1, i \ne j}^{n} |ha_{ij}| + |1 + ha_{jj}| - 1}{h} \\
&= \max_{1 \le j \le n} \lim_{h \to 0+} \frac{h \sum_{i=1, i \ne j}^{n} |a_{ij}| + ha_{jj}}{h} \\
&= \max_{1 \le j \le n} \left\{ a_{jj} + \sum_{i=1, i \ne j}^{n} |a_{ij}| \right\}.
\end{aligned}
$$

$\square$

Example 2.62 The trivial solution of the vector equation

$$x' = \begin{pmatrix} -3.3 & .3 & 3 & -.4 \\ 1 & -2 & 1 & 1.3 \\ -1.2 & .4 & -5 & .2 \\ -1 & .8 & .5 & -2 \end{pmatrix} x \qquad (2.15)$$

is globally asymptotically stable because

$$\mu_1(A) = -.1 < 0,$$

where A is the coefficient matrix in (2.15). Note that $\mu_\infty(A) = 1.3 > 0$. $\triangle$

2.5 Floquet Theory

Differential equations involving periodic functions play an important role in many applications. Let's consider the linear system

$$x' = A(t)x,$$

where the $n \times n$ matrix function A is a continuous, periodic function with smallest positive period ω. Such systems are called Floquet systems and the study of Floquet systems is called Floquet theory. A natural question is whether a Floquet system has a periodic solution with period ω. Although this is not neccessarily the case, it is possible to characterize all the solutions of such systems and to give conditions under which a periodic solution does exist. Fortunately, the periodic system turns out to be closely related to a linear system with constant coefficients, so the properties of these systems obtained in earlier sections can be applied. In particular, we can easily answer questions about the stability of periodic systems.

First we need some preliminary results about matrices.

Theorem 2.63 (Jordan Canonical Form) *If A is an $n \times n$ constant matrix, then there is a nonsingular $n \times n$ constant matrix P so that $A = PJP^{-1}$, where J is a block diagonal matrix of the form*

$$J = \begin{bmatrix} J_1 & 0 & \cdots & 0 \\ 0 & J_2 & \ddots & \vdots \\ \vdots & \ddots & \ddots & 0 \\ 0 & \cdots & 0 & J_k \end{bmatrix},$$

where either J_i is the 1×1 matrix $J_i = [\lambda_i]$ or

$$J_i = \begin{bmatrix} \lambda_i & 1 & 0 & \cdots & 0 \\ 0 & \lambda_i & 1 & \ddots & \vdots \\ \vdots & \ddots & \ddots & \ddots & \vdots \\ \vdots & \ddots & \ddots & \lambda_i & 1 \\ 0 & \cdots & 0 & 0 & \lambda_i \end{bmatrix},$$

$1 \le i \le k$, *and the λ_i's are the eigenvalues of A.*

Proof We will only prove this theorem for 2×2 matrices A. For a proof of the general result, see Horn and Johnson [**24**]. There are two cases:

Case 1: A has two linearly independent eigenvectors x^1 and x^2.

In this case we have eigenpairs λ_1, x^1 and λ_2, x^2 of A, where λ_1 and λ_2 might be the same. Let P be the matrix with columns x^1 and x^2, that is,

$$P = [x^1 \; x^2].$$

Then

$$\begin{aligned} AP &= A[x^1 \; x^2] \\ &= [Ax^1 \; Ax^2] \\ &= [\lambda_1 x^1 \; \lambda_2 x^2] \\ &= PJ, \end{aligned}$$

where
$$J = \begin{bmatrix} \lambda_1 & 0 \\ 0 & \lambda_2 \end{bmatrix}.$$

Hence
$$A = PJP^{-1},$$

where P is of the correct form.

  Case 2: Assume A has only one linearly independent eigenvector x^1.

  Let v be a vector that is independent of x^1. By the Cayley-Hamilton Theorem (Theorem 2.33),

$$(A - \lambda_1 I)(A - \lambda_1 I)v = 0,$$

so

$$(A - \lambda_1 I)v = cx^1,$$

for some $c \neq 0$. Define $x^2 = v/c$, so that

$$(A - \lambda_1 I)x^2 = x^1.$$

Set

$$P = [x^1 \ x^2].$$

Then

$$\begin{aligned} AP &= A[x^1 \ x^2] \\ &= [Ax^1 \ Ax^2] \\ &= [\lambda_1 x^1 \ \lambda_1 x^2 + x^1] \\ &= PJ, \end{aligned}$$

where

$$J = \begin{bmatrix} \lambda_1 & 1 \\ 0 & \lambda_1 \end{bmatrix}.$$

Hence

$$A = PJP^{-1},$$

where P is of the correct form. $\qquad\square$

Theorem 2.64 (Log of a Matrix) *If C is an $n \times n$ nonsingular matrix, then there is a matrix B such that*

$$e^B = C.$$

Proof We will just prove this theorem for 2×2 matrices. Let μ_1, μ_2 be the eigenvalues of C. Since C is nonsingular $\mu_1, \mu_2 \neq 0$. First we prove the result for two special cases.

  Case 1. Assume

$$C = \begin{bmatrix} \mu_1 & 0 \\ 0 & \mu_2 \end{bmatrix}.$$

In this case we seek a diagonal matrix

$$B = \begin{bmatrix} b_1 & 0 \\ 0 & b_2 \end{bmatrix},$$

so that $e^B = C$. That is, we want to choose b_1 and b_2 so that

$$e^B = \begin{bmatrix} e^{b_1} & 0 \\ 0 & e^{b_2} \end{bmatrix} = \begin{bmatrix} \mu_1 & 0 \\ 0 & \mu_2 \end{bmatrix}.$$

Hence we can just take

$$B = \begin{bmatrix} \ln \mu_1 & 0 \\ 0 & \ln \mu_2 \end{bmatrix}.$$

Case 2. Assume

$$C = \begin{bmatrix} \mu_1 & 1 \\ 0 & \mu_1 \end{bmatrix}.$$

In this case we seek a matrix B of the form

$$B = \begin{bmatrix} a_1 & a_2 \\ 0 & a_1 \end{bmatrix},$$

so that $e^B = C$. That is, we want to choose a_1 and a_2 so that

$$e^B = \begin{bmatrix} e^{a_1} & a_2 e^{a_1} \\ 0 & e^{a_1} \end{bmatrix} = \begin{bmatrix} \mu_1 & 1 \\ 0 & \mu_1 \end{bmatrix}.$$

Hence we can just take

$$B = \begin{bmatrix} \ln \mu_1 & \frac{1}{\mu_1} \\ 0 & \ln \mu_1 \end{bmatrix}.$$

Case 3. C is an arbritary 2×2 nonsingular constant matrix. By the Jordan canonical form theorem (Theorem 2.63) there is a nonsingular matrix P such that $C = PJP^{-1}$, where

$$J = \begin{bmatrix} \mu_1 & 0 \\ 0 & \mu_2 \end{bmatrix} \quad \text{or} \quad J = \begin{bmatrix} \mu_1 & 1 \\ 0 & \mu_1 \end{bmatrix}.$$

By the previous two cases there is a matrix B_1 so that

$$e^{B_1} = J.$$

Let

$$B := PB_1P^{-1};$$

then, using part (viii) in Theorem 2.39,

$$e^B = e^{PB_1P^{-1}} = Pe^{B_1}P^{-1} = C.$$

$\square$

Example 2.65 Find a log of the matrix

$$C = \begin{bmatrix} 2 & 1 \\ 3 & 4 \end{bmatrix}.$$

The characteristic equation for C is

$$\lambda^2 - 6\lambda + 5 = 0$$

and so the eigenvalues are $\lambda_1 = 1$, $\lambda_2 = 5$. The Jordan canonical form (see Theorem 2.63) of C is

$$J = \begin{bmatrix} 1 & 0 \\ 0 & 5 \end{bmatrix}.$$

Eigenpairs of C are

$$1, \ \begin{bmatrix} 1 \\ -1 \end{bmatrix}, \quad \text{and} \quad 5, \ \begin{bmatrix} 1 \\ 3 \end{bmatrix}.$$

From the proof of Theorem 2.64, if we let

$$P := \begin{bmatrix} 1 & 1 \\ -1 & 3 \end{bmatrix},$$

then

$$PB_1 P^{-1}$$

is a log of C provided B_1 is a log of J. Note that by the proof of Theorem 2.64,

$$B_1 = \begin{bmatrix} 0 & 0 \\ 0 & \ln 5 \end{bmatrix}$$

is a log of J. Hence a log of C is given by

$$
\begin{aligned}
B &= PB_1 P^{-1} \\
&= \begin{bmatrix} 1 & 1 \\ -1 & 3 \end{bmatrix} \begin{bmatrix} 0 & 0 \\ 0 & \ln 5 \end{bmatrix} \begin{bmatrix} \frac{3}{4} & -\frac{1}{4} \\ \frac{1}{4} & \frac{1}{4} \end{bmatrix} \\
&= \begin{bmatrix} \frac{1}{4}\ln 5 & \frac{1}{4}\ln 5 \\ \frac{3}{4}\ln 5 & \frac{3}{4}\ln 5 \end{bmatrix}.
\end{aligned}
$$

$\triangle$

Before we state and prove Floquet's theorem (Theorem 2.67) we give a motivating example.

Example 2.66 Consider the scalar differential equation

$$x' = (\sin^2 t) \, x.$$

A general solution of this differential equation is

$$\phi(t) = ce^{\frac{1}{2}t - \frac{1}{4}\sin(2t)}.$$

Note that even though the coefficient function in our differential equation is periodic with minimum period π, the only period π solution of our differential equation is the trivial solution. But note that all nontrivial solutions are of the form

$$\phi(t) = p(t)e^{bt},$$

where $p(t) = ce^{-\frac{1}{4}\sin(2t)} \neq 0$, for all $t \in \mathbb{R}$, is a continuously differentiable function on $\mathbb{R}$ that is periodic with period π (which is the minimum positive period of the coefficient function in our differential equation) and $b = \frac{1}{2}$ is a constant (1×1 matrix). $\triangle$

Floquet's theorem shows that any fundamental matrix for the Floquet system $x' = A(t)x$ can be written in a form like shown in Example 2.66.

Theorem 2.67 (Floquet's Theorem) *If Φ is a fundamental matrix for the Floquet system $x' = A(t)x$, where the matrix function A is continuous on $\mathbb{R}$ and has minimum positive period ω, then the matrix function Ψ defined by $\Psi(t) := \Phi(t + \omega)$, $t \in \mathbb{R}$ is also a fundamental matrix. Furthermore there is a nonsingular, continuously differentiable $n \times n$ matrix function P which is periodic with period ω and an $n \times n$ constant matrix B (possibly complex) so that*

$$\Phi(t) = P(t)e^{Bt},$$

for all $t \in \mathbb{R}$.

Proof Assume Φ is a fundamental matrix for the Floquet system $x' = A(t)x$. Define the matrix function Ψ by

$$\Psi(t) = \Phi(t + \omega),$$

for $t \in \mathbb{R}$. Then

$$
\begin{aligned}
\Psi'(t) &= \Phi'(t + \omega) \\
&= A(t + \omega)\Phi(t + \omega) \\
&= A(t)\Psi(t).
\end{aligned}
$$

Since $\det \Psi(t) \neq 0$ for all $t \in \mathbb{R}$, Ψ is a fundamental matrix of $x' = A(t)x$. Hence the first statement of the theorem holds.

Since Φ and Ψ are fundamental matrices for $x' = A(t)x$, Theorem 2.31 implies that there is a nonsingular constant matrix C so that

$$\Phi(t + \omega) = \Phi(t)C, \quad \text{for } t \in \mathbb{R}.$$

By Theorem 2.64 there is a matrix B such that

$$e^{B\omega} = C.$$

Then define the matrix function P by

$$P(t) = \Phi(t)e^{-Bt},$$

for $t \in \mathbb{R}$. Obviously, P is a continuously differentiable, nonsingular matrix function on $\mathbb{R}$. To see that P is periodic with period ω consider

$$
\begin{aligned}
P(t + \omega) &= \Phi(t + \omega)e^{-Bt - B\omega} \\
&= \Phi(t)Ce^{-B\omega}e^{-Bt} \\
&= \Phi(t)e^{-Bt} \\
&= P(t).
\end{aligned}
$$

Finally note that

$$\Phi(t) = P(t)e^{Bt},$$

for all $t \in \mathbb{R}$.   □

Definition 2.68 Let Φ be a fundamental matrix for the Floquet system $x' = A(t)x$. Then the eigenvalues μ of

$$C := \Phi^{-1}(0)\Phi(\omega)$$

are called the *Floquet multipliers* of the Floquet system $x' = A(t)x$.

Fundamental matrices for $x' = A(t)x$ are not unique, so we wonder if Floquet multipliers are well defined in Definition 2.68. To see that Floquet multipliers are well defined, let Φ and Ψ be fundamental matrices for the Floquet system $x' = A(t)x$ and let

$$C := \Phi^{-1}(0)\Phi(\omega)$$

and let

$$D := \Psi^{-1}(0)\Psi(\omega).$$

We want to show that C and D have the same eigenvalues. Since Φ and Ψ are fundamental matrices of $x' = A(t)x$, Theorem 2.31 yields a nonsingular constant matrix M such that

$$\Psi(t) = \Phi(t)M$$

for all $t \in \mathbb{R}$. It follows that

$$
\begin{aligned}
D &= \Psi^{-1}(0)\Psi(\omega) \\
&= M^{-1}\Phi^{-1}(0)\Phi(\omega)M \\
&= M^{-1}CM.
\end{aligned}
$$

Therefore, C and D are similar matrices (see Exercise 2.14) and hence have the same eigenvalues. Hence Floquet multipliers are well defined.

Example 2.69 Find the Floquet multipliers for the scalar differential equation

$$x' = (\sin^2 t)\, x.$$

In Example 2.66 we saw that a nontrivial solution of this differential equation is

$$\phi(t) = e^{\frac{1}{2}t - \frac{1}{4}\sin(2t)}.$$

Hence

$$c := \phi^{-1}(0)\phi(\pi) = e^{\frac{\pi}{2}}$$

and so $\mu = e^{\frac{\pi}{2}}$ is the Floquet multiplier for this differential equation.   △

Example 2.70 Find the Floquet multipliers for the Floquet system

$$x' = \begin{bmatrix} 1 & 1 \\ 0 & \frac{(\cos t + \sin t)}{(2 + \sin t - \cos t)} \end{bmatrix} x.$$

Solving this equation first for x_2 and then x_1, we get that

$$x_2(t) = \beta(2 + \sin t - \cos t),$$
$$x_1(t) = \alpha e^t - \beta(2 + \sin t),$$

for $t \in \mathbb{R}$. It follows that a fundamental matrix for our Floquet system is

$$\Phi(t) = \begin{bmatrix} -2 - \sin t & e^t \\ 2 + \sin t - \cos t & 0 \end{bmatrix}.$$

Since

$$C = \Phi^{-1}(0)\Phi(2\pi) = \begin{bmatrix} 1 & 0 \\ 0 & e^{2\pi} \end{bmatrix},$$

the Floquet multipliers are $\mu_1 = 1$ and $\mu_2 = e^{2\pi}$. $\triangle$

Theorem 2.71 Let $\Phi(t) = P(t)e^{Bt}$ be as in Floquet's theorem (Theorem 2.67). Then x is a solution of the Floquet system $x' = A(t)x$ iff the vector function y defined by $y(t) = P^{-1}(t)x(t)$, $t \in \mathbb{R}$ is a solution of $y' = By$.

Proof Assume x is a solution of the Floquet system $x' = A(t)x$. Then

$$x(t) = \Phi(t)x_0,$$

for some $n \times 1$ constant vector x_0. Let $y(t) = P^{-1}(t)x(t)$. Then

$$\begin{aligned} y(t) &= P^{-1}(t)\Phi(t)x_0 \\ &= P^{-1}(t)P(t)e^{Bt}x_0 \\ &= e^{Bt}x_0, \end{aligned}$$

which is a solution of

$$y' = By.$$

Conversely, assume y is a solution of $y' = By$ and set

$$x(t) = P(t)y(t).$$

Since y is a solution of $y' = By$, there is an $n \times 1$ constant vector y_0 such that

$$y(t) = e^{Bt}y_0.$$

It follows that

$$\begin{aligned} x(t) &= P(t)y(t) \\ &= P(t)e^{Bt}y_0 \\ &= \Phi(t)y_0, \end{aligned}$$

which is a solution of the Floquet system $x' = A(t)x$. $\square$

Theorem 2.72 Let $\mu_1, \mu_2, \cdots, \mu_n$ be the Floquet multipliers of the Floquet system $x' = A(t)x$. Then the trivial solution is
   (i) globally asymptotically stable on $[0, \infty)$ iff $|\mu_i| < 1$, $1 \leq i \leq n$;
   (ii) stable on $[0, \infty)$ provided $|\mu_i| \leq 1$, $1 \leq i \leq n$ and whenever $|\mu_i| = 1$, μ_i is a simple eigenvalue;
   (iii) unstable on $[0,\infty)$ provided there is an i_0, $1 \leq i_0 \leq n$, such that $|\mu_{i_0}| > 1$.

Proof We will just prove this theorem for the two-dimensional case. Let $\Phi(t) = P(t)e^{Bt}$ and C be as in Floquet's theorem. Recall that in the proof of Floquet's theorem, B was picked so that

$$e^{B\omega} = C.$$

By the Jordan canonical form theorem (Theorem 2.63) there are matrices M and J so that

$$B = MJM^{-1},$$

where either

$$J = \begin{bmatrix} \rho_1 & 0 \\ 0 & \rho_2 \end{bmatrix}, \quad \text{or} \quad J = \begin{bmatrix} \rho_1 & 1 \\ 0 & \rho_1 \end{bmatrix},$$

where ρ_1, ρ_2 are the eigenvalues of B. It follows that

$$\begin{aligned} C &= e^{B\omega} \\ &= e^{MJM^{-1}\omega} \\ &= Me^{J\omega}M^{-1} \\ &= MKM^{-1}, \end{aligned}$$

where either

$$K = \begin{bmatrix} e^{\rho_1\omega} & 0 \\ 0 & e^{\rho_2\omega} \end{bmatrix}, \quad \text{or} \quad K = \begin{bmatrix} e^{\rho_1\omega} & \omega e^{\rho_1\omega} \\ 0 & e^{\rho_1\omega} \end{bmatrix}.$$

Since the eigenvalues of K are the same (see Exercise 2.14) as the eigenvalues of C, we get that the Floquet multipliers are

$$\mu_i = e^{\rho_i\omega},$$

$i = 1, 2$, where it is possible that $\rho_1 = \rho_2$. Since

$$|\mu_i| = e^{Re(\rho_i)\omega},$$

we have that

$$\begin{aligned} |\mu_i| &< 1 \quad \text{iff} \quad Re(\rho_i) < 0 \\ |\mu_i| &= 1 \quad \text{iff} \quad Re(\rho_i) = 0 \\ |\mu_i| &> 1 \quad \text{iff} \quad Re(\rho_i) > 0. \end{aligned}$$

By Theorem 2.71 the equation

$$x(t) = P(t)y(t)$$

gives a one-to-one correspondence between solutions of the Floquet system $x' = A(t)x$ and $y' = By$. Note that there is a constant $Q_1 > 0$ so that

$$\|x(t)\| = \|P(t)y(t)\| \le \|P(t)\|\|y(t)\| \le Q_1\|y(t)\|,$$

for $t \in \mathbb{R}$ and since $y(t) = P^{-1}(t)x(t)$ there is a constant $Q_2 > 0$ such that

$$\|y(t)\| = \|P^{-1}(t)x(t)\| \le \|P^{-1}(t)\|\|x(t)\| \le Q_2\|x(t)\|,$$

for $t \in \mathbb{R}$. The conclusions of this theorem then follow from Theorem 2.49.
□

Theorem 2.73 *The number μ_0 is a Floquet multiplier of the Floquet system $x' = A(t)x$ iff there is a nontrivial solution x such that*

$$x(t + \omega) = \mu_0 x(t),$$

for all $t \in \mathbb{R}$. Consequently, the Floquet system has a nontrivial periodic solution of period ω if and only if $\mu_0 = 1$ is a Floquet multiplier.

Proof First assume μ_0 is a Floquet multiplier of the Floquet system $x' = A(t)x$. Then μ_0 is an eigenvalue of

$$C := \Phi^{-1}(0)\Phi(\omega),$$

where Φ is a fundamental matrix of $x' = A(t)x$. Let x_0 be an eigenvector corresponding to μ_0 and define the vector function x by

$$x(t) = \Phi(t)x_0, \quad t \in \mathbb{R}.$$

Then x is a nontrivial solution of $x' = A(t)x$ and

$$
\begin{aligned}
x(t + \omega) &= \Phi(t + \omega)x_0 \\
&= \Phi(t)Cx_0 \\
&= \Phi(t)\mu_0 x_0 \\
&= \mu_0 x(t),
\end{aligned}
$$

for all $t \in \mathbb{R}$.

Conversely, assume there is a nontrivial solution x such that

$$x(t + \omega) = \mu_0 x(t),$$

for all $t \in \mathbb{R}$. Let Ψ be a fundamental matrix of our Floquet system, then

$$x(t) = \Psi(t)y_0,$$

for all $t \in \mathbb{R}$ for some nontrivial vector y_0. By Floquet's theorem the matrix function $\Psi(\cdot + \omega)$ is also a fundamental matrix. Hence $x(t + \omega) = \mu_0 x(t)$, so $\Psi(t + \omega)y_0 = \mu_0\Psi(t)y_0$ and therefore

$$\Psi(t)Dy_0 = \Psi(t)\mu_0 y_0,$$

where $D := \Psi^{-1}(0)\Psi(\omega)$. It follows that

$$Dy_0 = \mu_0 y_0,$$

and so μ_0 is an eigenvalue of D and hence is a Floquet multiplier of our Floquet system. □

Theorem 2.74 *Assume $\mu_1, \mu_2, \cdots, \mu_n$ are the Floquet multipliers of the Floquet system $x' = A(t)x$. Then*

$$\mu_1\mu_2\cdots\mu_n = e^{\int_0^\omega \mathrm{tr}[A(t)]\,dt}.$$

Proof Let Φ be the solution of the matrix IVP

$$X' = A(t)X, \quad X(0) = I.$$

Then Φ is a fundamental matrix for $x' = A(t)x$ and

$$C := \Phi^{-1}(0)\Phi(\omega) = \Phi(\omega).$$

Then we get that

$$
\begin{aligned}
\mu_1\mu_2\cdots\mu_n &= \det C \\
&= \det \Phi(\omega) \\
&= e^{\int_0^\omega \operatorname{tr}[A(t)]\,dt} \det \Phi(0) \\
&= e^{\int_0^\omega \operatorname{tr}[A(t)]\,dt},
\end{aligned}
$$

where we have used Liouville's theorem (Theorem 2.23), which is the desired result.

$\square$

Example 2.75 (Hill's Equation) Consider the scalar differential equation (Hill's equation)

$$y'' + q(t)y = 0,$$

where we assume that q is a continuous periodic function on $\mathbb{R}$ with minimum positive period ω. G. W. Hill [21] considered equations of this form when he studied planetary motion. There are many applications of Hill's equation in mechanics, astronomy, and electrical engineering. For a more thorough study of Hill's equation than is given here, see [35]. Writing Hill's equation as a system in the standard way, we get the Floquet system

$$x' = \begin{bmatrix} 0 & 1 \\ -q(t) & 0 \end{bmatrix} x.$$

By the Floquet multipliers of Hill's equation we mean the Floquet multipliers of the preceding Floquet system. It follows from Theorem 2.74 that the Floquet multipliers of Hill's equation satisfy

$$\mu_1\mu_2 = 1.$$

$\triangle$

Example 2.76 (Mathieu's Equation) A special case of Hill's equation is Mathieu's equation,

$$y'' + (\alpha + \beta \cos t)y = 0,$$

where α and β are real parameters. We will assume that $\beta \neq 0$. Note that the Floquet multipliers of Mathieu's equation depend on α and β. From Theorem 2.74 the Floquet multipliers of Mathieu's equation satisfy

$$\mu_1(\alpha, \beta)\mu_2(\alpha, \beta) = 1.$$

Let

$$\gamma(\alpha, \beta) := \mu_1(\alpha, \beta) + \mu_2(\alpha, \beta).$$

Then the Floquet multipliers of Mathieu's equation satisfy the quadratic equation

$$\mu^2 - \gamma(\alpha, \beta)\mu + 1 = 0.$$

In particular,

$$\mu_{1,2} = \frac{\gamma \pm \sqrt{\gamma^2 - 4}}{2}.$$

There are five cases to consider:

Case 1: $\gamma > 2$.

In this case the Floquet multipliers satisfy

$$0 < \mu_2 < 1 < \mu_1.$$

It then follows from Theorem 2.72 that the trivial solution of Mathieu's equation is unstable on $[0, \infty)$ in this case. Using Exercise 2.70, we can show in this case that there is a general solution of Mathieu's equation of the form

$$y(t) = c_1 e^{\sigma t} p_1(t) + c_2 e^{-\sigma t} p_2(t),$$

where $\sigma > 0$ and p_i, $i = 1, 2$, are continuously differentiable functions on $\mathbb{R}$ which are periodic with period 2π.

Case 2: $\gamma = 2$.

In this case

$$\mu_1 = \mu_2 = 1.$$

It follows, using Exercise 2.70, that there is a nontrivial solution of period 2π. It has been proved in [**35**] that there is a second linearly independent solution that is unbounded. In particular, the trivial solution is unstable on $[0, \infty)$.

Case 3: $-2 < \gamma < 2$.

In this case the Floquet multipliers are not real and $\mu_2 = \overline{\mu}_1$ In this case, using Exercise 2.70, there is a general solution of the form

$$y(t) = c_1 e^{i\sigma t} p_1(t) + c_2 e^{-i\sigma t} p_2(t),$$

where $\sigma > 0$ and the p_i, $i = 1, 2$ are continuously differentiable functions on $\mathbb{R}$ that are periodic with period 2π. In this case it then follows from Theorem 2.72 that the trivial solution is stable on $[0, \infty)$.

Case 4: $\gamma = -2$.

In this case

$$\mu_1 = \mu_2 = -1.$$

It follows from Exercise 2.67 that there is a nontrivial solution that is periodic with period 4π. It has been shown in [**35**] that there is a second linearly independent solution that is unbounded. In particular, the trivial solution is unstable on $[0, \infty)$.

Case 5: $\gamma < -2$.

In this case the Floquet multipliers satisfy

$$\mu_2 < -1 < \mu_1 < 0.$$

In this case it follows from Theorem 2.72 that the trivial solution is unstable on $[0, \infty)$.   △

A very interesting fact concerning Mathieu's equation is that if $\beta > 0$ is fixed, then there are infinitely many intervals of α values, where the trivial solution of Mathieu's equation is alternately stable and unstable on $[0, \infty)$ (see [**35**]).

2.6 Exercises

2.1 Show that the characteristic equation for the constant coefficient scalar differential equation $y'' + ay' + by = 0$ is the same as the characteristic equation for the companion matrix of this differential equation.

2.2 Let $\mathbb{A}$ be the set of all continuous scalar functions on an interval I and define $M : \mathbb{A} \to \mathbb{A}$ by

$$Mx(t) = \int_a^t x(s) \, ds,$$

for $t \in I$, where a is a fixed point in I. Prove that M is a linear operator.

2.3 Determine in each case if the constant vectors are linearly dependent or linearly independent. Prove your answer.

(i)
$$\psi_1 = \begin{bmatrix} -4 \\ 4 \\ 1 \end{bmatrix}, \quad \psi_2 = \begin{bmatrix} 1 \\ -1 \\ -2 \end{bmatrix}, \quad \psi_3 = \begin{bmatrix} 2 \\ -2 \\ 1 \end{bmatrix}$$

(ii)
$$\psi_1 = \begin{bmatrix} 2 \\ 1 \\ -1 \end{bmatrix}, \quad \psi_2 = \begin{bmatrix} -1 \\ 3 \\ 2 \end{bmatrix}, \quad \psi_3 = \begin{bmatrix} 1 \\ -2 \\ 1 \end{bmatrix}$$

(iii)
$$\psi_1 = \begin{bmatrix} 2 \\ 1 \\ -1 \\ -2 \end{bmatrix}, \quad \psi_2 = \begin{bmatrix} -1 \\ 2 \\ 1 \\ 1 \end{bmatrix}, \quad \psi_3 = \begin{bmatrix} 1 \\ 13 \\ 2 \\ -1 \end{bmatrix}$$

2.4 Determine if the scalar functions ϕ_1, ϕ_2 defined by $\phi_1(t) = \ln(t)$, $\phi_2(t) = \ln(t^2)$ for $t \in (0, \infty)$ are linearly dependent or linearly independent on $(0, \infty)$. Prove your answer.

2.5 Determine if the given functions are linearly dependent or linearly independent on the given interval I. Prove your answers.

(i) $x_1(t) = 4\sin t$, $x_2(t) = 7\sin(-t)$, $I = \mathbb{R}$
(ii) $x_1(t) = 2\sin(2t)$, $x_2(t) = -3\cos(2t)$, $x_3(t) = 8$, $I = \mathbb{R}$
(iii) $x_1(t) = e^{3t}t$, $x_2(t) = e^{-2t}$, $I = \mathbb{R}$
(iv) $x_1(t) = e^{3t}t$, $x_2(t) = e^{3t+4}$, $I = \mathbb{R}$

(v) $x_1(t) = \sin t$, $x_2(t) = \cos t$, $x_3(t) = \sin(t + \frac{\pi}{4})$, $I = \mathbb{R}$

2.6 Prove that if $x_1, x_2, \cdots, x_k$ are k functions defined on I and if one of them is identically zero on I, then $x_1, x_2, \cdots, x_k$ are linearly dependent on I.

2.7 Prove that two functions x, y defined on I are linearly dependent on I iff one of them is a constant times the other.

2.8 Determine if the scalar functions ϕ_1, ϕ_2, ϕ_3 defined by $\phi_1(t) = 3$, $\phi_2(t) = 3\sin^2 t$, $\phi_3(t) = 4\cos^2 t$ for $t \in \mathbb{R}$ are linearly dependent or linearly independent on $\mathbb{R}$. Prove your answer.

2.9 Determine if the scalar functions ϕ_1, ϕ_2 defined by $\phi_1(t) = t^2 + 1$, $\phi_2(t) = 2t^2 + 3t - 7$ for $t \in \mathbb{R}$ are linearly independent or linearly dependent on $\mathbb{R}$. Verify your answer.

2.10 Determine if the scalar functions $\phi_1, \phi_2, \phi_3, \phi_4$ defined by $\phi_1(t) = \sin^2 t$, $\phi_2(t) = \cos^2 t$, $\phi_3(t) = \tan^2 t$, $\phi_4(t) = \sec^2 t$ for $t \in (-\frac{\pi}{2}, \frac{\pi}{2})$ are linearly dependent or linearly independent on $(-\frac{\pi}{2}, \frac{\pi}{2})$. Prove your answer.

2.11 Find four two dimensional vector functions that are linearly independent on $\mathbb{R}$ and prove that they are linearly independent on $\mathbb{R}$.

2.12 Find two scalar functions that are linearly independent on $\mathbb{R}$, but linearly dependent on $(0, \infty)$. Prove your answer.

2.13 Find the inverse of each of the following matrices:

(i) $A = \begin{bmatrix} 2 & -6 \\ -1 & 4 \end{bmatrix}$

(ii) $B = \begin{bmatrix} 2 & -2 & 2 \\ 0 & 1 & 0 \\ 4 & 0 & 3 \end{bmatrix}$

2.14 Two $n \times n$ matrices A, B are said to be *similar* if there is a nonsingular $n \times n$ matrix M such that $A = M^{-1}BM$. Prove that similar matrices have the same eigenvalues.

2.15 Show that if an $n \times n$ matrix has eigenvalues $\lambda_1, \lambda_2, \cdots, \lambda_n$ (not necessarily distinct) with linearly independent eigenvectors

$$x_1, x_2 \cdots, x_n,$$

respectively, then a general solution of $x' = Ax$ is $x(t) = c_1 e^{\lambda_1 t} x_1 + \cdots + c_n e^{\lambda_n t} x_n$.

2.16 Show that if a square matrix has all zeros either above or below the main diagonal, then the numbers down the diagonal are the eigenvalues of the matrix.

2.17 If λ_0 is an eigenvalue of A, find an eigenvalue of
(i) A^T (the transpose of A)

(ii) A^n, where n is a positive integer

(iii) A^{-1}, provided $\det(A) \neq 0$

Be sure to verify your answers.

2.18 Show that any $n+1$ solutions of (2.3) on (c, d) are linearly dependent on (c, d).

2.19 Show that the characteristic equation for any 2×2 constant matrix A is

$$\lambda^2 - \text{tr}[A]\lambda + \det(A) = 0,$$

where $\text{tr}[A]$ is defined in Definition 2.22. Use this result to find the characteristic equation for each of the following matrices:

(i) $A = \begin{bmatrix} 1 & 4 \\ -2 & 3 \end{bmatrix}$

(ii) $A = \begin{bmatrix} 1 & 2 \\ 3 & 4 \end{bmatrix}$

(iii) $A = \begin{bmatrix} 3 & 2 \\ 2 & 0 \end{bmatrix}$

2.20 Using Theorem 2.14, solve the following differential equations:

(i) $x' = \begin{bmatrix} 2 & 1 \\ 3 & 4 \end{bmatrix} x$

(ii) $x' = \begin{bmatrix} 2 & 2 \\ 2 & -1 \end{bmatrix} x$

(iii) $x' = \begin{bmatrix} 3 & 2 & 4 \\ 2 & 0 & 2 \\ 4 & 2 & 3 \end{bmatrix} x$

2.21 Solve the IVP

$$x' = \begin{bmatrix} 1 & 3 \\ 3 & 1 \end{bmatrix} x, \quad x(0) = \begin{bmatrix} -2 \\ 1 \end{bmatrix} x.$$

2.22 Work each of the following:

(i) Find eigenpairs for the matrix $A = \begin{bmatrix} -1 & -6 \\ 1 & 4 \end{bmatrix} x$

(ii) Use your answer in (i) to find two linearly independent solutions (prove that they are linearly independent) of $x' + Ax$ on $\mathbb{R}$.

(iii) Use your answer in (ii) to find a fundamental matrix for $x' = Ax$.

(iv) Use your answer in (iii) and Theorem 2.31 to find e^{At}.

2.23 (Complex Eigenvalues) Using Theorem 2.14, solve the following differential equations:

(i) $x' = \begin{bmatrix} 0 & 9 \\ -9 & 0 \end{bmatrix} x$

(ii) $x' = \begin{bmatrix} 1 & 1 \\ -1 & 1 \end{bmatrix} x$

(iii) $x' = \begin{bmatrix} 1 & 2 & 1 \\ -2 & 1 & 0 \\ 0 & 0 & 3 \end{bmatrix} x$

2.24 Prove the last statement in Corollary 2.24.

2.25 Using Theorem 2.14, solve the following differential equations:

(i) $x' = \begin{bmatrix} -2 & 5 \\ -2 & 4 \end{bmatrix} x$

(ii) $x' = \begin{bmatrix} 1 & 1 \\ -1 & 1 \end{bmatrix} x$

(iii) $x' = \begin{bmatrix} 3 & 0 & 4 \\ 0 & 2 & 0 \\ 0 & 0 & -3 \end{bmatrix} x$

2.26 Use Theorem 2.14 to find a fundamental matrix for

$$x' = \begin{bmatrix} 4 & 0 & 0 \\ 0 & 5 & 1 \\ 0 & 1 & 5 \end{bmatrix} x.$$

2.27 Show that the matrix function Φ defined by

$$\Phi(t) := \begin{bmatrix} t^2 & t^3 \\ 2t & 3t^2 \end{bmatrix},$$

for $t \in (0, \infty)$ is a fundamental matrix for the vector differential equation

$$x' = \begin{bmatrix} 0 & 1 \\ -\frac{6}{t^2} & \frac{4}{t} \end{bmatrix} x.$$

2.28 Show that the matrix function Φ defined by

$$\Phi(t) := \begin{bmatrix} e^{-t} & 2 \\ 1 & e^t \end{bmatrix},$$

for $t \in \mathbb{R}$ is a fundamental matrix for the vector differential equation

$$x' = \begin{bmatrix} 1 & -2e^{-t} \\ e^t & -1 \end{bmatrix} x.$$

Find the solution satisfying the initial condition

$$x(0) = \begin{bmatrix} 1 \\ -1 \end{bmatrix}.$$

2.29 Verify the Cayley-Hamilton Theorem (Theorem 2.33) directly for the general 2×2 matrix

$$A = \begin{bmatrix} a & b \\ c & d \end{bmatrix}.$$

2.30 Verify the Cayley-Hamilton Theorem (Theorem 2.33) directly for the matrix

$$A = \begin{bmatrix} 1 & 0 & 2 \\ 2 & 1 & -1 \\ 1 & -1 & 2 \end{bmatrix}.$$

2.31 Use the Putzer algorithm (Theorem 2.35) to find e^{At} for each of the following:

(i) $A = \begin{bmatrix} 2 & 1 \\ -9 & -4 \end{bmatrix}$

(ii) $A = \begin{bmatrix} 10 & 4 \\ -9 & -2 \end{bmatrix}$

(iii) $A = \begin{bmatrix} 2 & -2 & 2 \\ 0 & 1 & 1 \\ -4 & 8 & 3 \end{bmatrix}$

(iv) $A = \begin{bmatrix} 4 & 0 & 0 \\ 0 & 5 & 1 \\ 0 & 1 & 5 \end{bmatrix}$

2.32 (Multiple Eigenvalue) Use the Putzer algorithm (Theorem 2.35) to find e^{At} given that

$$A = \begin{bmatrix} -5 & 2 \\ -2 & -1 \end{bmatrix}.$$

2.33 (Complex Eigenvalues) Use the Putzer algorithm (Theorem 2.35) to find e^{At} for each of the following:

(i) $A = \begin{bmatrix} 1 & 2 \\ -5 & -1 \end{bmatrix}$

(ii) $A = \begin{bmatrix} 2 & 3 \\ -3 & 2 \end{bmatrix}$

(iii) $A = \begin{bmatrix} 0 & 2 \\ -1 & 2 \end{bmatrix}$

2.34 Use the Putzer algorithm (Theorem 2.35) to help you solve each of the following differential equations:

(i) $x' = \begin{bmatrix} 1 & 1 \\ 1 & 1 \end{bmatrix} x$

(ii) $x' = \begin{bmatrix} 2 & 1 \\ -1 & 4 \end{bmatrix} x$

(iii) $x' = \begin{bmatrix} -1 & -6 \\ 1 & 4 \end{bmatrix} x$

(iv) $x' = \begin{bmatrix} 1 & 1 & 0 \\ 0 & 1 & 0 \\ 0 & 1 & 1 \end{bmatrix} x$

2.35 Solve each of the following differential equations:

(i) $x' = \begin{bmatrix} -4 & 3 \\ -2 & 3 \end{bmatrix} x$

(ii) $x' = \begin{bmatrix} -4 & 1 \\ 2 & -3 \end{bmatrix} x$

(iii) $x' = \begin{bmatrix} 2 & -2 & 2 \\ 0 & 1 & 1 \\ -4 & 8 & 3 \end{bmatrix} x$

2.36 Work each of the following:

   (i) Solve the differential equations for the vibrating system in Example 2.2 in case there is no friction: $m_1 = m_2 = k_1 = k_2 = k_3 = 1$ and $c = 0$.

   (ii) Decide whether the trivial solution is stable in this case, and discuss the implications of your answer for the vibrating system.

   (iii) Find the solution that satisfies the initial conditions $u(0) = 1$, $u'(0) = v(0) = v'(0) = 0$. Also, sketch a graph of the solution.

2.37 Work each of the following:

   (i) Solve the differential equations for the vibrating system in Example 2.2 for the parameter values $m_1 = m_2 = k_1 = k_2 = k_3 = 1$, $c = 2$. (*Note*: The eigenvalues are not distinct in this case.)

   (ii) Show that the trivial solution is globally asymptotically stable.

2.38 Find 2×2 matrices A and B such that

$$e^A e^B \neq e^{A+B}.$$

2.39 Show that if A and B are $n \times n$ constant matrices and $AB = BA$, then

$$e^{At} B = B e^{At},$$

for $t \in \mathbb{R}$. Also, show that $e^A B = B e^A$.

2.40 Use the Putzer algorithm (Theorem 2.35) to find

$$e^{\begin{bmatrix} 0 & 1 \\ -1 & 0 \end{bmatrix} t}.$$

Repeat the same problem using Theorem 2.39 part (*vii*). Use your answer and Theorem 2.39 part (*iii*) to prove the addition formulas for the trigonometric sine and cosine functions:

$$\sin(t + s) = \sin(t)\cos(s) + \sin(s)\cos(t)$$
$$\cos(t + s) = \cos(t)\cos(s) - \sin(t)\sin(s).$$

2.41 Show that

$$
e^{\begin{bmatrix} \lambda_1 & 0 & 0 & \cdots & 0 \\ 0 & \lambda_2 & 0 & \ddots & 0 \\ 0 & 0 & \lambda_3 & \ddots & 0 \\ \vdots & \ddots & \ddots & \ddots & 0 \\ 0 & 0 & \cdots & 0 & \lambda_n \end{bmatrix}} = \begin{bmatrix} e^{\lambda_1} & 0 & 0 & \cdots & 0 \\ 0 & e^{\lambda_2} & 0 & \ddots & 0 \\ 0 & 0 & e^{\lambda_3} & \ddots & 0 \\ \vdots & \ddots & \ddots & \ddots & 0 \\ 0 & 0 & \cdots & 0 & e^{\lambda_n} \end{bmatrix}.
$$

2.42 Use the variation of constants formula to solve each of the following IVPs:

(i) $x' = \begin{bmatrix} 1 & 0 \\ 0 & 2 \end{bmatrix} x + \begin{bmatrix} 1 \\ 2 \end{bmatrix}$, $\quad x(0) = \begin{bmatrix} 1 \\ 1 \end{bmatrix}$

(ii) $x' = \begin{bmatrix} 2 & 0 \\ 0 & 3 \end{bmatrix} x + \begin{bmatrix} e^{2t} \\ e^{3t} \end{bmatrix}$, $\quad x(0) = \begin{bmatrix} 1 \\ 2 \end{bmatrix}$

(iii) $x' = \begin{bmatrix} 2 & 1 \\ 0 & 2 \end{bmatrix} x + \begin{bmatrix} e^{2t} \\ te^{2t} \end{bmatrix}$, $\quad x(0) = \begin{bmatrix} -1 \\ 1 \end{bmatrix}$

(iv) $x' = \begin{bmatrix} 0 & 0 & 0 \\ 0 & 2 & 1 \\ 0 & 0 & 2 \end{bmatrix} x + \begin{bmatrix} t \\ 0 \\ 0 \end{bmatrix}$, $\quad x(0) = \begin{bmatrix} 1 \\ 1 \\ 1 \end{bmatrix}$

2.43 Use the variation of constants formula to solve each of the following IVP's

(i) $x' = \begin{bmatrix} 0 & 1 \\ -1 & 0 \end{bmatrix} x + \begin{bmatrix} 0 \\ t \end{bmatrix}$, $\quad x(0) = \begin{bmatrix} 0 \\ 1 \end{bmatrix}$

(ii) $x' = \begin{bmatrix} -2 & 0 \\ 0 & 4 \end{bmatrix} x + \begin{bmatrix} 1 \\ e^{-t} \end{bmatrix}$, $\quad x(0) = \begin{bmatrix} -1 \\ 2 \end{bmatrix}$

2.44 Use Theorem 2.40 and Theorem 2.42 to help you solve the IVP

$$
x' = \begin{bmatrix} 2t & 0 \\ 0 & 3 \end{bmatrix} x + \begin{bmatrix} t \\ 1 \end{bmatrix}, \quad x(0) = \begin{bmatrix} 0 \\ 2 \end{bmatrix}.
$$

2.45 Let $\epsilon > 0$. Graph $\{x \in \mathbb{R}^2 : \|x\| < \epsilon\}$ when

(i) $\| \cdot \|_2$ is the Euclidean norm (l_2 norm

(ii) $\| \cdot \|_\infty$ is the maximum norm (l_∞ norm)

(iii) $\| \cdot \|_1$ is the traffic norm (l_1 norm)

2.46 Prove: If A has an eigenvalue with positive real part, then there is a solution x of $x' = Ax$ so that $\|x(t)\| \to \infty$ as $t \to \infty$.

2.47 Determine the stability of the trivial solution for each of the following:

(i) $x' = \begin{bmatrix} -1 & 3 \\ 2 & -2 \end{bmatrix} x$

(ii) $x' = \begin{bmatrix} 5 & 8 \\ -7 & -10 \end{bmatrix} x$

(iii) $x' = \begin{bmatrix} 0 & 2 & 1 \\ -2 & 0 & -1 \\ 0 & 0 & -1 \end{bmatrix} x$

2.48 Determine the stability of the trivial solution for each of the following:

(i) $x' = \begin{bmatrix} -1 & -6 \\ 0 & 4 \end{bmatrix} x$

(ii) $x' = \begin{bmatrix} 0 & -9 \\ 4 & 0 \end{bmatrix} x$

(iii) $x' = \begin{bmatrix} 0 & 1 \\ -2 & -3 \end{bmatrix} x$

2.49 Show that the limit in Definition 2.58 exists by showing, for $0 < \theta < 1, h > 0$, that $\|I + \theta h A\| \le \theta \|I + hA\| + (1 - \theta)$, which implies that

$$\frac{\|I + \theta h A\| - 1}{\theta h} \le \frac{\|I + hA\| - 1}{h}$$

and by showing that $\frac{\|I + hA\| - 1}{h}$ is bounded below by $-\|A\|$.

2.50 Show that a matrix norm induced by a vector norm on $\mathbb{R}^n$ is a norm on M_n.

2.51 Show that the matrix norm on M_n induced by the vector norm $\| \cdot \|$ is given by

$$\|A\| = \max_{\|x\|=1} \|Ax\|.$$

2.52 Show that if $A, B \in M_n$, then $\|AB\| \le \|A\| \cdot \|B\|$.

2.53 Prove Theorem 2.56.

2.54 Find the matrix norm of the matrix

$$A = \begin{bmatrix} -1 & 1 \\ 2 & -2 \end{bmatrix}$$

corresponding to the maximum norm $\| \cdot \|_\infty$, the traffic norm $\| \cdot \|_1$, and the Euclidean norm $\| \cdot \|_2$, respectively. Also, find $\mu_\infty(A), \mu_1(A), \mu_2(A)$.

2.55 Find the matrix norm of the matrix

$$A = \begin{bmatrix} -2 & -1 & 1 \\ -1 & -3 & 2 \\ 0 & -1 & -5 \end{bmatrix}$$

corresponding the maximum norm $\| \cdot \|_\infty$ and the traffic norm $\| \cdot \|_1$, respectively. Also, find $\mu_\infty(A)$ and $\mu_1(A)$.

2.56 Determine if the trivial solution of the vector equation

$$x' = \begin{pmatrix} -2.2 & .3 & 1 & -.4 \\ 1.5 & -3 & -1 & .3 \\ -1.2 & .8 & -3 & .7 \\ -.1 & .3 & .3 & -1 \end{pmatrix} x$$

is globally asymptotically stable or not.

2.57 Determine if the trivial solution of the vector equation

$$x' = \begin{pmatrix} -2 & 1.7 & .1 & .1 \\ 1.8 & -4 & 1 & 1 \\ 0 & 1 & -2 & .5 \\ .4 & 1 & .5 & -2 \end{pmatrix} x$$

is globally asymptotically stable or not.

2.58 Work each of the following:

    (i) Find the matrix norm of

$$A = \begin{pmatrix} -6 & 2 & -3 \\ 2.5 & -7 & 4 \\ -2 & 1 & -8 \end{pmatrix}$$

    (ii) Find the matrix norm of A corresponding to the traffic norm $\| \cdot \|_1$.

    (iii) Find the Lozinski measure $\mu_1(A)$.

    (iv) What can you say about the stability of the trivial solution of $x' = Ax$?

2.59 Prove part (ii) of Theorem 2.61.

2.60 Find a log of each of the following matrices:

    (i) $A = \begin{bmatrix} 1 & 0 \\ 0 & e \end{bmatrix}$

    (ii) $B = \begin{bmatrix} 2 & 0 \\ 0 & -1 \end{bmatrix}$

    (iii) $C = \begin{bmatrix} 4 & 2 \\ -1 & 1 \end{bmatrix}$

    (iv) $D = \begin{bmatrix} 3 & 1 \\ -1 & 1 \end{bmatrix}$

2.61 Find the Floquet multipliers for each of the following scalar equations:

    (i) $x' = (2\sin(3t))\, x$

    (ii) $x' = (\cos^2 t)\, x$

    (iii) $x' = (-1 + \sin(4t))\, x$

2.62 Assume the scalar function a is continuous on $\mathbb{R}$ and $a(t+\omega) = a(t)$ for $t \in \mathbb{R}$, where $\omega > 0$. Prove directly by solving the scalar differential equation $x' = a(t)x$ that every nontrivial solution is of the form $x(t) = p(t)e^{rt}$ for $t \in \mathbb{R}$, where $p(t) \neq 0$ for $t \in \mathbb{R}$ is a continuously differentiable function on $\mathbb{R}$ that is periodic with period ω and r is the average value of $a(t)$ on $[0, \omega]$, [i.e. $r = \frac{1}{\omega} \int_0^\omega a(t)\, dt$]. Show that $\mu = e^{r\omega}$ is the Floquet multiplier for $x' = a(t)x$. In particular, show that all solutions of $x' = a(t)x$ are periodic with period ω iff $\int_0^\omega a(t)\, dt = 0$.

2.63 Find the Floquet multipliers for each of the following Floquet systems:

(i) $x' = \begin{bmatrix} 3 & 0 \\ 0 & \sin^2 t \end{bmatrix} x$

(ii) $x' = \begin{bmatrix} -1 + \cos t & 0 \\ \cos t & -1 \end{bmatrix} x$

(iii) $x' = \begin{bmatrix} -1 & 0 \\ \sin t & -1 \end{bmatrix} x$

(iv) $x' = \begin{bmatrix} -3 + 2\sin t & 0 \\ 0 & -1 \end{bmatrix} x$

2.64 Determine the stability of the trivial solution for each of the differential equations in Exercise 2.61 by looking at the Floquet multipliers that you found in Exercise 2.61.

2.65 Find the Floquet multipliers for each of the following Floquet systems and by looking at the Floquet multipliers determine the stability of the trivial solution on $[0, \infty)$:

(i) $x' = \begin{bmatrix} -3 + 2\sin t & 0 \\ 0 & -1 \end{bmatrix} x$

(ii) $x' = \begin{bmatrix} \cos(2\pi t) + 1 & 0 \\ \cos(2\pi t) & 1 \end{bmatrix} x$

(iii) $x' = \begin{bmatrix} -2 & 0 \\ \sin(2t) & -2 \end{bmatrix} x$

2.66 Show that

$$\Phi(t) = \begin{bmatrix} e^t(\cos t - \sin t) & e^{-t}(\cos t + \sin t) \\ e^t(\cos t + \sin t) & e^{-t}(-\cos t + \sin t) \end{bmatrix}$$

is a fundamental matrix for the Floquet system

$$x' = \begin{bmatrix} -\sin(2t) & \cos(2t) - 1 \\ \cos(2t) + 1 & \sin(2t) \end{bmatrix} x.$$

Find the Floquet multipliers for this Floquet system. What do the Floquet multipliers tell you about the stability of the trivial solution?

2.67 Show that if $\mu = -1$ is a Floquet multiplier for the Floquet system $x' = A(t)x$, then there is a nontrivial periodic solution with period 2ω.

2.68 Without finding the Floquet multipliers, find the product of the Floquet multipliers of the Floquet system

$$x' = \begin{bmatrix} 2 & \sin^2 t \\ \cos^2 t & \sin t \end{bmatrix} x.$$

2.69 Show that

$$x(t) = \begin{bmatrix} -e^{\frac{t}{2}} \cos t \\ e^{\frac{t}{2}} \sin t \end{bmatrix}$$

is a solution of the Floquet system

$$x' = \begin{bmatrix} -1 + \left(\frac{3}{2}\right)\cos^2 t & 1 - \left(\frac{3}{2}\right)\cos t \sin t \\ -1 - \left(\frac{3}{2}\right)\sin t \cos t & -1 + \left(\frac{3}{2}\right)\sin^2 t \end{bmatrix} x.$$

Using just the preceding solution, find a Floquet multiplier for the proceding system. Without solving the system, find the other Floquet multiplier. What can you say about the stability of the trivial solution? Show that for all t the coefficient matrix in the proceding Floquet system has eigenvalues with negative real parts. This example is due to Markus and Yamabe [**36**].

2.70 (Floquet Exponents) Show that if μ_0 is a Floquet multiplier for the Floquet system $x' = A(t)x$, then there is a number ρ_0 (called a Floquet exponent) such that there is a nontrivial solution $x_0(t)$ of the Floquet system $x' = A(t)x$ of the form

$$x_0(t) = e^{\rho_0 t} p_0(t),$$

where p_0 is a continuously differentiable function on $\mathbb{R}$ that is periodic with period ω.

Chapter 3

Autonomous Systems

3.1 Introduction

In Example 1.13, we discussed the logistic model of population growth:

$$N' = rN\left(1 - \frac{N}{K}\right).$$

Now suppose there is a population P of predators that prey on the organisms in the population N. If we assume that interactions between predators and prey decrease the prey population and increase the predator population at rates proportional to the product of the populations, then a reasonable model might be

$$\begin{aligned} N' &= rN\left(1 - \frac{N}{K}\right) - aNP, \\ P' &= bNP - cP, \end{aligned}$$

where a, b, c, K, and r are positive parameters. Note that we have assumed that in the absence of prey, the predator population satisfies $P' = -cP$ and so dies out.

The predator-prey equations constitute an *autonomous* system, that is, a collection of equations that do not explicitly contain the independent variable. More generally, autonomous systems have the form

$$\begin{aligned} x_1' &= f_1(x_1, x_2, \cdots, x_n), \\ x_2' &= f_2(x_1, x_2, \cdots, x_n), \\ &\vdots \\ x_n' &= f_n(x_1, x_2, \cdots, x_n). \end{aligned}$$

An equivalent vector equation is obtained by choosing

$$x = \begin{bmatrix} x_1 \\ x_2 \\ \vdots \\ x_n \end{bmatrix}, \quad f = \begin{bmatrix} f_1 \\ f_2 \\ \vdots \\ f_n \end{bmatrix},$$

so that we have

$$x' = f(x). \tag{3.1}$$

W.G. Kelley and A.C. Peterson, *The Theory of Differential Equations:*
Classical and Qualitative, Universitext 278, DOI 10.1007/978-1-4419-5783-2_3,
© Springer Science+Business Media, LLC 2010

The following theorem gives conditions under which initial value problems for (3.1) have unique solutions (see Chapter 8 for a proof).

Theorem 3.1 (Existence-Uniqueness Theorem) *Assume $f : \mathbb{R}^n \to \mathbb{R}^n$ is continuous. Then for each $t_0 \in \mathbb{R}$ and $x_0 \in \mathbb{R}^n$, the initial value problem (IVP)*
$$x' = f(x), \quad x(t_0) = x_0,$$
has a solution x. Furthermore, x has a maximal interval of existence (α, ω), where $-\infty \leq \alpha < t_0 < \omega \leq \infty$. If $\alpha > -\infty$, then $\lim_{t \to \alpha+} \|x(t)\| = \infty$, and if $\omega < \infty$, then $\lim_{t \to \omega-} \|x(t)\| = \infty$. If, in addition, f has continuous first-order partial derivatives with respect to $x_1, x_2, \cdots, x_n$ on $\mathbb{R}^n$, then the above IVP has a unique solution.

We shall assume throughout this chapter that the conditions of this theorem are satisfied. As in Chapter 2, we will use the notation $\phi(t, x)$ to represent the unique solution of (3.1) that has the value x at $t = 0$. The last part of the theorem assures us that solutions either exist for all t or go off to infinity (in norm) in finite time. The assumption that f is defined for all x is not essential, but the conclusions are somewhat more complicated if this condition is not satisfied. See Chapter 8 for the more general case.

The following theorem states that solutions change continuously with respect to their starting point. A more general result is proved in Chapter 8.

Theorem 3.2 *For each t in its maximal interval of existence, $\phi(t, x)$ is continuous as a function of x.*

Theorem 3.4 contains the distinguishing characteristics of an autonomous system. First, we need a definition.

Definition 3.3 *If x is a solution of (3.1) on its maximal interval of existence (α, ω), then $\{x(t) : \alpha < t < \omega\} \subset \mathbb{R}^n$ is called an* orbit *or* trajectory *of (3.1).*

Theorem 3.4 *The following hold:*

   (i) *Assume $x(t)$ satisfies (3.1) on an interval (c, d). Then for any constant h, $x(t - h)$ satisfies (3.1) on the interval $(c + h, d + h)$.*
   (ii) *If two orbits of (3.1) have a common point, then the orbits are identical.*

Proof Assume x is a solution of (3.1) on (c, d) and let
$$u(t) := x(t - h)$$
for $t \in (c + h, d + h)$. Then
$$\begin{aligned} u'(t) &= x'(t - h) \\ &= f(x(t - h)) \\ &= f(u(t)), \end{aligned}$$

for $t \in (c+h, d+h)$, and part (i) is established.

Now assume that x and y are solutions of (3.1) whose orbits have a common point, that is, $x(t_1) = y(t_2)$ for some numbers t_1 and t_2. By part (i), the function w defined by $w(t) := y(t - t_1 + t_2)$ is a solution of (3.1). Since $w(t_1) = x(t_1)$, we have by Theorem 3.1 that $w(t) = x(t)$ for all t in the maximal interval of existence, so $y(t - t_1 + t_2) = x(t)$ for all such t, and the orbits for x and y must be identical.   □

Note that the orbits of (3.1) *partition* $\mathbb{R}^n$, that is, each point of $\mathbb{R}^n$ lies on exactly one orbit of (3.1).

Example 3.5 Since for each positive constant α, $x(t) = \alpha \cos t$, $y(t) = -\alpha \sin t$ is a solution of the planar system

$$x' = y, \tag{3.2}$$
$$y' = -x, \tag{3.3}$$

the circles $x^2 + y^2 = \alpha^2$, $\alpha > 0$, are orbits for this system. The only other orbit for the system is the origin. Figure 1 shows several orbits of (3.2), (3.3) (the arrows in Figure 1 will be explained later).   △

Definition 3.6 We say that x_0 is an *equilibrium point* of (3.1) provided

$$f(x_0) = 0.$$

An equilibrium point x_0 is a one-point orbit since $x(t) = x_0$ is a constant solution of (3.1).

Example 3.7 Find the equilibrium points for the predator-prey model:

$$N' = rN(1 - N/K) - aNP,$$
$$P' = bNP - cP.$$

First, set $P' = P(bN - c) = 0$. There are two possibilities: $P = 0$ or $N = c/b$. If $P = 0$, then setting the first equation equal to 0 yields

$$rN(1 - N/K) = 0,$$

so $N = 0$ or $N = K$. In the other case, subsitute $N = c/b$ into the first equation and set the equation equal to zero:

$$r\frac{c}{b}\left(1 - \frac{c}{bK}\right) - \frac{ac}{b}P = 0,$$

so

$$P = \frac{r}{a}\left(1 - \frac{c}{bK}\right).$$

Consequently, the predator-prey model has three equilibrium points:

$$(0,0), \quad (K,0), \quad \text{and} \quad \left(\frac{c}{b}, \frac{r}{a}\left(1 - \frac{c}{bK}\right)\right).$$

Note that if $c = bK$, the second and third equilibrium points are the same.   △

3.2 Phase Plane Diagrams

In this section, we will concentrate on the important special case of two-dimensional autonomous systems:

$$x' = f(x, y), \tag{3.4}$$
$$y' = g(x, y). \tag{3.5}$$

(We assume throughout that $f : \mathbb{R}^2 \to \mathbb{R}$ and $g : \mathbb{R}^2 \to \mathbb{R}$ are continuous along with their first-order derivatives.) Graphing several representative orbits of (3.4), (3.5) and assigning an appropriate direction is what we call the *phase plane diagram* of (3.4), (3.5). If x, y is a solution of (3.4), (3.5), then we can think of $(x(t), y(t))$ as the position of an object in the xy-plane at time t, of $x'(t) = f(x(t), y(t))$ as the velocity of the object in the x-direction at time t, and of $y'(t) = g(x(t), y(t))$ as the velocity of the object in the y-direction at time t.

Example 3.8 Draw the phase plane diagram for the system (3.2), (3.3).

From Example 3.5 we know that the circles $x^2 + y^2 = \alpha^2$, $\alpha \in \mathbb{R}$ are orbits for this system. Note that if an object is at a point on a circle in the upper half plane, then its velocity in the x direction by equation (3.2) is

$$x'(t) = y(t) > 0.$$

Hence the phase plane diagram is given in Figure 1.   △

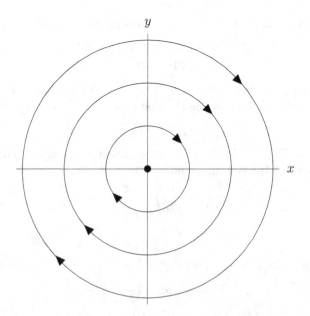

FIGURE 1. Phase plane diagram for $x' = y$, $y' = -x$.

A possible way to find the phase plane diagram for the planar system (3.4), (3.5) is to assume x, y is a solution of (3.4), (3.5). Then for those values of t such that $f(x(t), y(t)) \neq 0$, we have

$$\frac{dy}{dx}(t) = \frac{\frac{dy}{dt}(t)}{\frac{dx}{dt}(t)} = \frac{g(x(t), y(t))}{f(x(t), y(t))}.$$

Hence to find the orbits we want to solve the differential equation

$$\frac{dy}{dx} = \frac{g(x, y)}{f(x, y)}.$$

In particular, we get the orbits of (3.2), (3.3) by solving

$$\frac{dy}{dx} = \frac{g(x, y)}{f(x, y)} = \frac{-x}{y}.$$

Solving this equation by the method of separation of variables, we once again get the circles

$$x^2 + y^2 = \alpha^2.$$

Example 3.9 Draw the phase plane diagram for the differential equation

$$x'' = x - x^3.$$

When we say the phase plane diagram for this scalar differential equation we mean, unless otherwise stated, the phase plane diagram for this scalar equation written as a system in the standard way. (There are many different ways that we can write this scalar equation as an equivalent system.) Namely, assume x is a solution and define y by $y(t) = x'(t)$. Then (x, y) solves the system

$$\begin{aligned} x' &= y, \\ y' &= x - x^3. \end{aligned}$$

To find the orbits we want to solve the differential equation

$$\frac{dy}{dx} = \frac{g(x, y)}{f(x, y)} = \frac{x - x^3}{y}.$$

Solving this differential equation by the method of separating variables, we get the orbits

$$y^2 = \alpha + x^2 - \frac{x^4}{2}.$$

A phase plane diagram that contains some of these curves is given in Figure 2. Note that if an object is in the upper half-plane, then

$$x'(t) = y(t) > 0,$$

so the velocity of the object in the x-direction at a point in the upper half-plane is positive, and we can add the arrows in the phase plane diagram. Also, the orbits that cross the x-axis are orthogonal to the x-axis (why is this?). The equilibrium points are $(-1, 0)$, $(0, 0)$, and $(1, 0)$. There are two orbits that tend to the equilibrium point $(0, 0)$ as $t \to \pm\infty$. These are known as *homoclinic orbits*. Use Theorem 3.1 to convince yourself that any solution of (3.4), (3.5) corresponding to a homoclinic orbit has maximal interval of existence $(-\infty, \infty)$. The curve with equation $y^2 = x^2 - \frac{x^4}{2}$ is called the *separatrix* because because it separates two distinct types of orbits from each other.

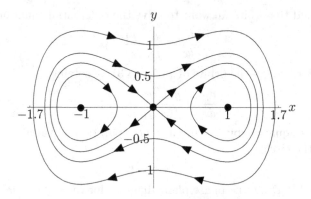

FIGURE 2. Phase plane diagram for $x'' = x - x^3$.

△

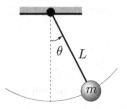

FIGURE 3. The simple pendulum.

Example 3.10 (The Pendulum Problem) A weight with mass m is suspended at the end of a rod of length L. Let θ be measured in the positive sense as indicated in Figure 3. Since the rod is rigid the mass travels along the arc of a circle with radius L. Ignoring air resistance and the mass of the rod, we get by applying Newton's second law ($F = ma$) in the direction tangential to the circle that

$$mL\theta''(t) = -mg\sin\theta(t),$$

where g is the acceleration due to gravity. Simplifying, we have

$$\theta''(t) + \frac{g}{L}\sin\theta(t) = 0.$$

Now assume that θ is a solution of this last equation, and set

$$x(t) = \theta\left(\sqrt{\frac{L}{g}}t\right).$$

Then x is a solution of what we will call the *pendulum equation*

$$x'' + \sin x = 0. \tag{3.6}$$

Note that the phase plane diagram for the pendulum equation (3.6) is now easier to draw since there are no parameters. Also, if we considered the pendulum problem on say the moon, then the acceleration due to gravity g would be different but the pendulum equation (3.6) would be the same. Writing equation (3.6) in the standard way as an equivalent system, we have

$$\begin{aligned} x' &= y, & (3.7) \\ y' &= -\sin x. & (3.8) \end{aligned}$$

To find the orbits we want to solve the differential equation

$$\frac{dy}{dx} = \frac{g(x,y)}{f(x,y)} = \frac{-\sin x}{y}.$$

Again by the method of separating variables, we find that the orbits are the curves

$$y^2 = 2\cos x + \alpha. \tag{3.9}$$

The phase diagam for the pendulum equation is given in Figure 4. Note that

$$y^2 = 2\cos x + 2$$

is the equation of the separatrix of (3.7), (3.8). Study the phase plane diagram and explain physically what is going on for each of the various types of orbits graphed in Figure 4. One such orbit is the orbit in the upper half-plane that goes from the equilibrium point $(-\pi, 0)$ to the equilibrium point $(\pi, 0)$. This type of an orbit is called a *heteroclinic orbit*. Use Theorem 3.1 to convince yourself that any solution of (3.4), (3.5) corresponding to a heteroclinic orbit has maximal interval of existence $(-\infty, \infty)$. $\triangle$

The last three examples are special cases of *Hamiltonian systems*. More generally, any two-dimensional system of the form

$$\begin{aligned} x' &= \frac{\partial h}{\partial y}, \\ y' &= -\frac{\partial h}{\partial x}, \end{aligned}$$

where $h : \mathbb{R}^2 \to \mathbb{R}$ and its first-order partial derivatives are continuous is called a Hamiltonian system. For example, in the pendulum problem

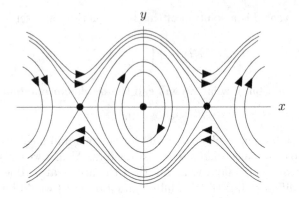

FIGURE 4. Phase plane diagram for the pendulum equation.

$h(x, y) = y^2/2 - \cos x$. We call such a function h a Hamiltonian function for the system (3.7), (3.8). Since

$$\frac{d}{dt} h(x(t), y(t))$$

$$= \frac{\partial h}{\partial x}(x(t), y(t)) x'(t) + \frac{\partial h}{\partial y}(x(t), y(t)) y'(t)$$

$$= \frac{\partial h}{\partial x}(x(t), y(t)) \frac{\partial h}{\partial y}(x(t), y(t)) - \frac{\partial h}{\partial y}(x(t), y(t)) \frac{\partial h}{\partial x}(x(t), y(t))$$

$$= 0,$$

h is constant on solution curves. Hence orbits are of the type $h(x, y) = C$. In mechanics, Hamiltonian systems often model problems in which energy is conserved, and such a function h represents the total energy of the mechanical system. Note that a necessary condition for the system (3.4), (3.5) to be a Hamiltonian system is that

$$\frac{\partial f}{\partial x} = -\frac{\partial g}{\partial y}.$$

Example 3.11 Find all Hamiltonian functions (energy functions) for the system

$$x' = f(x, y) = 3y^2 - 2x^3 y,$$
$$y' = g(x, y) = 3x^2 y^2 - 2x.$$

To find all Hamiltonian functions for this system, consider

$$\frac{\partial h}{\partial y}(x, y) = f(x, y) = 3y^2 - 2x^3 y.$$

It follows that

$$h(x, y) = y^3 - x^3 y^2 + k(x).$$

Then using
$$\frac{\partial h}{\partial x}(x, y) = -g(x, y)$$
we get that
$$-3x^2y^2 + k'(x) = 2x - 3x^2y^2.$$
It follows that
$$k'(x) = 2x$$
and hence
$$k(x) = x^2 + \alpha$$
and finally
$$h(x, y) = y^3 - x^3y^2 + x^2 + \alpha$$
are the Hamiltonian functions.

$\triangle$

The system in the next example is not a Hamiltonian system.

Example 3.12 Consider again the predator-prey model
$$N' = rN(1 - N/K) - aNP,$$
$$P' = bNP - cP.$$

In this case, we cannot compute equations for the orbits using separation of variables. However, we can get an idea of the nature of the orbits by considering the *nullclines*, that is, the curves on which one of the components of velocity is zero. The velocity (or direction) vectors are given by
$$\begin{bmatrix} N' \\ P' \end{bmatrix},$$
so we get a vertical direction vector if
$$N' = rN(1 - N/K) - aNP$$
$$= N[r(1 - N/K) - aP] = 0.$$
The lines $N = 0$ and $r(1 - N/K) - aP = 0$ are the N nullclines. Similarly, we get the P nullclines, on which the direction vectors are horizontal, by setting $P' = 0$: $P = 0$ and $N = c/b$. Figure 5 shows the N and P nullclines and small line segments indicating vertical or horizontal directions for the case that $K > c/b$. Note that equilibrium points occur wherever an N nullcline intersects a P nullcline.

We get the actual directions (up or down, left or right) by taking the signs of the components of the direction vectors into account. For example, on the N nullcline, $r(1 - N/K) - aP = 0$, the direction will be upward if the second component is positive: $P' = P(bN - c) > 0$, that is, if $P > 0$ and $N > c/b$ or if $P < 0$ and $N < c/b$. See Figure 6 for the direction of motion on all nullclines. Of course, the first quadrant is the only relevant quadrant for predator-prey interactions. From Figure 6, we see that the direction of motion of the orbits in this quadrant is

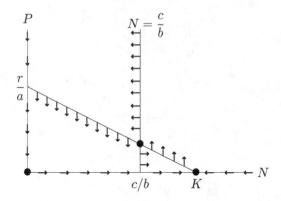

FIGURE 5. Nullclines for the predator-prey model.

counterclockwise about an equilibrium point. We will determine later in this chapter that these orbits actually spiral in toward the equilibrium as $t \to \infty$. For more practice with the important concept of nullclines, see Exercises 3.13, 3.14, and 3.15. $\triangle$

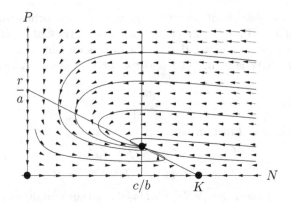

FIGURE 6. Direction vectors for the predator-prey model.

3.3 Phase Plane Diagrams for Linear Systems

In this section we study phase plane diagrams for the linear planar system

$$
\begin{aligned}
x' &= ax + by & (3.10) \\
y' &= cx + dy, & (3.11)
\end{aligned}
$$

where we assume that the coefficient matrix

$$
A := \begin{bmatrix} a & b \\ c & d \end{bmatrix}
$$

is nonsingular, that is,

$$\det(A) \neq 0.$$

In this case $(0,0)$ is the only equilibrium point for (3.10), (3.11) and zero is not an eigenvalue of A. We will see that the phase plane diagrams depend on the eigenvalues λ_1, λ_2 of the matrix A. The discussion will be divided into several cases.

Case 1: $\lambda_1 > \lambda_2 > 0$ *(unstable node).* Draw the phase plane diagram for the system

$$x' = \frac{3}{2}x + \frac{1}{2}y, \tag{3.12}$$

$$y' = \frac{1}{2}x + \frac{3}{2}y. \tag{3.13}$$

The characteristic equation is

$$\lambda^2 - 3\lambda + 2 = 0,$$

so $\lambda_1 = 2 > \lambda_2 = 1 > 0$, which is the case we are considering. Eigenpairs are

$$2, \begin{bmatrix} 1 \\ 1 \end{bmatrix} \quad \text{and} \quad 1, \begin{bmatrix} 1 \\ -1 \end{bmatrix}.$$

From these eigenpairs we get from Theorem 2.14 that

$$\begin{bmatrix} x(t) \\ y(t) \end{bmatrix} = e^{2t} \begin{bmatrix} 1 \\ 1 \end{bmatrix}, \quad \begin{bmatrix} x(t) \\ y(t) \end{bmatrix} = e^{t} \begin{bmatrix} 1 \\ -1 \end{bmatrix}$$

are solutions of the vector equation corresponding to (3.12), (3.13). A general solution of the vector equation corresponding to (3.12), (3.13) is

$$\begin{bmatrix} x(t) \\ y(t) \end{bmatrix} = c_1 e^{2t} \begin{bmatrix} 1 \\ 1 \end{bmatrix} + c_2 e^{t} \begin{bmatrix} 1 \\ -1 \end{bmatrix}.$$

The solutions

$$\begin{bmatrix} x(t) \\ y(t) \end{bmatrix} = c_1 e^{2t} \begin{bmatrix} 1 \\ 1 \end{bmatrix}$$

for $c_1 > 0$ give the orbit in Figure 7 that is the open ray emanating from the origin with polar angle $\frac{\pi}{4}$, while the solutions

$$\begin{bmatrix} x(t) \\ y(t) \end{bmatrix} = c_1 e^{2t} \begin{bmatrix} 1 \\ 1 \end{bmatrix}$$

for $c_1 < 0$ give the orbit in Figure 7 that is the open ray emanating from the origin with polar angle $-\frac{3\pi}{4}$. Similarly, the solutions

$$\begin{bmatrix} x(t) \\ y(t) \end{bmatrix} = c_2 e^{t} \begin{bmatrix} 1 \\ -1 \end{bmatrix}$$

for $c_2 > 0$ give the orbit in Figure 7 that is the open ray emanating from the origin with polar angle $-\frac{\pi}{4}$. Finally, the solutions

$$\begin{bmatrix} x(t) \\ y(t) \end{bmatrix} = c_2 e^{t} \begin{bmatrix} 1 \\ -1 \end{bmatrix}$$

for $c_2 < 0$ give the orbit in Figure 7 that is the open ray emanating from the origin with polar angle $\frac{3\pi}{4}$. Next assume $c_1 \neq 0$, $c_2 \neq 0$ and note that

$$
\begin{aligned}
\begin{bmatrix} x(t) \\ y(t) \end{bmatrix}
&= c_1 e^{2t} \begin{bmatrix} 1 \\ 1 \end{bmatrix} + c_2 e^{t} \begin{bmatrix} 1 \\ -1 \end{bmatrix} \\
&= e^{t} \left\{ c_1 e^{t} \begin{bmatrix} 1 \\ 1 \end{bmatrix} + c_2 \begin{bmatrix} 1 \\ -1 \end{bmatrix} \right\} \\
&\approx c_2 e^{t} \begin{bmatrix} 1 \\ -1 \end{bmatrix},
\end{aligned}
$$

for large negative t. So in backward time (as $t \to -\infty$) these solutions approach the origin in the direction of the eigenvector $\begin{bmatrix} 1 \\ -1 \end{bmatrix}$, if $c_2 > 0$, and in the direction of the vector $- \begin{bmatrix} 1 \\ -1 \end{bmatrix}$, if $c_2 < 0$. Also note that

$$
\begin{aligned}
\begin{bmatrix} x(t) \\ y(t) \end{bmatrix}
&= c_1 e^{2t} \begin{bmatrix} 1 \\ 1 \end{bmatrix} + c_2 e^{t} \begin{bmatrix} 1 \\ -1 \end{bmatrix} \\
&= e^{2t} \left\{ c_1 \begin{bmatrix} 1 \\ 1 \end{bmatrix} + c_2 e^{-t} \begin{bmatrix} 1 \\ -1 \end{bmatrix} \right\} \\
&\approx c_1 e^{2t} \begin{bmatrix} 1 \\ 1 \end{bmatrix},
\end{aligned}
$$

for large t. With this in mind one can draw the other orbits for the planar system (3.12), (3.13) in Figure 7. In this case we say the origin is an *unstable node*. For an unstable node there are essentially two ways to approach the origin in backward time. Two orbits approach the origin in backward time in the direction of the eigenvector corresponding to the smaller eigenvalue λ_2, and all other nontrivial orbits approach the origin in backwards time in the direction of the eigenvector corresponding to the larger eigenvalue λ_1.

Case 2: $\lambda_1 < \lambda_2 < 0$ *(stable node).* Draw the phase plane diagram for the system

$$
x' = -\frac{3}{2}x - \frac{1}{2}y, \tag{3.14}
$$

$$
y' = -\frac{1}{2}x - \frac{3}{2}y. \tag{3.15}
$$

The characteristic equation is

$$
\lambda^2 + 3\lambda + 2 = 0,
$$

so the eigenvalues are $\lambda_1 = -2$, $\lambda_2 = -1$. Note that $\lambda_1 < \lambda_2 < 0$, which is the stable node case. Let x, y be a solution of the system (3.12), (3.13) and set

$$
u(t) = x(-t), \quad v(t) = y(-t);
$$

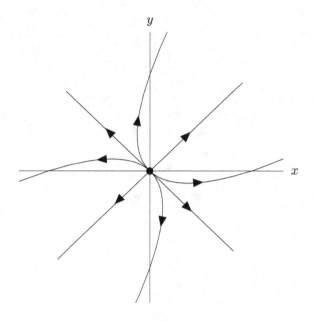

FIGURE 7. Unstable node.

then

$$
\begin{aligned}
u'(t) &= -x'(-t) \\
&= \frac{3}{2}x(-t) + \frac{1}{2}y(-t) \\
&= \frac{3}{2}u(t) + \frac{1}{2}v(t),
\end{aligned}
$$

and

$$
\begin{aligned}
v'(t) &= -y'(-t) \\
&= \frac{1}{2}x(-t) + \frac{3}{2}y(-t) \\
&= \frac{1}{2}u(t) + \frac{3}{2}v(t).
\end{aligned}
$$

Hence we have shown that if x, y is a solution of the system (3.12), (3.13), then the pair of functions defined by $x(-t)$, $y(-t)$ is a solution of (3.14), (3.15). This implies that to get the phase plane diagram for (3.14), (3.15) all we have to do is reverse the arrows in the phase plane diagram for (3.12), (3.13). See Figure 8.

   Case 3: $\lambda_1 = \lambda_2 \neq 0$ *(degenerate node).*

   Subcase 1: Two linearly independent eigenvectors corresponding to $\lambda_1 = \lambda_2$.

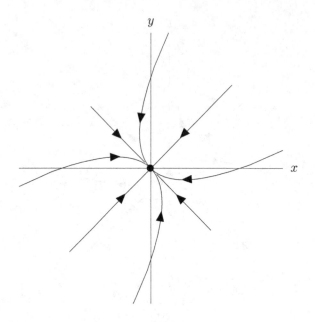

FIGURE 8. Stable node.

Draw the phase plane diagram for the system

$$x' = 3x, \tag{3.16}$$
$$y' = 3y. \tag{3.17}$$

The eigenvalues in this case are $\lambda_1 = \lambda_2 = 3$. It is easy to check that there are two linearly independent eigenvectors corresponding to $\lambda = 3$. To get the phase plane diagram, consider the differential equation

$$\frac{dy}{dx} = \frac{3y}{3x} = \frac{y}{x}.$$

Solving this differential equation, we get that

$$y = \alpha x$$

are the integral curves. Using $x' = 3x > 0$ if $x > 0$ and $x' = 3x < 0$ if $x < 0$, we get the phase plane diagram in Figure 9. Note that the origin is unstable in this case.

For the system

$$x' = -3x, \tag{3.18}$$
$$y' = -3y, \tag{3.19}$$

the eigenvalues are $\lambda_1 = \lambda_2 = -3$. Again there are two linearly independent eigenvectors corresponding to $\lambda = -3$. It can be shown that the phase plane diagram for the system (3.18), (3.19) is the phase plane diagram in Figure 9 with the arrows reversed.

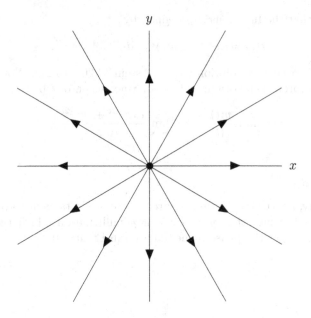

FIGURE 9. Degenerate node.

Subcase 2: Only one linearly independent eigenvector corresponding to $\lambda_1 = \lambda_2$.
Draw the phase plane diagram for the system

$$\begin{aligned} x' &= -2x, & (3.20) \\ y' &= x - 2y. & (3.21) \end{aligned}$$

In this case the eigenvalues are $\lambda_1 = \lambda_2 = -2$ and there is only one linearly independent eigenvector. An eigenvector corresponding to $\lambda = -2$ is the vector $\begin{bmatrix} 0 \\ 1 \end{bmatrix}$. It follows that

$$\begin{bmatrix} x(t) \\ y(t) \end{bmatrix} = c_1 e^{-2t} \begin{bmatrix} 0 \\ 1 \end{bmatrix}$$

are solutions. Then the positive y-axis and the negative y-axis are orbits, and the arrows associated with these orbits point toward the origin (see Figure 10). A general solution of $x' = -2x$ is defined by $x(t) = \alpha e^{-2t}$, $t \in \mathbb{R}$. Substituting this in the differential equation $y' = x - 2y$, we get the differential equation

$$y' + 2y = \alpha e^{-2t}.$$

A general solution y of this last equation is given by

$$y(t) = \alpha t e^{-2t} + \beta e^{-2t}, \quad t \in \mathbb{R}.$$

It follows that the functions x, y defined by

$$x(t) = \alpha e^{-2t}, \quad y(t) = \alpha t e^{-2t} + \beta e^{-2t},$$

for $t \in \mathbb{R}$ is a general solution to the system (3.20), (3.21). Note that all solutions approach the origin as $t \to \infty$. Since when $\alpha \neq 0$

$$\lim_{t \to \infty} \frac{y(t)}{x(t)} = \lim_{t \to \infty} \frac{\alpha t e^{-2t} + \beta e^{-2t}}{\alpha e^{-2t}}$$
$$= \lim_{t \to \infty} \frac{\alpha t + \beta}{\alpha}$$
$$= \infty,$$

respectively, nontrivial solutions approach the origin tangent to the y-axis. Also, using the fact that $y = \frac{1}{2}x$ is a y nullcline on which the flow is horizontal, we get the phase plane diagram in Figure 10.

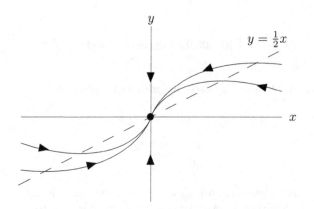

FIGURE 10. Another degenerate node.

For the system

$$x' = 2x, \qquad (3.22)$$
$$y' = -x + 2y, \qquad (3.23)$$

the eigenvalues are $\lambda_1 = \lambda_2 = 2$ and there is only one linearly independent eigenvector. Hence we again are in this subcase 2. The phase plane diagram for (3.22), (3.23) is obtained from the phase plane diagram in Figure 10 with the arrows reversed.

Case 4: $\lambda_1 < 0 < \lambda_2$ *(saddle point).* Draw the phase plane diagram for the system

$$x' = -x + 3y, \tag{3.24}$$
$$y' = 3x - y. \tag{3.25}$$

In this case the characteristic equation is

$$\lambda^2 + 2\lambda - 8 = 0,$$

so the eigenvalues are $\lambda_1 = -4$, $\lambda_2 = 2$. So we are in the case $\lambda_1 < 0 < \lambda_2$ which we are calling the saddle-point case. Eigenpairs are

$$-4, \begin{bmatrix} -1 \\ 1 \end{bmatrix} \quad \text{and} \quad 2, \begin{bmatrix} 1 \\ 1 \end{bmatrix}.$$

From these eigenpairs we get, using Theorem 2.14, that

$$\begin{bmatrix} x(t) \\ y(t) \end{bmatrix} = e^{-4t} \begin{bmatrix} -1 \\ 1 \end{bmatrix}, \quad \begin{bmatrix} x(t) \\ y(t) \end{bmatrix} = e^{2t} \begin{bmatrix} 1 \\ 1 \end{bmatrix}$$

are solutions of the vector equation corresponding to (3.24), (3.25). Hence a general solution of the vector equation corresponding to (3.24), (3.25) is

$$\begin{bmatrix} x(t) \\ y(t) \end{bmatrix} = c_1 e^{-4t} \begin{bmatrix} -1 \\ 1 \end{bmatrix} + c_2 e^{2t} \begin{bmatrix} 1 \\ 1 \end{bmatrix}.$$

The solutions

$$\begin{bmatrix} x(t) \\ y(t) \end{bmatrix} = c_1 e^{-4t} \begin{bmatrix} -1 \\ 1 \end{bmatrix},$$

for $c_1 > 0$ give the orbit in Figure 11 that is the open ray emanating from the origin with polar angle $\frac{3\pi}{4}$. Since these solutions approach the origin as $t \to \infty$, the arrow points toward the origin. For $c_1 < 0$ these solutions give the orbit in Figure 11 that is the open ray emanating from the origin with polar angle $-\frac{\pi}{4}$. Again the arrow on this ray points toward the origin. The line $y = -x$ is called the *stable manifold* for the system (3.24), (3.25). The solutions

$$\begin{bmatrix} x(t) \\ y(t) \end{bmatrix} = c_2 e^{2t} \begin{bmatrix} 1 \\ 1 \end{bmatrix}$$

imply that an object that is on the line $y = x$ stays on that line and, if $x \neq 0$, then the object moves away from the origin (see Figure 11). The line $y = x$ is called the *unstable manifold* of the system (3.24), (3.25). To see how an object moves when it is not on the stable or unstable manifold, consider the solutions

$$\begin{bmatrix} x(t) \\ y(t) \end{bmatrix} = c_1 e^{-4t} \begin{bmatrix} -1 \\ 1 \end{bmatrix} + c_2 e^{2t} \begin{bmatrix} 1 \\ 1 \end{bmatrix},$$

where $c_1 \neq 0$, $c_2 \neq 0$. Note that

$$\begin{bmatrix} x(t) \\ y(t) \end{bmatrix} \approx c_1 e^{-4t} \begin{bmatrix} -1 \\ 1 \end{bmatrix},$$

for t a large negative number. Also,

$$\begin{bmatrix} x(t) \\ y(t) \end{bmatrix} \approx c_2 e^{2t} \begin{bmatrix} 1 \\ 1 \end{bmatrix},$$

for t a large positive number.

To draw a more accurate phase plane diagram of (3.24), (3.25), we consider the nullclines. The x nullcline is

$$-x + 3y = 0 \quad \text{or} \quad y = \frac{x}{3},$$

and the y nullcline is

$$3x - y = 0 \quad \text{or} \quad y = 3x.$$

The nullclines are drawn in Figure 11 as dashed lines. Note that any orbit crossing the nullcline $y = \frac{1}{3}x$ is vertical at the crossing point and any orbit crossing the nullcline $y = 3x$ is horizontal at the crossing point.

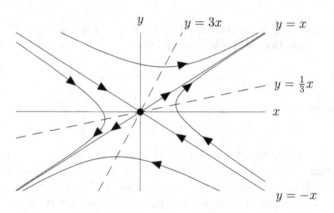

FIGURE 11. Saddle point.

Finally, we consider the complex cases. Consider the system

$$x' = ax + by, \qquad (3.26)$$
$$y' = -bx + ay, \qquad (3.27)$$

where $b \neq 0$. The characteristic equation for this system is

$$\lambda^2 - 2a\lambda + (a^2 + b^2) = 0$$

and so the eigenvalues are the complex numbers $a \pm i|b|$. Let x, y be a solution of (3.26), (3.27), and let $r(t)$, $\theta(t)$ be the polar coordinates of $x(t)$, $y(t)$; that is,

$$x(t) = r(t) \cos \theta(t), \quad y(t) = r(t) \sin \theta(t).$$

Since

$$r^2(t) = x^2(t) + y^2(t)$$

we get that
$$2r(t)r'(t) = 2x(t)x'(t) + 2y(t)y'(t).$$
Hence
$$\begin{aligned} r'(t) &= \frac{x(t)x'(t) + y(t)y'(t)}{r(t)} \\ &= \frac{x(t)[ax(t) + by(t)] + y(t)[-bx(t) + ay(t)]}{r(t)} \\ &= \frac{a[x^2(t) + y^2(t)]}{r(t)} \\ &= ar(t). \end{aligned}$$

Again assume x, y is a solution of (3.26), (3.27) and consider
$$\tan \theta(t) = \frac{y(t)}{x(t)}.$$

Differentiating both sides, we get
$$\begin{aligned} \sec^2 \theta(t)\theta'(t) &= \frac{x(t)y'(t) - x'(t)y(t)}{x^2(t)} \\ &= \frac{x(t)[-bx(t) + ay(t)] - [ax(t) + by(t)]y(t)}{x^2(t)} \\ &= -b\frac{x^2(t) + y^2(t)}{r^2(t)\cos^2(t)} \\ &= -\frac{b}{\cos^2(t)}. \end{aligned}$$

Therefore,
$$\theta'(t) = -b.$$

Hence if x, y is a solution of (3.26), (3.27) and $r(t)$, $\theta(t)$ are the polar coordinates of $x(t)$, $y(t)$, then r, θ solves the system
$$\begin{aligned} r' &= ar, & (3.28) \\ \theta' &= -b. & (3.29) \end{aligned}$$

Case 5: $\lambda_1 = i\beta$, $\lambda_2 = -i\beta$, *where* $\beta > 0$ *(center).* Draw the phase plane diagram for the system
$$\begin{aligned} x' &= by, & (3.30) \\ y' &= -bx, & (3.31) \end{aligned}$$

where $b \neq 0$. In this case the system (3.30), (3.31) is obtained from the system (3.26), (3.27) by setting $a = 0$. From before the eigenvalues are $\lambda_1 = i|b|$, $\lambda_2 = -i|b|$, so we are in the center case. If x, y is a solution of (3.26), (3.27) and $r(t)$, $\theta(t)$ are the polar coordinates of $x(t)$, $y(t)$, then r, θ solves the system
$$\begin{aligned} r' &= 0, & (3.32) \\ \theta' &= -b. & (3.33) \end{aligned}$$

Equation (3.32) implies that orbits stay the same distance from the origin, and equation (3.33) implies that the object has constant angular velocity $-b$. In Figure 12 we draw the phase plane diagram when $b < 0$ (if $b > 0$ the arrows are pointed in the opposite direction.) In this case the equilibrium point $(0,0)$ is called a *center*.

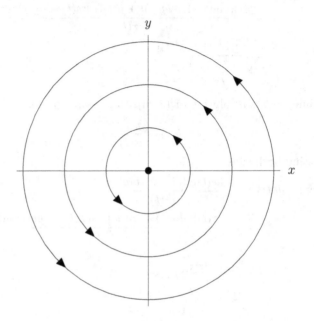

Figure 12. A center.

Case 6: $\lambda_1 = \alpha + i\beta$, $\lambda_2 = \alpha - i\beta$, *where* $\alpha \neq 0$, $\beta > 0$ *(spiral point).* Draw the phase plane diagram for the system

$$x' = ax + by, \tag{3.34}$$
$$y' = -bx + ay, \tag{3.35}$$

where $a \neq 0$ and $b \neq 0$. From before the eigenvalues are $\lambda_1 = a + i|b|$, $\lambda_2 = a - i|b|$, so we are in the spiral point case. If x, y is a solution of (3.26), (3.27) and $r(t)$, $\theta(t)$ are the polar coordinates of $x(t)$, $y(t)$, then r, θ solves the system (3.28), (3.29). Equation (3.28) implies that if an object is not at the origin and $a < 0$, then as time increases it is getting closer to the origin. Equation (3.29) implies that if an object is not at the origin, then its angular velocity is $-b$. In Figure 13 we draw the phase plane diagram when $a < 0$ and $b > 0$. In this case if $\alpha < 0$ the origin is called a *stable spiral point*. If $\alpha > 0$, then the object spirals away from the origin and we say the origin is an *unstable spiral point*.

Now that we have considered all the cases, let's look at an example.

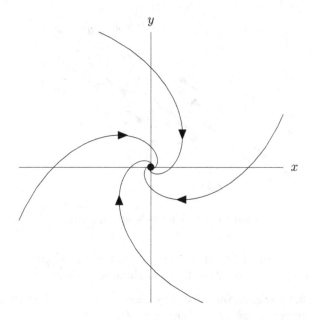

FIGURE 13. A stable spiral point.

Example 3.13 Draw the phase plane diagram for the system

$$x' = 4x - 5y, \qquad (3.36)$$
$$y' = 2x - 2y. \qquad (3.37)$$

The eigenvalues are $\lambda_1 = 1 + i$, $\lambda_2 = 1 - i$. This is Case 6. Since λ_1 and λ_2 have positive real parts, we know that the origin is an unstable spiral point. Orbits either spiral in the counterclockwise direction or in the clockwise direction. To determine which, way note that

$$x'|_{x=0} = -5y.$$

It follows from this equation that when an orbit passes through a point on the positive y-axis, x is decreasing. We now know that orbits spiral in the counterclockwise direction. Since the nullclines are $y = \frac{4}{5}x$ and $y = x$, we get Figure 14.   △

3.4 Stability of Nonlinear Systems

In this section, we will analyze the behavior of the orbits of the system of n equations in n unknowns:

$$x' = f(x) \qquad (3.38)$$

near an equilibrium point x_0. We assume throughout that $f : \mathbb{R}^n \to \mathbb{R}^n$ is continuous and has continuous first-order partial derivatives with respect to the components of x. Again, we will use the notation $\phi(\cdot, x_0)$ to denote

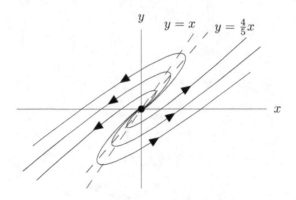

FIGURE 14. Unstable spiral point.

the unique solution of (3.38) so that $\phi(0, x_0) = x_0$, where $x_0 \in \mathbb{R}^n$. An important property of orbits is given in the next theorem.

Theorem 3.14 (Semigroup Property) *Let $x_0 \in \mathbb{R}^n$ and let (α, ω) be the maximal interval of existence of $\phi(t, x_0)$. Then*

$$\phi(t + \tau, x_0) = \phi(t, \phi(\tau, x_0)),$$

if $t, \tau, t + \tau \in (\alpha, \omega)$.

Proof Let $x_0 \in \mathbb{R}^n$ and assume $\phi(t, x_0)$ has maximal interval of existence (α, ω). Assume $t_1, t_2, t_1 + t_2 \in (\alpha, \omega)$. By Theorem 3.4 the function ψ defined by

$$\psi(t) = \phi(t + t_1, x_0)$$

is a solution of $x' = f(x)$ on $(\alpha - t_1, \omega - t_1)$. But $\psi(\cdot)$, $\phi(\cdot, \phi(t_1, x_0))$ are solutions of $x' = f(x)$ satisfying the same initial conditions at $t = 0$. Since solutions of IVPs are unique,

$$\psi(t) = \phi(t + t_1, x_0) = \phi(t, \phi(t_1, x_0)).$$

Letting $t = t_2$, we get the desired result

$$\phi(t_2 + t_1, x_0) = \phi(t_2, \phi(t_1, x_0)).$$

$\square$

The concepts of stability and asymptotic stability were defined for linear systems in Chapter 2. The next definition extends these ideas to nonlinear systems.

Definition 3.15 Let $x_0 \in \mathbb{R}^n$ and $r > 0$.

(i) The open ball centered at x_0 with radius r is the set

$$B(x_0, r) \equiv \{x \in \mathbb{R}^n : ||x - x_0|| < r\}.$$

(ii) Let x_0 be an equilibrium point for (3.38). Then x_0 is *stable* if for each ball $B(x_0, r)$, there is a ball $B(x_0, s)$ (here $s \leq r$) so that if $x \in B(x_0, s)$, then $\phi(t, x)$ remains in $B(x_0, r)$ for $t \geq 0$.

(iii) If, in addition to the conditions in part (ii), there is a ball $B(x_0, p)$ so that for each $x \in B(x_0, p)$, $\phi(t, x) \to x_0$ as $t \to \infty$, then x_0 is *asymptotically stable*.

Roughly speaking, an equilibrium point x_0 is stable if solutions of (3.38) starting near x_0 do not wander too far away from x_0 in future time. Asymptotic stability requires that x_0 is stable, and solutions near x_0 must approach it as a limit as $t \to \infty$.

One of the most useful methods for investigating stability is due to A. M. Liapunov [32]. Before we can state Liapunov's stability theorem (Theorem 3.18), we give a few definitions.

Definition 3.16 If $V : \mathbb{R}^n \to \mathbb{R}$ has partial derivatives with respect to each component of x, then we define the *gradient* of V to be the $1 \times n$ matrix function (or we could think of it as an n-dimensional vector function)

$$\text{grad } V(x) = \left[\begin{array}{cccc} \frac{\partial V}{\partial x_1}(x) & \frac{\partial V}{\partial x_2}(x) & \cdots & \frac{\partial V}{\partial x_n}(x) \end{array} \right].$$

Definition 3.17 Let x_0 be an equilibrium point for (3.38). A continuously differentiable function V defined on an open set $U \subset \mathbb{R}^n$ with $x_0 \in U$ is called a *Liapunov function* for (3.38) on U provided $V(x_0) = 0$, $V(x) > 0$ for $x \neq x_0$, $x \in U$, and

$$\text{grad } V(x) \cdot f(x) \leq 0, \tag{3.39}$$

for $x \in U$. If the inequality (3.39) is strict for $x \in U$, $x \neq x_0$, then V is called a *strict Liapunov function* for (3.38) on U.

Note that (3.39) implies that if $x \in U$, then

$$\frac{d}{dt} V(\phi(t, x)) = \text{grad } V(\phi(t, x)) \cdot f(\phi(t, x)) \leq 0,$$

as long as $\phi(t, x)$ remains in U, so V is decreasing along orbits as long as they stay in U. Here is Liapunov's stability theorem:

Theorem 3.18 (Liapunov's Stability Theorem) *If V is a Liapunov function for (3.38) on an open set U containing an equilibrium point x_0, then x_0 is stable. If V is a strict Liapunov function, then x_0 is asymptotically stable.*

Proof Assume V is a Liapunov function for (3.38) on an open set U containing an equilibrium point x_0. Pick $r > 0$ sufficiently small so that $B(x_0, r) \subset U$ and define

$$m \equiv \min\{V(x) : |x - x_0| = r\} > 0.$$

Now

$$W \equiv \{x : V(x) < \frac{m}{2}\} \cap B(x_0, r)$$

is open and contains x_0. Choose $s > 0$ so that $B(x_0, s) \subset W$. For $x \in B(x_0, s)$,

$$V(\phi(t, x)) < \frac{m}{2},$$

as long as $\phi(t, x)$ remains in W since $V(\phi(t, x))$ is decreasing. Thus $\phi(t, x)$ cannot intersect the boundary of $B(x_0, r)$ for $t \geq 0$, so $\phi(t, x)$ remains in $B(x_0, r)$ for $t \geq 0$, and x_0 is stable.

Now suppose V is a strict Liapunov function, but x_0 is not asymptotically stable. Then there is an $x \in B(x_0, s)$ so that $\phi(t, x)$ does not go to x_0 as $t \to \infty$. Since the orbit is bounded, there is an $x_1 \neq x_0$ and a sequence $t_k \to \infty$ so that $\phi(t_k, x) \to x_1$ as $k \to \infty$. Note that by the semigroup property for orbits (Theorem 3.14),

$$\phi(t_k + 1, x) = \phi(1, \phi(t_k, x)).$$

By Theorem 3.2, as $k \to \infty$,

$$V(\phi(t_k + 1), x)) = V(\phi(1, \phi(t_k, x))) \to V(\phi(1, x_1)) < V(x_1),$$

so there is an integer N for which

$$V(\phi(t_N + 1, x)) < V(x_1).$$

Choose k so that $t_k > t_N + 1$. Then

$$V(x_1) \leq V(\phi(t_k, x)) < V(\phi(t_N + 1, x)),$$

a contradiction. We conclude that x_0 is asymptotically stable. $\square$

The main difficulty in applying Liapunov's theorem is that Liapunov functions are often hard to find. In the next elementary example, we can find one easily.

Example 3.19 The origin (0,0) is an equilibrium point for the system

$$\begin{aligned} x' &= -x - xy^2, \\ y' &= -y + 3x^2y. \end{aligned}$$

Let's try the simplest possible type of Liapunov function $V(x, y) = ax^2 + by^2$, where a and b are positive constants that we must determine. Note that V is zero at (0,0) and positive elsewhere. Now

$$\begin{aligned} \text{grad } V(x, y) \cdot f(x, y) &= [2ax \quad 2by] \cdot \begin{bmatrix} -x - xy^2 \\ -y + 3x^2y \end{bmatrix} \\ &= -2ax^2 - 2ax^2y^2 - 2by^2 + 6bx^2y^2 \\ &= -6x^2 - 2y^2 < 0, \end{aligned}$$

unless $x = y = 0$, where we choose $a = 3$, $b = 1$ in order to eliminate two of the terms. Thus $V(x, y) = 3x^2 + y^2$ defines a strict Liapunov function for this system on $\mathbb{R}^2$, so the equilibrium (0,0) is asymptotically stable. $\triangle$

Example 3.20 Let h be a real-valued function defined on $\mathbb{R}^n$ with continuous first-order partial derivatives. Then the system

$$x' = -\operatorname{grad} h(x) \tag{3.40}$$

is called a *gradient system*. Note that equilibrium points for the system (3.40) are critical points for h. Suppose that h has an isolated critical point at x_0 and h has a local minimum at x_0. Then there is an open set U that contains no critical points of h other than x_0, and $h(x_0) < h(x)$, for all $x \in U$, $x \neq x_0$. Define $V : U \to \mathbb{R}$ by $V(x) = h(x) - h(x_0)$, for $x \in U$. Then $V(x_0) = 0$, $V(x) > 0$ for $x \in U$, $x \neq x_0$, and

$$\operatorname{grad} V(x) \cdot [-\operatorname{grad} h(x)] = -\|\operatorname{grad} h(x)\|^2 < 0,$$

for $x \in U$, $x \neq x_0$. Since V is a strict Liapunov function for (3.40) on U, the equilibrium point x_0 is asymptotically stable. (Compare this example to the scalar case given in Theorem 1.17.) $\triangle$

Actually, Liapunov functions can be used to deduce additional information about solutions. Let's start with a couple of definitions.

Definition 3.21 A set S is said to be *positively invariant* for system (3.38) if for each $x_0 \in S$, $\phi(t, x_0) \in S$, for $t \in [0, \omega)$.

Definition 3.22 The ω-*limit set* for an orbit $\phi(t, x_0)$ with right maximal interval of existence $[0, \infty)$ is defined to be

$$\{z \in \mathbb{R}^n : \text{ there is a sequence } t_k \to \infty \text{ so that } \phi(t_k, x_0) \to z\}.$$

The α-*limit set* for an orbit $\phi(t, x_0)$ with left maximal interval of existence $(-\infty, 0]$ is defined to be

$$\{z \in \mathbb{R}^n : \text{ there is a sequence } t_k \to -\infty \text{ so that } \phi(t_k, x_0) \to z\}.$$

Theorem 3.23 *If V is a Liapunov function for (3.38) on a bounded open set U, then for any constant $c > 0$ such that the set $\{x \in U : V(x) \leq c\}$ is closed in $\mathbb{R}^n$, this set is positively invariant.*

Proof Fix a constant $c > 0$ such that $S \equiv \{x \in U : V(x) \leq c\}$ is a closed set in $\mathbb{R}^n$. For each $x \in S$, $V(\phi(t, x)) \leq V(x) \leq c$ for $t \geq 0$, as long as the solution exists, and since S is bounded, Theorem 3.1 implies that the solution exists for all $t \geq 0$. Consequently, $\phi(t, x)$ remains in S for all $t \geq 0$, so S is positively invariant. $\square$

As a result of this theorem we get that for Example 3.19, the elliptical regions $\{(x, y) : 3x^2 + y^2 \leq c\}$, $c > 0$, are positively invariant sets.

The next theorem shows that, under the right conditions, the existence of a Liapunov function implies that solutions in a certain set must converge to an equilibrium point. Note that the Liapunov function is not assumed to be strict!

Theorem 3.24 (LaSalle Invariance Theorem) *Let V be a Liapunov function for (3.38) on a bounded open set U containing an equilibrium point x_0. If $c > 0$ is a constant so that $S := \{x : V(x) \leq c\}$ is a closed set in $\mathbb{R}^n$, and if there is no $x \neq x_0$ in S for which $V(\phi(t, x))$ is constant for $t \geq 0$, then for all $x \in S$, $\phi(t, x) \to x_0$ as $t \to \infty$.*

Proof Let $x \in S$ and define W to be the ω-limit set for the orbit $\phi(t, x)$. Then W is a closed, nonempty subset of S. First, we claim V is constant on W. If not, choose $t_k \to \infty$, $s_i \to \infty$, so that $\phi(t_k, x) \to z_1$, $\phi(s_i, x) \to z_2$, and $a := V(z_1) < b := V(z_2)$. By the continuity of V, there is an N so that if $k > N$, then $V(\phi(t_k, x)) < (a + b)/2$. If $s_i > t_k$, then $V(\phi(s_i, x)) \leq V(\phi(t_k, x)) < (a + b)/2$ for $k > N$, which contradicts $V(\phi(s_i, x)) \to b$.

Next, we claim that W is positively invariant. Let $z \in W$. Then $\phi(t_k, x) \to z$ for some $t_k \to \infty$. For any $s \geq 0$, we get, using the semigroup property (Theorem 3.14),

$$\phi(t_k + s, x) = \phi(s, \phi(t_k, x)) \to \phi(s, z),$$

as $k \to \infty$, by the continuity of solutions on initial conditions (Theorem 3.2), so $\phi(s, z) \in W$ for $s \geq 0$, and hence W is positively invariant.

Thus for any $z \in W$, $\phi(t, z) \in W$ for $t \geq 0$, so $V(\phi(t, z))$ is constant for $t \geq 0$. The hypotheses of the theorem imply that $z = x_0$, and we conclude that W contains only the equilibrium point x_0. Now if there were an $x \in S$ for which $\phi(t, x)$ did not go to x_0 as $t \to \infty$, then there would have to be a sequence $t_k \to \infty$ so that $\phi(t_k, x) \to x_1 \neq x_0$, in contradiction of $W = \{x_0\}$. □

Example 3.25 (Pendulum Problem with Friction) Now consider the pendulum with friction:

$$\theta'' + r\theta' + \frac{g}{L} \sin \theta = 0.$$

(The frictionless pendulum was discussed in Example 3.10.) Here we are assuming that the force of friction (and/or air resistance) is proportional to the angular velocity. Let's write the equation in system form by setting $x = \theta$, $y = \theta'$:

$$\begin{aligned} x' &= y, \\ y' &= -ry - \frac{g}{L} \sin x. \end{aligned}$$

We choose a Liapunov function that is essentially (see Exercise 3.18) the total energy of the system:

$$V(x, y) = \frac{y^2}{2} + \frac{g}{L}(1 - \cos x),$$

for $(x, y) \in U = \{(x, y) : -\pi/2 < x < \pi/2, -\sqrt{2} < y < \sqrt{2}\}$. This is a reasonable choice since the presence of friction should cause the energy of the pendulum to diminish as time progresses. Note that $V(0,0) = 0$ and $V(x,y) > 0$ for all other $(x,y) \in U$. Also,

$$\text{grad } V(x,y) \cdot f(x,y)$$

$$= \left[\frac{g}{L}\sin x \quad y\right] \cdot \left[\begin{array}{c} y \\ -ry - \frac{g}{L}\sin x \end{array}\right]$$

$$= -ry^2 \leq 0,$$

so V is a Liapunov function on U. Let $0 < c < \min\{1, \frac{g}{L}\}$; then $S := \{(x,y) \in U : V(x,y) \leq c\}$ is a closed set in $\mathbb{R}^2$. Now V is constant only on solutions that lie on the x-axis, but the only such solution in U is the equilibrium $(0,0)$ (why?) Then Theorem 3.24 implies that all orbits in S go to $(0,0)$ as $t \to \infty$. What does this conclusion say about the motion of the pendulum? △

3.5 Linearization of Nonlinear Systems

In this section we will reexamine the behavior of solutions of

$$x' = f(x), \tag{3.41}$$

near an equilibrium x_0 by relating (3.41) directly to a linear system. Specifically, assume f has the form

$$f(x) = A(x - x_0) + g(x), \tag{3.42}$$

where A is an $n \times n$ constant matrix, and g satisfies

$$\lim_{x \to x_0} \frac{\|g(x)\|}{\|x - x_0\|} = 0. \tag{3.43}$$

Conditions (3.42) and (3.43) are equivalent to the condition that f is differentiable at x_0. This is certainly true if f has continuous first-order partials at x_0, as we are assuming in this chapter. The matrix A is the Jacobian matrix of f at x_0 given by

$$A = \left[\begin{array}{cccc} \frac{\partial f_1}{\partial x_1} & \frac{\partial f_1}{\partial x_2} & \cdots & \frac{\partial f_1}{\partial x_n} \\ \frac{\partial f_2}{\partial x_1} & \frac{\partial f_2}{\partial x_2} & \cdots & \frac{\partial f_2}{\partial x_n} \\ \vdots & \vdots & & \vdots \\ \frac{\partial f_n}{\partial x_1} & \frac{\partial f_n}{\partial x_2} & \cdots & \frac{\partial f_n}{\partial x_n} \end{array}\right],$$

where $f_1, f_2, \cdots, f_n$ are the components of f, and all partial derivatives are computed at x_0. The main theme of this section is the following principle:

If all the eigenvalues of A have nonzero real parts, then the behavior of the orbits of (3.41) near x_0 is similar to the behavior of the orbits of the linearized system

$$x' = Ax,$$

near the origin.

In particular, we have the following stability theorem.

Theorem 3.26 *Let f be defined by (3.42) and g satisfy property (3.43).*

    (i) *If all eigenvalues of A have negative real parts, then the equilibrium x_0 is asymptotically stable.*

    (ii) *If some eigenvalue of A has positive real part, then x_0 is unstable.*

Proof (i) Choose a constant $\delta > 0$ so that $\Re(\lambda) < -\delta < 0$, for each eigenvalue λ of A. Similar to the proof of Theorem 2.49, there is a constant $C > 0$ so that

$$||e^{At}x|| \leq Ce^{-\delta t}||x||,$$

for $t \geq 0$ and all $x \in \mathbb{R}^n$. Let A^T be the transpose of A. Since A and A^T have the same eigenvalues, there is a constant $D > 0$ so that

$$||e^{tA^T}|| \leq De^{-\delta t}||x||,$$

for $t \geq 0$. Define

$$Bx \equiv \int_0^\infty e^{tA^T} e^{tA} x \, dt.$$

Since

$$||e^{tA^T} e^{tA} x|| \leq CDe^{-\delta t}||x||,$$

for $t \geq 0$, the integral converges by the comparison test for improper integrals. Then B is a well-defined $n \times n$ matrix. Note that B is symmetric since

$$
\begin{aligned}
B^T &= \int_0^\infty \left(e^{tA}\right)^T \left(e^{tA^T}\right)^T dt \\
&= \int_0^\infty e^{tA^T} e^{tA} \, dt \\
&= B.
\end{aligned}
$$

Define a trial Liapunov function V by

$$
\begin{aligned}
V(x) &= (x - x_0)^T B(x - x_0) \\
&= \int_0^\infty (x - x_0)^T e^{tA^T} e^{tA} (x - x_0) \, dt \\
&= \int_0^\infty ||e^{tA}(x - x_0)||^2 \, dt
\end{aligned}
$$

for $x \in \mathbb{R}^n$. Then $V(x_0) = 0$ and $V(x) > 0$ for $x \neq x_0$.

Finally, we claim that $\operatorname{grad} V(x) \cdot f(x) < 0$ for $||x - x_0||$ small and positive. Since

$$\operatorname{grad} V(x) = 2B(x - x_0),$$

we have by the Cauchy-Schwarz inequality

$$
\begin{aligned}
&\operatorname{grad} V(x) \cdot f(x) \\
&= 2(x - x_0)^T B[A(x - x_0) + g(x)] \\
&\leq 2(x - x_0)^T BA(x - x_0) + 2||x - x_0|| \, ||B|| \, ||g(x)||.
\end{aligned}
\tag{3.44}
$$

In order to calculate the first term in the last expression, we use the fact from Chapter 2 that e^{tA} and e^{tA^T} satisfy the linear systems for the matrices A and A^T, respectively:

$$
\begin{aligned}
2x^T BAx &= x^T A^T Bx + x^T BAx \\
&= \int_0^\infty x^T \left[A^T e^{tA^T} e^{tA} + e^{tA^T} e^{tA} A \right] x \, dt \\
&= \int_0^\infty x^T \left[\left(e^{tA^T} \right)' e^{tA} + e^{tA^T} \left(e^{tA} \right)' \right] x \, dt \\
&= \int_0^\infty x^T \left[e^{tA^T} e^{tA} \right]' x \, dt \\
&= -\|x\|^2,
\end{aligned}
$$

by the fundamental theorem of calculus. Returning to our calculation (3.44), we have

$$
\text{grad } V(x) \cdot f(x) \leq -\|x - x_0\|^2 + 2\|x - x_0\|^2 \, \|B\| \, \frac{\|g(x)\|}{\|x - x_0\|}.
$$

Finally, (3.43) implies that

$$
\text{grad } V(x) \cdot f(x) < 0,
$$

for $\|x - x_0\|$ small and positive. Since V is a strict Liapunov function on a small ball containing x_0, we know from Theorem 3.18 that x_0 is asymptotically stable.

(ii) See Hirsch and Smale [**23**].   □

Before looking at examples, we consider the special case of a two-dimensional system. Then A is a 2×2 matrix, and its characteristic equation is

$$
\lambda^2 - (\text{tr } A)\lambda + \det A = 0.
$$

(See Exercise 2.19.) The eigenvalues are

$$
\lambda = \frac{\text{tr } A \pm \sqrt{(\text{tr } A)^2 - 4 \det A}}{2}.
$$

Now consider two cases.

Case 1: $(\text{tr } A)^2 - 4 \det A \geq 0$.

In this case, the eigenvalues are real. In order that they both be negative, it is necessary that $\text{tr } A < 0$. If, in addition, $\det A \leq 0$, then one eigenvalue will be negative and the other will be nonnegative. Thus it is also necessary to have $\det A > 0$.

Case 2: $(\text{tr } A)^2 - 4 \det A < 0$.

Now the eigenvalues are complex, and det $A > 0$. The real part of each eigenvalue is $\frac{1}{2}$tr A, so again the eigenvalues will have negative real part if and only if tr $A < 0$.

To summarize, in the two-dimensional case both eigenvalues of A have negative real parts if and only if tr $A < 0$ and det $A > 0$.

Example 3.27 Consider again the predator-prey model

$$
\begin{aligned}
N' &= rN(1 - N/K) - aNP, \\
P' &= bNP - cP.
\end{aligned}
$$

(See Examples 3.7 and 3.12.) One of the equilibrium points was

$$
(N_0, P_0) = \left(\frac{c}{b}, \frac{r}{a} \left(1 - \frac{c}{bK} \right) \right).
$$

The Jacobian matrix evaluated at this equilibrium is

$$
\begin{aligned}
A &= \begin{bmatrix} r - \frac{2rN_0}{K} - aP_0 & -aN_0 \\ bP_0 & bN_0 - c \end{bmatrix} \\
&= \begin{bmatrix} -\frac{rc}{bK} & -aN_0 \\ bP_0 & 0 \end{bmatrix}.
\end{aligned}
$$

Then tr $A = -rc/bK < 0$ and det $A = abN_0P_0$. We conclude that the equilibrium is asymptotically stable if (N_0, P_0) is in the first quadrant. In terms of the parameters, the stability requirement is $bK > c$.

Another equilibrium point for this system is $(K, 0)$. We can show that this equilibrium is stable for $bK < c$ (see Exercise 3.31). Also, note that this equilibrium coincides with the first one if $bK = c$. Consequently, we have a two-dimensional example of a transcritical bifurcation (see Example 1.24). What are the implications of this bifurcation for the predator and prey populations?   △

Example 3.28 Given that the radius $x(t)$ of a vapor bubble in a liquid at time t satisfies the differential equation

$$
xx'' + \frac{3}{2}(x')^2 = 1 - \frac{1}{x},
$$

determine the stability of the bubble which at $t = 0$ has radius 1 and is at rest.

Writing the preceding differential equation as a vector equation in the standard way, that is, letting

$$
z = \begin{bmatrix} x \\ y \end{bmatrix},
$$

where $y = x'$, we get the vector equation

$$
z' = \begin{bmatrix} y \\ \frac{1}{x} - \frac{1}{x^2} - \frac{3y^2}{2x} \end{bmatrix}.
$$

Note that the bubble which at $t = 0$ has radius 1 and is at rest corresponds to the equilibrium point $(1, 0)$ for the preceding system. In this case

$$\left[\begin{array}{cc} f_x(x, y) & f_y(x, y) \\ g_x(x, y) & g_y(x, y) \end{array} \right] = \left[\begin{array}{cc} 0 & 1 \\ -\frac{1}{x^2} + \frac{2}{x^3} + \frac{3y^2}{2x^2} & \frac{3y}{x} \end{array} \right].$$

Hence

$$\left[\begin{array}{cc} f_x(1, 0) & f_y(1, 0) \\ g_x(1, 0) & g_y(1, 0) \end{array} \right] = \left[\begin{array}{cc} 0 & 1 \\ 1 & 0 \end{array} \right].$$

Since the determinant of this matrix equals -1, the equilibrium point $(1, 0)$ is unstable. That is, the bubble with radius 1 at rest is unstable.

$\triangle$

Part (ii) of the stability theorem deserves more comment. If the 2×2 constant matrix A has a positive eigenvalue and a negative eigenvalue, then the equilibrium x_0 is called a *saddle point*. In Section 3.3, we saw that saddles for two-dimensional linear systems have the property that there are two orbits that approach the saddle as $t \to \infty$ along a line (the *stable manifold*), and two orbits that converge to the saddle as $t \to -\infty$ along a different line (the *unstable manifold*). Here is a similar result for nonlinear systems.

Theorem 3.29 *In the case of two equations in two unknowns, assume A has eigenvalues λ_1, λ_2 with $\lambda_1 < 0 < \lambda_2$. Then there are two orbits of (3.41) that go to x_0 as $t \to \infty$ along a smooth curve tangent at x_0 to the eigenvectors for λ_1 and two orbits that go to x_0 as $t \to -\infty$ along a smooth curve tangent at x_0 to the eigenvectors of λ_2.*

See Hubbard and West [**27**] for a proof.

Example 3.30 For the pendulum with friction (see Example 3.25), the system is

$$\begin{aligned} x' &= y, \\ y' &= -ry - \frac{g}{L}\sin x, \end{aligned}$$

with equilibria $(n\pi, 0)$ for all integers n. The Jacobian matrix at $(n\pi, 0)$ is

$$A = \left[\begin{array}{cc} 0 & 1 \\ \frac{g}{L}(-1)^{n+1} & -r \end{array} \right]$$

with eigenvalues

$$\lambda = \frac{-r \pm \sqrt{r^2 + (-1)^{n+1}4g/L}}{2}.$$

If n is even, then there are two eigenvalues with negative real parts, and the equilibrium point is asymptotically stable. If n is odd, then

$$\lambda_1 \equiv \frac{-r - \sqrt{r^2 + 4g/L}}{2} < 0 < \lambda_2 \equiv \frac{-r + \sqrt{r^2 + 4g/L}}{2},$$

so these equilibrium points are saddles. Since

$$(A - \lambda I)\begin{bmatrix} x \\ y \end{bmatrix} = \begin{bmatrix} -\lambda & 1 \\ -\frac{g}{L}(-1)^{n+1} & -r - \lambda \end{bmatrix}\begin{bmatrix} x \\ y \end{bmatrix},$$

the eigenvectors (when drawn as position vectors) corresponding to each eigenvalue λ_i lie on the line $y = \lambda_i x$, for $i = 1, 2$. Figure 15 shows the flow near $(\pi, 0)$.   △

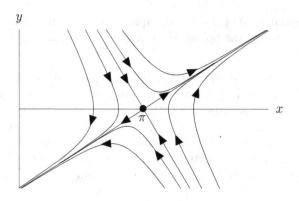

FIGURE 15. Orbits near $(\pi, 0)$.

Example 3.31 In Section 3.2, we encountered some Hamiltonian systems of the form

$$x' = \frac{\partial h}{\partial y},$$
$$y' = -\frac{\partial h}{\partial x}$$

and showed that the orbits are of the type $h(x, y) = C$, where C is constant. Note that an equilibrium point (x_0, y_0) is a critical point for h. The Jacobian matrix is

$$A = \begin{bmatrix} \frac{\partial^2 h}{\partial x \partial y} & \frac{\partial^2 h}{\partial y^2} \\ -\frac{\partial^2 h}{\partial x^2} & -\frac{\partial^2 h}{\partial x \partial y} \end{bmatrix}.$$

Now $\text{tr}[A] = 0$, and

$$\det A = \frac{\partial^2 h}{\partial x^2}\frac{\partial^2 h}{\partial y^2} - \left(\frac{\partial^2 h}{\partial x \partial y}\right)^2.$$

By the second derivative test for functions of two variables, if $\det A > 0$, then h has a local max or min at (x_0, y_0), so such an equilibrium point is the center of a family of closed curves of the form $h(x, y) = C$. On the other hand, if $\det A < 0$, then (x_0, y_0) is a saddle point. The frictionless pendulum studied in Section 3.2 is an elementary example of a Hamiltonian system with both centers and saddles.   △

Let's mention just one more characteristic of nonlinear systems that is inherited from the linearized system. If the 2×2 matrix A has complex eigenvalues with nonzero real part, then we know from Section 3.2 that the solutions of $x' = Ax$ spiral around the origin. Specifically, if the real part is positive, orbits spiral outward away from the origin as t increases, while in the asymptotically stable case, orbits spiral into the origin. For the corresponding nonlinear system $x' = f(x)$, the orbits near the equilibrium exhibit similar behavior.

Example 3.32 In Example 3.30, we found the eigenvalues

$$\lambda = \frac{-r \pm \sqrt{r^2 + (-1)^{n+1}4g}}{2}.$$

For n even, the eigenvalues are complex if $r^2 - 4g < 0$, that is, if friction is relatively small. Consequently, the phase plane diagram in the vicinity of the equilibrium points $(-\pi, 0)$, $(0, 0)$, and $(\pi, 0)$ must be of the general form shown in Figure 16. What are the possible motions of the pendulum? $\triangle$

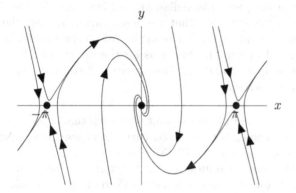

FIGURE 16. Phase plane diagram for the pendulum with friction.

Example 3.33 For the predator-prey model considered in Example 3.27, the eigenvalues for the Jacobian matrix at (P_0, N_0) are computed from

$$\det(A - \lambda I) = \lambda^2 + \frac{rc}{bK}\lambda + abN_0P_0 = 0.$$

These eigenvalues are complex if the discriminate is negative:

$$\frac{r^2c^2}{b^2K^2} - 4abN_0P_0 < 0,$$

or

$$\frac{rc}{b^2K^2} < 4\left(1 - \frac{c}{bK}\right).$$

In this case, orbits will spiral into the stable equilibrium at (N_0, P_0); see Figure 17. $\triangle$

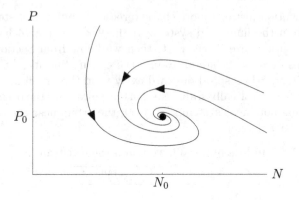

FIGURE 17. A phase plane diagram for the predator-prey model.

3.6 Existence and Nonexistence of Periodic Solutions

Looking at the phase plane diagram in Figure 3.10 for the simple pendulum problem, we see orbits that are closed curves between the separatrix and the critical point $(0,0)$. These orbits (also called *cycles* in Definition 3.34) correspond to periodic solutions for the system of differential equations for the simple pendulum. More generally, consider the system of n equations in n unknowns

$$x' = f(x). \tag{3.45}$$

Assume that this equation has a solution x such that $x(0) = x(\omega)$ (that is, the solution corresponds to a closed curve in phase space). Now $u(t) := x(t + \omega)$ is also a solution of (3.45), and $u(0) = x(\omega) = x(0)$. From the uniqueness theorem for solutions of initial value problems, it follows that $x(t + \omega) = x(t)$ for all t, so closed curves do represent periodic solutions.

Definition 3.34 A *cycle* is a nonconstant periodic solution of (3.45). If a cycle is the ω-limit set or the α-limit set of a distinct orbit, then the cycle is called a *limit cycle*. If a limit cycle is the ω-limit set of every nearby orbit, then the limit cycle is said to be *stable*.

In practice, cycles occur in physical, biological, or chemical systems that exhibit self-sustained oscillations (that is, oscillations that occur without an outside stimulus). For the simple pendulum, these cycles are not isolated since different initial positions of the pendulum lead to different periodic motions. However, it is more common in applications to have an isolated limit cycle, for which small perturbations yield solutions that converge to the stable limit cycle. Our first example is of this type.

Example 3.35 Show that the system

$$x' = x + y - x\left(x^2 + y^2\right), \tag{3.46}$$

$$y' = -x + y - y\left(x^2 + y^2\right) \tag{3.47}$$

has a limit cycle.

The expression $x^2 + y^2$ in these differential equations suggests that we find the equivalent system in polar coordinates. Let x, y be a solution of this system and let r, θ be the corresponding polar coordinates. Differentiating both sides of the equation

$$r^2(t) = x^2(t) + y^2(t),$$

we get that

$$2r(t)r'(t) = 2x(t)x'(t) + 2y(t)y'(t).$$

It follows that

$$
\begin{aligned}
&r(t)r'(t) \\
=\ &x(t)\left[x(t) + y(t) - x(t)\left(x^2(t) + y^2(t)\right)\right] \\
+\ &y(t)\left[-x(t) + y(t) - y(t)\left(x^2(t) + y^2(t)\right)\right] \\
=\ &x^2(t) + y^2(t) - x^2(t)\left(x^2(t) + y^2(t)\right) - y^2(t)\left(x^2(t) + y^2(t)\right) \\
=\ &x^2(t) + y^2(t) - \left(x^2(t) + y^2(t)\right)^2 \\
=\ &r^2(t) - r^4(t).
\end{aligned}
$$

Hence r solves the differential equation

$$r' = r(1 - r^2).$$

Next, differentiating both sides of the equation

$$\tan\theta(t) = \frac{y(t)}{x(t)},$$

we get that

$$
\begin{aligned}
\sec^2\left(\theta(t)\right)\theta'(t) &= \frac{x(t)y'(t) - y(t)x'(t)}{x^2(t)} \\
&= \frac{-x^2(t) - y^2(t)}{x^2(t)} \\
&= \frac{-r^2(t)}{r^2(t)\cos^2\theta(t)} \\
&= \frac{-1}{\cos^2\theta(t)}.
\end{aligned}
$$

Therefore, θ solves the differential equation

$$\theta' = -1.$$

Then the equivalent polar coordinate system is

$$
\begin{aligned}
r' &= r(1 - r^2), \\
\theta' &= -1.
\end{aligned}
$$

From this we get the phase plane diagram in Figure 18. Note that the circle $r = 1$ is a stable limit cycle. $\triangle$

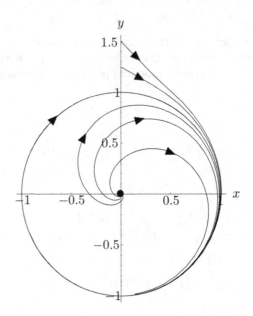

FIGURE 18. A periodic orbit.

The preceding example was very simple since we were able to use polar coordinates to find the periodic solution, which turned out to be a circle in the phase plane! In general, the problem of finding cycles is difficult. The most fundamental result in this area, the Poincaré-Bendixson theorem, works only in two-dimensions, and we will mostly limit our discussion to the two-dimensional case for the remainder of this section.

Theorem 3.36 (Poincaré-Bendixson) *Consider equation* (3.45) *for the case $n = 2$. If $\phi(t, x)$ is a bounded orbit for $t \geq 0$ and W is its ω-limit set, then either W is a cycle, or for each $y \in W$, the ω-limit set of $\phi(t, y)$ is a set of one or more equilibrium points.*

A complete proof of this theorem is surprisingly technical (see Hirsch and Smale [**23**]). We give just a rough sketch in order that the reader can get an idea of why it is true.

Suppose that W contains a y that is not an equilibrium point. Choose z in the ω-limit set of y. If z is not an equilibrium point, we can define a *local section S* at z to be a curve (minus the endpoints) containing z so that no solution of (3.45) is tangent to S (see Figure 19). Now we can show that there are times t_1 and t_2 so that $\phi(t_1, y)$, $\phi(t_2, y)$ are in $S \cap W$. If $\phi(t_1, y) = \phi(t_2, y)$, then the orbit $\phi(t, y)$ is closed and is a cycle in W. Otherwise, there are sequences $r_i \to \infty$, $s_j \to \infty$, so that $\phi(r_i, x) \in S$, $\phi(r_i, x) \to \phi(t_1, y)$ and $\phi(s_j, x) \in S$, $\phi(s_j, x) \to \phi(t_2, y)$ as i and j go to infinity. However, this impossible since the trajectory $\phi(t, x)$ can cross S in only one direction. (If it crossed in both directions, by continuity, the vector field would have to be tangent to S at some point in between.) Consequently, either y lies on a cycle or the ω-limit set of $\phi(t, y)$ consists only of equilibrium points. In the case that y lies on a cycle, it can be shown that this cycle is the entire ω-limit set.

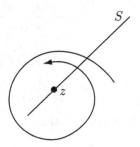

FIGURE 19. A local section at z.

The next example illustrates the use of this theorem in a geometrically simple case.

Example 3.37 Show that the system

$$x' = x + y - x(x^2 + 2y^2), \tag{3.48}$$
$$y' = -x + y - y(x^2 + 2y^2) \tag{3.49}$$

has a limit cycle.

Let x, y be a solution of this system and let r, θ be the corresponding polar coordinates. Differentiating both sides of the equation

$$r^2(t) = x^2(t) + y^2(t),$$

we get that

$$2r(t)r'(t) = 2x(t)x'(t) + 2y(t)y'(t).$$

It follows that

$$
\begin{aligned}
&r(t)r'(t) \\
=\ & x(t)\left[x(t) + y(t) - x(t)\left(x^2(t) + 2y^2(t)\right)\right] \\
+\ & y(t)\left[-x(t) + y(t) - y(t)\left(x^2(t) + 2y^2(t)\right)\right] \\
=\ & x^2(t) + y^2(t) - x^2(t)\left(x^2(t) + 2y^2(t)\right) - y^2(t)\left(x^2(t) + 2y^2(t)\right) \\
=\ & x^2(t) + y^2(t) - \left(x^2(t) + y^2(t)\right)\left(x^2(t) + 2y^2(t)\right) \\
=\ & r^2(t) - r^4(t)\left(1 + \sin^2(t)\right).
\end{aligned}
$$

Hence $r(t)$ solves the differential equation

$$r' = r - r^3\left[1 + \sin^2\theta\right].$$

Next, differentiating both sides of the equation

$$\tan\theta(t) = \frac{y(t)}{x(t)},$$

we get that

$$
\begin{aligned}
\sec^2(\theta(t))\,\theta'(t) &= \frac{x(t)y'(t) - y(t)x'(t)}{x^2(t)} \\
&= \cdots \\
&= \frac{-x^2(t) - y^2(t)}{x^2(t)} \\
&= \frac{-r^2(t)}{r^2(t)\cos^2\theta(t)} \\
&= \frac{-1}{\cos^2\theta(t)}.
\end{aligned}
$$

Therefore, θ solves the differential equation

$$
\theta' = -1.
$$

Hence the equivalent polar coordinate system is

$$
\begin{aligned}
r' &= r - r^3\left[1 + \sin^2\theta\right], \\
\theta' &= -1.
\end{aligned}
$$

Note that

$$
\begin{aligned}
r'|_{r=\frac{1}{2}} &= \frac{1}{2} - \frac{1}{8}\left[1 + \sin^2\theta\right] \\
&\geq \frac{1}{2} - \frac{1}{4} > 0.
\end{aligned}
$$

This calculation implies that all orbits that intersect the circle $r = \frac{1}{2}$ enter into the exterior of the circle $r = \frac{1}{2}$ as t increases. Next, note that

$$
\begin{aligned}
r'|_{r=1.1} &= 1.1 - 1.331\left[1 + \sin^2\theta\right] \\
&\leq 1.1 - 1.331 < 0.
\end{aligned}
$$

This implies that all orbits that intersect the circle $r = 1.1$ enter into the interior of the circle $r = 1.1$ as t increases. Thus the annular region

$$
D := \left\{(x,y): \frac{1}{2} < r < 1.1\right\}
$$

is positively invariant. Since D contains no equilibrium points of the system (3.48), (3.49) we have by the Poincaré-Bendixson theorem (Theorem 3.36) that the system (3.48), (3.49) has at least one limit cycle in D.

$\triangle$

In the preceding example, the invariant annular region has the property that all orbits that intersect the boundary enter into the region. In some cases, a portion of the boundary may be invariant, and the question arises whether this set can be an ω-limit set. The next result is useful in answering this question.

Theorem 3.38 *Let D be a closed, positively invariant set in $\mathbb{R}^2$ bounded by a simple closed curve C. Suppose A is a positively invariant, proper subset of C. If A is an ω-limit set for some point in D, then A is a set of equilibrium points.*

Proof Choose a point x that is in C but not in A. Suppose that A is the ω-limit set for the orbit $\phi(t, y)$, $y \in D$. Assume that there is a point $z \in A$ that is not an equlibrium point. Then the point $\phi(1, z)$ is also not an equilibrium point, and the flow at $\phi(1, z)$ is tangent to C at $\phi(1, z)$. Thus there is a local section S at $\phi(1, z)$ (see the sketch of the proof of the Poincaré-Bendixson theorem). Now the portion of S inside C is a subset of a curve E that connects $\phi(1, z)$ to x (see Figure 20). Now $\phi(t, y)$ must cross E infinitely many times, but it can cross S only in one direction. Then it must cross E in infinitely many points that are not in S. It follows that the ω-limit set of $\phi(t, y)$ must contain a point that is not in A, and we have a contradiction. Thus all the points in A are equilibrium points. $\qquad\square$

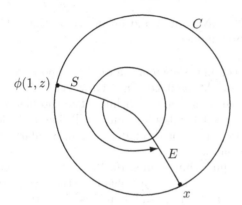

FIGURE 20. Flow through an extended local section.

Example 3.39 Now consider the scaled predator-prey model

$$x' = x(1 - x) - \frac{9xy}{10x + 1},$$

$$y' = by\left(1 - \frac{10y}{10x + 1}\right),$$

where $b > 0$. Compare to the simpler model studied earlier (Example 3.7).

The predation term in the first equation limits the number of prey that a predator can harvest when x is large. For each fixed level of prey x, the second equation is logistic such that the carrying capacity of the predator population increases as a function of x. The x nullclines of the system are $x = 0$ and

$$y = \frac{1}{9}(1 - x)(10x + 1),$$

and the y nullclines are $y = x + .1$ and $y = 0$ (see Figure 21).

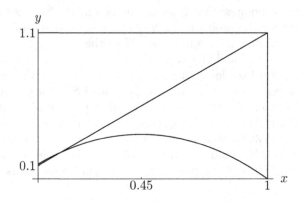

FIGURE 21.   Nullclines and an invariant region for the predator-prey model.

We claim the rectangle shown in Figure 21 is a positively invariant set. For example, on the right-hand side ($x = 1$), we have $x' < 0$ since this line segment is above the parabolic x nullcline. The top has inward flow for similar reasons. The left-hand side is an x nullcline on which the flow is vertical, so no orbits can escape through this side. Similarly, the flow along the bottom is horizontal.

There are four equilibrium points: $(0,0)$, $(1,0)$, $(0,.1)$ and $(.1,.2)$. The reader can check by calculating the Jacobian matrix that $(0,0)$ is a repelling point, while $(1,0)$ and $(0,.1)$ are saddle points. Now the Jacobian matrix at $(.1,.2)$ is

$$A = \begin{bmatrix} .35 & -.45 \\ b & -b \end{bmatrix}.$$

Then $\det A = .1b > 0$ and $\operatorname{tr}[A] = -b + .35 < 0$ if $b > .35$. Thus if $b > .35$, the equilibrium $(.1,.2)$ is asymptotically stable, and we expect that solutions originating in the interior of the invariant rectangle will converge to the equilibrium.

On the other hand, if $0 < b < .35$, this equilibrium is a repelling point, so no other orbit can have this point in its ω-limit set. The same is true of the origin. Now let P be any point in the interior of the invariant rectangle and let W be its ω-limit set. Suppose that W does not contain a cycle. Then by the Poincaré-Bendixson theorem, W must consist of equilibrium points and orbits converging to equilibrium points. The remaining two equilibrium points are saddles on the left and bottom boundaries of the rectangle. The only orbits that converge to them are on these boundaries, so the previous theorem implies that W cannot be a subset of the boundary of the rectangle. Consequently, the ω-limit set does contain a limit cycle. The flow pattern indicates that such a cycle must encircle the equilibrium $(.1, .2)$ in a counterclockwise direction. A graph of a solution converging to a cycle for the case $b = .3$ is given in Figure 22. $\triangle$

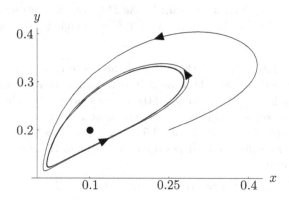

FIGURE 22.   A periodic solution of the predator-prey model.

Another approach to the question of periodic solutions is to try to show that none exist.

Example 3.40 Consider a gradient system

$$x' = -\operatorname{grad} h(x).$$

(Here we can take $x \in \mathbb{R}^n$.) Let x_0 be a point that is not an equilibrium point. For $a < b$, we have

$$
\begin{aligned}
h(\phi(b, x_0)) - h(\phi(a, x_0)) &= \int_a^b \frac{d}{dt} h(\phi(t, x_0)) \, dt \\
&= \int_a^b \operatorname{grad} h(\phi(t, x_0)) \cdot \phi'(t, x_0) \, dt \\
&= -\int_a^b \|\operatorname{grad} h(\phi(t, x_0))\|^2 \, dt < 0.
\end{aligned}
$$

It follows that $\phi(b, x_0) \neq \phi(a, x_0)$, so x_0 does not lie on a cycle. Consequently, gradient systems have no cycles. $\triangle$

Since for the remainder of this section we are considering only two-dimensional problems, we recast equation (3.45) in the form

$$
\begin{aligned}
x' &= f(x, y), & (3.50) \\
y' &= g(x, y). & (3.51)
\end{aligned}
$$

The next theorem has implications for the existence and location of periodic solutions.

Theorem 3.41 *Inside any cycle there has to be an equilibrium point of* (3.50), (3.51).

Proof We give a sketch of the proof here (for more details see Hirsch and Smale [**23**]). Let C be a simple closed curve in the plane $\mathbb{R}^2$ and assume $F(x, y)$ is a continuous vector field on and inside of C and that $F(x, y) \neq 0$ on C. Let $P = (x, y)$ be a point on C. As P moves once around the curve C the angle θ of inclination of the vector $F(x, y)$ changes by the amount

$$\Delta\theta = 2n\pi$$

where n is an integer called the index of the curve C with respect to the vector field (see Example 3.42 below). If we deform C continuously without passing through any critical points of the vector field $F(x, y)$, then the index changes continuously and hence must remain constant.

Now assume C is a cycle of (3.50), (3.51) and assume $x(t)$, $y(t)$ is a corresponding solution with period $T > 0$. Then the curve C is given parametrically by

$$C: \ x = x(t), \ y = y(t), \quad 0 \le t \le T.$$

Also let

$$F(x, y) = \left[\begin{array}{c} f(x, y) \\ g(x, y) \end{array} \right]$$

be the vector field associated with (3.50), (3.51). Then as t goes from 0 to T, the point $P(t) = (x(t), y(t))$ moves around the curve C one time. But

$$P'(t) = (x'(t), y'(t)) = (f(x(t)), g(y(t))) = F(x(t), y(t))$$

is the tangent vector to the curve C and hence the index of the curve C with respect to the vector field $F(x, y)$ is ± 1 (+1 if the curve C is positively oriented and -1 if the curve is negatively oriented). If there are no critical points inside C, then for any point (x_0, y_0) interior to C we can continuously deform C into a circle about (x_0, y_0) with an arbitrarily small radius with center at (x_0, y_0). On C_0 the vector field is nearly constant (in particular $f(x, y) \neq 0$ inside and on C_0). Therefore the index of the vector field $F(x, y)$ on C_0 must be zero, which is a contradiction. Hence there has to be at least one critical point of (3.50), (3.51) inside of C. □

Example 3.42 Find the index of the mapping

$$F(x, y) = \left[\begin{array}{c} -2xy \\ y^2 - x^2 \end{array} \right]$$

with respect to the simple closed curve C, $x^2 + y^2 = 1$, positively oriented. A parametric representation of this curve C is given by

$$x = \cos t, \quad y = \sin t, \quad 0 \le t \le 2\pi.$$

It follows that

$$F(x(t), y(t)) = \left[\begin{array}{c} -2\cos t \sin t \\ \sin^2 t - \cos^2 t \end{array} \right] = \left[\begin{array}{c} -\sin(2t) \\ -\cos(2t) \end{array} \right].$$

Hence the index of the mapping F with respect to C is $n = -2$. $\triangle$

We apply Theorem 3.41 in the following example.

Example 3.43 Show that the system

$$\begin{aligned} x' &= y + yx^2, \\ y' &= xy + 2 \end{aligned}$$

has no periodic solutions.

Since this system does not have any equilibrium points, we have by Theorem 3.41 that this system does not have any periodic solutions.

$\triangle$

Note that Theorem 3.41 also implies that the cycles whose existence was established in the earlier examples must encircle the equilibrium.

Before we state our next theorem, we need a couple of definitions.

Definition 3.44 We define the *divergence* of the vector field

$$F(x, y) = \begin{bmatrix} f(x, y) \\ g(x, y) \end{bmatrix}$$

by

$$\text{div } F(x, y) = f_x(x, y) + g_y(x, y).$$

Definition 3.45 A domain $D \subset \mathbb{R}^2$ is said to be a *simply connected domain* provided it is connected and for any simple closed curve C in D the interior of C is a subset of D.

Theorem 3.46 (Bendixson-Dulac Theorem) *Assume there is a continuously differentiable function $\alpha(\cdot, \cdot)$ on a simply connected domain $D \subset \mathbb{R}^2$ such that the*

$$\text{div } [\alpha(x, y)F(x, y)]$$

is either always positive or always negative on D. Then the system (3.50), (3.51) *does not have a cycle in D.*

Proof Assume that the system (3.50), (3.51) does have a cycle. Then there is a nonconstant periodic solution x, y in D. Assume this periodic solution has minimum period ω, let C be the corresponding simple closed curve oriented by increasing t, $0 \le t \le \omega$, and let E be the region interior to C. Since div $[\alpha(x, y)F(x, y)]$ is of one sign on $E \subset D$, we have that the double integral

$$\int_E \text{div } [\alpha(x, y)F(x, y)] \, dA \ne 0.$$

On the other hand, we get, using Green's theorem,

$$\int_E \operatorname{div}\left[\alpha(x,y)F(x,y)\right] dA$$

$$= \int_E \left\{ \frac{\partial}{\partial x}\left[\alpha(x,y)f(x,y)\right] + \frac{\partial}{\partial y}\left[\alpha(x,y)g(x,y)\right] \right\} dA$$

$$= \pm \int_C \left[\alpha(x,y)f(x,y)\, dy - \alpha(x,y)g(x,y)\, dx\right]$$

$$= \pm \int_0^\omega \alpha(x(t),y(t))\left[f(x(t),y(t))y'(t) - g(x(t),y(t))x'(t)\right] dt$$

$$= \pm \int_0^\omega \alpha(x(t),y(t))[f(x(t),y(t))g(x(t),y(t))$$
$$-g(x(t),y(t))f(x(t),y(t))]dt$$

$$= 0,$$

which is a contradiction.   □

Example 3.47 Show that the system

$$\begin{aligned} x' &= -x + 3y + y^2 - 3x^3 + x^4, \\ y' &= 2x - 3y - 4x^3y - y^7 \end{aligned}$$

has no cycles.

Note that

$$\begin{aligned} \operatorname{div} F(x,y) &= \operatorname{div}\begin{bmatrix} f(x,y) \\ g(x,y) \end{bmatrix} \\ &= \operatorname{div}\begin{bmatrix} -x + 3y + y^2 - 3x^3 + x^4 \\ 2x - 3y - 4x^3y - y^7 \end{bmatrix} \\ &= -1 - 9x^2 + 4x^3 - 3 - 4x^3 - 7y^6 \\ &= -4 - 9x^2 - 7y^6 \\ &< 0, \end{aligned}$$

on the simply connected domain $\mathbb{R}^2$. Hence by the Bendixson-Dulac theorem with $\alpha(x,y) = 1$ there are no cycles.   △

Example 3.48 Show that the system

$$\begin{aligned} x' &= y^3, \\ y' &= -x - y + x^2 + y^4 \end{aligned}$$

has no cycles.

First note that

$$\text{div } F(x,y) = \text{div} \begin{bmatrix} f(x,y) \\ g(x,y) \end{bmatrix}$$

$$= \text{div} \begin{bmatrix} y^3 \\ -x-y+x^2+y^4 \end{bmatrix}$$

$$= -1 + 4y^3,$$

which is not of one sign on $\mathbb{R}^2$. Hence we cannot apply the Bendixson-Dulac theorem with $\alpha(x,y) = 1$ to conclude that there are no cycles in R^2. We now set out to find a continuously differentiable function $\alpha(\cdot, \cdot)$ so that the

$$\text{div } [\alpha(x,y)F(x,y)]$$

is either always positive or always negative on $\mathbb{R}^2$. Consider

$$\text{div } [\alpha(x,y)F(x,y)]$$

$$= \text{div} \begin{bmatrix} \alpha(x,y)f(x,y) \\ \alpha(x,y)g(x,y) \end{bmatrix}$$

$$= \text{div} \begin{bmatrix} \alpha(x,y)[y^3] \\ \alpha(x,y)[-x-y+x^2+y^4] \end{bmatrix}$$

$$= \alpha_x(x,y)y^3 + \alpha_y(x,y)[-x-y+x^2+y^4] + \alpha(x,y)[-1+4y^3].$$

Studying this last equation, we see it might be beneficial if we assume that α is just a function of x. Then

$$\text{div } [\alpha(x,y)F(x,y)] = -\alpha(x) + [\alpha'(x) + 4\alpha(x)]y^3.$$

To eliminate the term involving y^3, we will choose $\alpha(x)$ so that

$$\alpha'(x) + 4\alpha(x) = 0,$$

so let

$$\alpha(x) = e^{-4x}.$$

Then α is a continuously differentiable function on $\mathbb{R}^2$ and

$$\text{div } [\alpha(x,y)F(x,y)] = -e^{-4x} < 0,$$

on $\mathbb{R}^2$. It follows from the Bendixson-Dulac theorem that there are no cycles in $\mathbb{R}^2$.   △

We next give a result (see [46]) concerning the existence of a unique limit cycle for Liénard's equation

$$x'' + f(x)x' + g(x) = 0. \tag{3.52}$$

Theorem 3.49 (Liénard's Theorem) *Assume that*

  (i) *$f,g : \mathbb{R} \to \mathbb{R}$ are continuously differentiable;*

(ii) f, g are even and odd functions respectively with $g(x) > 0$ on $(0, \infty)$.

(iii) $F(x) := \int_0^x f(t)\, dt$ has a positive zero a with $F(x) < 0$ for $0 < x < a$ and $F(x) > 0$ for $x > a$, F is nondecreasing for $x > a$, and $\lim_{x \to \infty} F(x) = \infty$. Then (3.52) has a unique cycle surrounding the origin and every nontrivial orbit approaches this cycly spirally.

We next apply Theorem 3.49 to the van der Pol equation which arises in the study of vacuum tubes.

Example 3.50 The van der Pol equation

$$x'' + \mu(x^2 - 1)x' + x = 0, \quad \mu > 0 \tag{3.53}$$

has a unique limit cycle. Note that the van der Pol equation (3.53) equation is a Liénard equation with $f(x) = \mu(x^2 - 1)$, $g(x) = x$ and

$$F(x) = \int_0^x \mu(s^2 - 1)ds = \mu\left(\frac{1}{3}x^3 - x\right).$$

Note that all the hypotheses of Theorem 3.49 hold, where $a = \sqrt{3}$. △

There is another important approach to the problem of finding limit cycles that avoids the difficulty of identifying a positively invariant region. The idea is to study the stability of an equilibrium as the value of some parameter λ varies in the system:

$$x' = f(x, y, \lambda), \tag{3.54}$$
$$y' = g(x, y, \lambda). \tag{3.55}$$

Suppose that the equilibrium is asymptotically stable for $\lambda < \lambda_0$ and that the eigenvalues of the Jacobian matrix evaluated at the equilibrium are complex so that solutions spiral in toward the equilibrium. If the equilibrium is unstable for $\lambda > \lambda_0$ and the eigenvalues remain complex, then it might happen that solutions would spiral toward a limit cycle for these values of λ. However, the following example shows that this is not necessarily the case.

Example 3.51 The system

$$x' = y,$$
$$y' = -x + \lambda y$$

has a single equilibrium $(0, 0)$, and the Jacobian matrix has eigenvalues $.5(\lambda \pm \sqrt{\lambda^2 - 4})$. For this simple linear system, it is easy to see that the orbits spiral inward toward $(0, 0)$ for $\lambda < 0$ and are unbounded spirals for $\lambda > 0$, so there is no limit cycle. Note that when $\lambda = 0$, the origin is a center surrounded by an infinite family of circular cycles. △

However, if we add the requirement that the equilibrium point is asymptotically stable at $\lambda = \lambda_0$, then we do get a limit cycle. This is the substance of the famous Hopf bifurcation theorem:

Theorem 3.52 (Hopf Bifurcation) *Suppose that* $(x_0(\lambda), y_0(\lambda))$ *is an equilibrium point for the system* (3.54), (3.55), *and the Jacobian matrix for the system evaluated at this equilibrium has eigenvalues* $\alpha(\lambda) \pm i\beta(\lambda)$. *Also, assume that at some* $\lambda = \lambda_0$, $\alpha(\lambda_0) = 0$, $\alpha'(\lambda_0) > 0$, $\beta(\lambda_0) \neq 0$, *and the equlibrium is asymptotically stable. Then there is a* λ_1 *such that* (3.54), (3.55) *has a limit cycle encircling the equilibrium for* $\lambda_0 < \lambda < \lambda_1$.

See Chicone [8] for a more complete description of the Hopf bifurcation and a proof.

Example 3.53 Let's consider a nonlinear system that is similar to the preceding example that had no limit cycle:

$$x' = y,$$
$$y' = -x + \lambda y - ay^3,$$

where $a > 0$. As in the earlier example, the equilibrium is $(0,0)$ and the eigenvalues are $.5(\lambda \pm \sqrt{\lambda^2 - 4})$. Let's show that $(0,0)$ is asymptotically stable when $\lambda = 0$. Define a trial Liapunov function $V(x,y) = x^2 + y^2$, and note that

$$\frac{d}{dt}V(\phi(t)) = 2x(t)x'(t) + 2y(t)y'(t)$$
$$= 2y^2(t)(\lambda - ay^2(t))$$
$$\leq 0,$$

when $\lambda = 0$, with equality only for $y = 0$. Since there is no solution other than the origin that lies on the x-axis, the LaSalle invariance theorem (Theorem 3.24) implies that the origin is asymptotically stable. Since the other hypotheses of the Hopf bifurcation theorem are clearly satisfied, we conclude that the system has a limit cycle for $0 < \lambda < \lambda_1$, for some positive constant λ_1.   △

The predator-prey model that was considered earlier in this section can also be analyzed by the Hopf bifurcation theorem:

Example 3.54 Recall that the equilibrium point $(.1, .2)$ for the system

$$x' = x(1 - x) - \frac{9xy}{10x + 1},$$
$$y' = by\left(1 - \frac{10y}{10x + 1}\right)$$

experiences a change in stability when $b = .35$. If we let $\lambda = .35 - b$, then the eigenvalues of the Jacobian matrix evaluated at $(.1, .2)$ are

$$\frac{\lambda \pm \sqrt{\lambda^2 - .4(.35 - \lambda)}}{2}.$$

At $\lambda = 0$, the eigenvalues are complex, the real part changes sign, and the derivative of the real part is $.5$. It is possible, but complicated, to show that there is a strict Liapunov function defined near $(.1, .2)$ when $\lambda = 0$.

Alternatively, we can use software for graphing solutions of differential equations to check the asymptotic stablility of $(.1, .2)$ when $\lambda = 0$. As was shown earlier by an application of the Poincaré-Bendixson theorem, there does exist a limit cycle for $\lambda > 0$, that is, for $b < .35$. $\triangle$

3.7 Three-Dimensional Systems

Autonomous systems of three equations in three unknowns can exhibit all the types of behavior we have seen for two-dimensional systems, plus much more complicated behavior. Since many biological, chemical, and physical systems have been observed to have complicated, even chaotic, properties, mathematical models having these properties are potentially useful in understanding such phenomena.

Let's begin with linear systems of three equations in three unknowns:

$$\begin{bmatrix} x \\ y \\ z \end{bmatrix}' = \begin{bmatrix} a_{11} & a_{12} & a_{13} \\ a_{21} & a_{22} & a_{23} \\ a_{31} & a_{32} & a_{33} \end{bmatrix} \begin{bmatrix} x \\ y \\ z \end{bmatrix}.$$

As in the two-dimensional case, the type of orbits that the system has depends on the eigenvalues of the matrix. Recall that if all eigenvalues have negative real parts, then the origin is globally asymptotically stable. There is a useful criterion for stability called the Routh-Hurwitz criterion that depends on the coefficients in the characteristic equation. Suppose that the characteristic equation for the matrix in the linear system is

$$\lambda^3 + b\lambda^2 + c\lambda + d = 0.$$

Then the origin is globally asymptotically stable if

$$b > 0, \ d > 0, \ \text{and} \ bc > d.$$

If one or more of these inequalities are reversed, then the origin is unstable. See Pielou [40] for a discussion and for criteria for higher dimensions.

We won't try to characterize all three-dimensional linear systems, but a couple of examples will serve to illustrate how the two-dimensional categories extend to three dimensions.

Example 3.55 First, consider the system

$$\begin{bmatrix} x \\ y \\ z \end{bmatrix}' = \begin{bmatrix} 0 & 1 & 0 \\ 0 & 0 & 1 \\ -34 & -21 & -4 \end{bmatrix} \begin{bmatrix} x \\ y \\ z \end{bmatrix}.$$

The characteristic equation is

$$\lambda^3 + 4\lambda^2 + 21\lambda + 34 = 0.$$

The Routh-Hurwitz criterion is certainly satisfied, so the origin is asymptotically stable. We can get more precise information by computing the

eigenpairs

$$-2, \begin{bmatrix} 1 \\ -2 \\ 4 \end{bmatrix}; \quad -1 \pm 4i, \quad \begin{bmatrix} -1 \\ 1 \\ 15 \end{bmatrix} \pm i \begin{bmatrix} 0 \\ 4 \\ -8 \end{bmatrix}.$$

In phase space, orbits spiral around the one-dimensional subspace spanned by an eigenvector for $\lambda = -2$ toward the two-dimensional subspace spanned by the real and imaginary parts of the eigenvectors for the complex eigenvalues. See Figure 23. △

FIGURE 23.   A stable spiral.

Example 3.56 The system

$$\begin{bmatrix} x \\ y \\ z \end{bmatrix}' = \begin{bmatrix} 0 & 1 & 0 \\ 0 & 0 & 1 \\ 2 & 1 & -2 \end{bmatrix} \begin{bmatrix} x \\ y \\ z \end{bmatrix}$$

has the characteristic equation

$$\lambda^3 + 2\lambda^2 - \lambda - 2 = 0.$$

Note that the origin is unstable since the last coefficient is negative. The eigenpairs are

$$1, \begin{bmatrix} 1 \\ 1 \\ 1 \end{bmatrix}; \quad -1, \begin{bmatrix} 1 \\ -1 \\ 1 \end{bmatrix}; \quad -2, \begin{bmatrix} 1 \\ -2 \\ 4 \end{bmatrix}.$$

In this case, most orbits converge to the stable plane containing the eigenvectors for the negative eigenvalues as $t \to -\infty$ and converge to the line containing the eigenvectors for the positive eigenvalue as $t \to \infty$ (see Figure 24). Here the origin is a saddle with a two-dimensional stable subspace and a one-dimensional unstable subspace. △

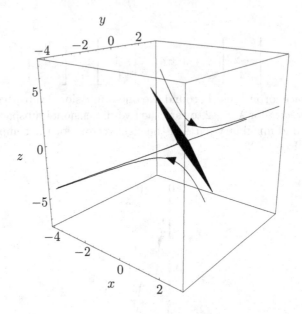

FIGURE 24.   A three-dimensional saddle.

The general autonomous system in three dimensions is

$$\begin{aligned}
x' &= f(x, y, z), \\
y' &= g(x, y, z), \\
z' &= h(x, y, z).
\end{aligned} \qquad (3.56)$$

In the next example, which involves a three-species food chain, the behavior of the solutions is simple and predictable.

Example 3.57 Suppose there are three species with population densities $x(t)$, $y(t)$, and $z(t)$ at time t, and the third species preys on the second, while the second species preys on the first. A simple model of this arrangement is

$$\begin{aligned}
x' &= ax(1 - x) - xy, \\
y' &= -by + cxy - dyz, \\
z' &= -kz + yz,
\end{aligned}$$

where all parameters are positive. The unique equilibrium in the first octant is

$$(\overline{x}, \overline{y}, \overline{z}) = \left(1 - \frac{k}{a}, k, \frac{c(a - k) - ab}{ad} \right),$$

whenever

$$1 > \frac{k}{a} + \frac{b}{c}.$$

Then the Jacobian matrix at this equilibrium is

$$\begin{bmatrix} -a\bar{x} & -\bar{x} & 0 \\ c\bar{y} & 0 & -dk \\ 0 & \bar{z} & 0 \end{bmatrix},$$

and the characteristic equation is

$$\lambda^3 + a\bar{x}\lambda^2 + k(d\,\bar{z} + c\bar{x})\lambda + adk\bar{x}\,\bar{z} = 0.$$

In order for the equilibrium to be asymptotically stable, the Routh- Hurwitz criterion requires

$$a\bar{x} > 0, \quad adk\bar{x}\,\bar{z} > 0, \quad ak\bar{x}\,(d\bar{z} + c\bar{x}) > adk\bar{x}\,\bar{z}.$$

Consequently, this equilibrium is asymptotically stable if it is in the first octant. What do you think would happen to the three species if the values of the parameters place this equilibrium in a different quadrant? △

In recent years, it has been shown that certain types of three-dimensional systems always have fairly simple dynamics. For example, let's assume that the system (3.56) models competition among three species. Then an increase in any one species tends to inhibit the growth of the other two. Specifically, we impose the condition that the partial derivatives f_y, f_z, g_x, g_z, h_x, and h_y are less than or equal to zero. If all these partial derivatives are greater than or equal to zero, then the system is said to be cooperative. The following theorem, which is similar to the Poincaré-Bendixson theorem, is due to Hirsch [18] (see also Smith [47]):

Theorem 3.58 *Let D be a bounded invariant set in $\mathbb{R}^3$ for the system (3.56), which is assumed to be either competitive or cooperative. Then for every orbit in D, the ω-limit set either contains an equilibrium or is a cycle.*

To study complicated behavior for system (3.56), there is no better place to begin than with the famous example due to Lorenz [34]. In 1963, he published a study of the following system that now bears his name:

$$\begin{aligned} x' &= \sigma(y - x), \\ y' &= rx - y - xz, \\ z' &= xy - bz, \end{aligned}$$

where σ, r, and b are positive parameters. He derived this system from a model of fluid convection in the atmosphere, and other researchers have found it to be relevant to a variety of phenomena from lasers to water wheels. He argued convincingly, though without rigorous proof, that for some parameter values the solutions exhibit extremely complicated behavior that might be chaotic, in some sense.

Let's see what we can deduce using the methods of this chapter. As usual, we first look for equilibrium points and quickly observe that the origin is one of them. To get the others, we have from the first equation that $y = x$, then from the third equation that $z = x^2/b$, and from the second equation that $x(r - 1 - x^2/b) = 0$. Now if $x = 0$, then we get the origin, so set $r - 1 - x^2/b = 0$ to obtain the two additional equilibrium points

$$P, Q \equiv (\pm\sqrt{b(r - 1)}, \pm\sqrt{b(r - 1)}, r - 1),$$

which exist for $r \geq 1$. Note that the three equilibrium points are all at the origin when $r = 1$, so there is a *pitchfork bifurcation* as r is increased through $r = 1$, in which a single equilibrium branches into three.

To test stability, we compute the Jacobian matrix:

$$\begin{bmatrix} -\sigma & \sigma & 0 \\ r - z & -1 & -x \\ y & x & -b \end{bmatrix}.$$

At the origin, this matrix has eigenvalues

$$-b \text{ and } \frac{-(\sigma + 1) \pm \sqrt{(\sigma + 1)^2 + 4\sigma(r - 1)}}{2},$$

so the origin is asymptotically stable for $0 < r < 1$, and for $r > 1$ it has two negative eigenvalues and one positive eigenvalue, like the linear system considered earlier in this section. At the equilibria P and Q, the characteristic equation is

$$\lambda^3 + (\sigma + b + 1)\lambda^2 + b(\sigma + r)\lambda + 2\sigma b(r - 1) = 0.$$

From the Routh-Hurwitz criterion, we see that P and Q are asymptotically stable if

$$1 < r < \frac{\sigma(\sigma + b + 3)}{\sigma - b - 1}, \text{ in case } \sigma > b + 1,$$
$$1 < r, \text{ in case } \sigma \leq b + 1.$$

Thus in the case that $\sigma \leq b + 1$, the Lorenz system has fairly simple dynamics since there is always at least one stable equilibrium point.

Next, we show that all orbits of the Lorenz system are bounded. Define

$$V(x, y, z) := rx^2 + \sigma y^2 + \sigma(z - 2r).$$

Then along any solution x, y, z of the Lorenz system, we have

$$\frac{d}{dt}V(x(t), y(t), z(t))$$
$$= 2rx(t)x'(t) + 2\sigma y(t)y'(t) + 2\sigma(z(t) - 2r)z'(t)$$
$$= 2r\sigma x(t)(y(t) - x(t)) + 2\sigma y(t)(rx(t) - y(t) - x(t)z(t))$$
$$\quad + 2\sigma(z(t) - 2r)(x(t)y(t) - bz(t))$$
$$= -2\sigma[rx^2(t) + y^2(t) + b(z(t) - r)^2 - r^2b]$$
$$< 0,$$

if $rx^2 + y^2 + b(z - r)^2 > r^2b$. Define

$$E_1 \equiv \{(x, y, z) : rx^2 + y^2 + b(z - r)^2 \leq 2r^2b\}$$

and
$$E_2 \equiv \{(x, y, z) : rx^2 + \sigma y^2 + \sigma(z - 2r)^2 \leq C\}.$$
Then for large enough values of C, E_1 is a subset of E_2, so that
$$\frac{d}{dt} V(x(t), y(t), z(t)) < 0,$$
on the boundary of E_2, so E_2 is invariant. Since every orbit must start inside E_2 for large C, every orbit must be bounded.

We now know that the ω-limit set for each orbit must be a nonempty, compact set, and some numerical experimentation is needed to study the nature of these sets. For the parameter values $r = 10$, $b = 8/3$, and $\sigma = 10$, we are in the part of the parameter domain where the equilibrium points P and Q are asymptotically stable, and Figure 25 shows a typical orbit spiralling in toward P.

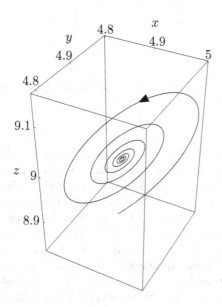

FIGURE 25. A stable spiral for the Lorenz system.

If r is increased to 21, while the other parameters are unchanged, we still have asymptotically stable equilibria, but orbits tend to roam around for some period of time before starting the final spiral (see Figure 26). This phenomenon is sometimes called *transient chaos* and seems to be due to the presence of an unstable homoclinic orbit (one that belongs to both the stable and unstable manifolds) at the origin.

By the time r reaches the value 28, all equilibrium points are unstable, and numerical computations reveal a very complicated ω-limit set with two distinct sections. Orbits seem to bounce back and forth between these sections rather randomly. (See Figure 27 and a graph of x versus t for

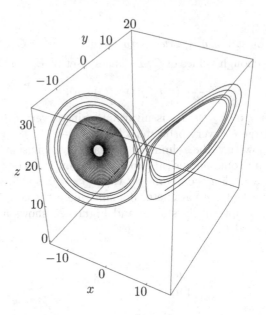

FIGURE 26.   Transient chaos in the Lorenz system.

$0 \le t \le 30$ in Figure 28.) This ω-limit set is not a closed orbit, and it appears to have dimension greater than 2. One of the basic attributes of chaotic behavior is *sensitive dependence on initial conditions*, which means (roughly) that for every orbit, there are orbits starting near it whose graphs are quite different. In Figure 29, x is plotted against t for orbits with initial conditions $(3, 4, 5)$ and $(3.01, 4, 5)$. Although the orbits are close for $0 \le t \le 11$, they are only occasionally similar for larger values of t.

Let's take a different approach to this so-called *strange attractor*. Fix $r = 28$ and $\sigma = 10$ and set $b = .5$. Figure 30 shows that there is a stable limit cycle with an approximate period of 3.5. If b is increased to .55, then the period of the cycle doubles to about 7 (see Figure 31). In Figure 32, we can see that at $b = .559$ the period has doubled again. These period doublings seem to continue as we continue to increase b, so this phenomenon has become known as the *period doubling route to chaos*, and it has been observed in many discrete and continuous models.

Once people started looking for chaotic behavior, they began to find it everywhere! There is category of problems, called nonlinear oscillation problems, for which complicated behavior is the norm. As a basic example (due to Ueda [**50**]), consider a simple electric circuit with linear resistance, nonlinear inductance, and an alternating current source:

$$y'' + ry' + y^3 = b\cos t,$$

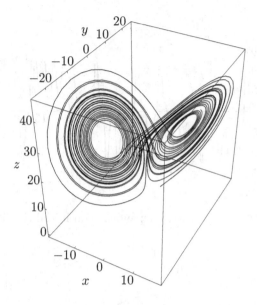

FIGURE 27. A strange attractor for the Lorenz system.

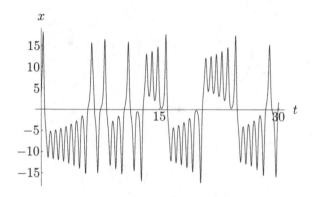

FIGURE 28. A graph of x for the strange attractor.

where y is the flux in the inductor. Of course, this is a second-order differential equation, but it is nonautonomous, so it tranforms into a three-dimensional autonomous system. Let $x = t$, $y = y$, and $z = y'$:

$$
\begin{aligned}
x' &= 1, \\
y' &= z, \\
z' &= b\cos x - y^3 - rz,
\end{aligned}
$$

where y is the flux in the inductor. For this example, we will be content to look at some numerical solutions of this system in the y, z-plane. First, choose $r = .1$, $b = 12$ and initial conditions $x = 0$, $y = 1.54$, $z = 0$.

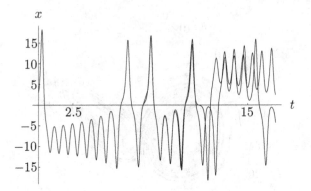

FIGURE 29.   Sensitive dependence on initial conditions.

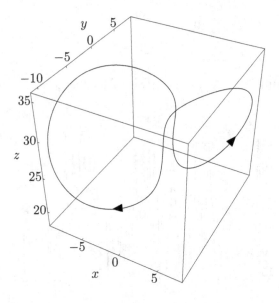

FIGURE 30.   A limit cycle for the Lorenz system.

Figure 33 shows a solution that is periodic in the y, z-plane (but not in three dimensions). A slight change in the initial value of y from 1.54 to 1.55 sends the solution spinning into what looks like a chaotic orbit (see Figure 34, in which the orbit is plotted for $300 \le t \le 400$). Figure 35 is a graph of y versus t for this solution on an initial interval. Note that the solution first appears to be settling into a periodic orbit, but around $t = 40$ the regular pattern is lost and an erratic pattern of oscillation emerges.

Alternatively, we can retain the original initial conditions, and change the value of the parameter b from 12 to 11.9 to obtain an intricate orbit (see Figure 36).

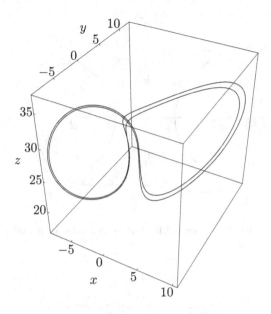

FIGURE 31. The first doubling of the period.

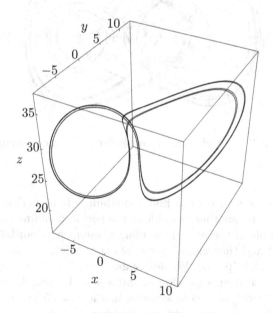

FIGURE 32. The second doubling of the period.

This system displays a wide variety of periodic solutions. For example, if we reduce the parameter b to 9, while keeping $r = .1$ and the original initial conditions, we obtain the elaborate periodic solution in Figure

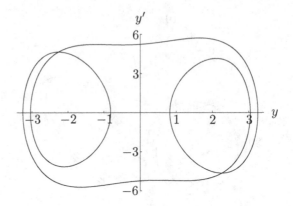

FIGURE 33.   An orbit that is periodic in y and y'.

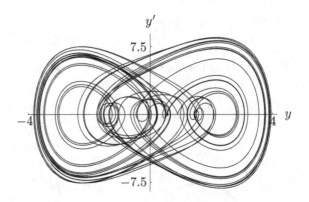

FIGURE 34.   A chaotic (?) orbit for a nonlinear oscillator.

37. Now increase b to 9.86 to find a doubling of the period (Figure 38). This appears to be another example of the period doubling route to chaos. Another example of the period doubling phenomenon can be obtained by setting $b = 12$ and then choosing $r = .34$ (periodic), $r = .33$ (period doubles), and $r = .318$ (period doubles again).

There has been an explosion of interest and research activity in the area of complicated nonlinear systems in the last 25 years, leading to a host of new ideas and techniques for studying these problems. A thorough discussion of these would fill several large volumes! Some of the important terms are *Smale horseshoes* (chaos producing mechanisms), *discrete dynamics* (nonlinear dynamics in discrete time), *fractals* (complicated sets with nonintegral dimension), *Liapunov exponents* (a measure of contraction of a set of orbits near a strange attractor), and *shadowing* (existence of real

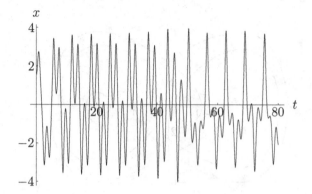

FIGURE 35. A plot of y versus t for a nonlinear oscillator.

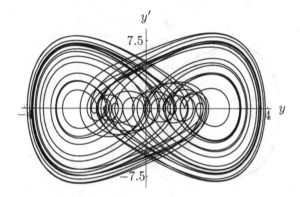

FIGURE 36. Another chaotic (?) orbit for a nonlinear oscillator.

orbits near computed orbits). Here is a partial list of the fundamental references on nonlinear dynamics: Guckenheimer and Holmes [17], Wiggins [53], Devaney [12], and Peitgen, Jürgens, and Saupe [39].

3.8 Differential Equations and *Mathematica*

In this section, we will indicate briefly how a computer algebra system can be used to find solutions of elementary differential equations and to generate information about solutions of differential equations that cannot be solved. Although we restrict the discussion to *Mathematica*, version 7.0, other popular computer algebra systems such as *Maple* and MATLAB have similar capabilities.

The basic command for solving differential equations is DSolve. For example, to solve the logistic differential equation (see Example 1.13), we enter

```
DSolve[x'[t]==r*x[t]*(1-x[t]/K), x[t], t]
```

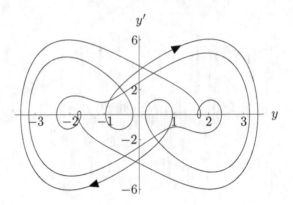

Figure 37. An elaborate periodic soution of a nonlinear oscillator.

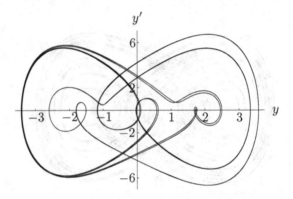

Figure 38. A doubling of the period.

Mathematica responds with

$$\left\{\left\{ x[t] \to \frac{e^{rt+KC[1]}K}{-1+e^{rt+KC[1]}} \right\}\right\}.$$

Here $C[1]$ stands for an arbitrary constant, and we have the general solution of the logistic equation. If we want the solution that satisfies an initial condition, we simply supply *Mathematica* with a list of the differential equation and the equation for the initial value:

```
DSolve[{x'[t]==r*x[t]*(1-x[t]/K), x[0]==x0}, x[t], t]
```

and we get

$$\left\{\left\{ x[t] \to \frac{e^{rt}Kx0}{K-x0+e^{rt}x0} \right\}\right\}.$$

Mathematica knows all the standard methods for solving first-order differential equations and can also solve systems of linear equations with constant coefficients. Let's ask *Mathematica* to solve the system in Example 2.17:

```
DSolve[{x'[t]==3*x[t]+y[t], y'[t]==-13*x[t]-3*y[t],
    x[0]==1, y[0]==2}, {x[t], y[t]}, t]
```

Since we included initial values, we obtain a unique solution

$$\{\{x[t] \to \frac{1}{2}(2\text{Cos}[2t] + 5\text{Sin}[2t]), y[t] \to \frac{1}{2}(4\text{Cos}[2t] - 19\text{Sin}[2t])\}\}$$

Also, many linear equations with variable coefficients can be solved in terms of special functions. For instance, if we type

```
DSolve[x''[t]==t^2 x[t], x[t], t]
```

the answer is

$$\left\{\left\{x[t] \to C[2]\text{ParabolicCylinderD}\left[-\frac{1}{2}, i\sqrt{2}t\right]\right.\right.$$
$$\left.\left. + C[1]\text{ParabolicCylinderD}\left[-\frac{1}{2}, \sqrt{2}t\right]\right\}\right\},$$

where `ParabolicCylinderD[v,z]` stands for the parabolic cylinder function $D_v(z)$.

Mathematica has a built-in package, called *VectorFieldPlots*, that will plot an array of direction arrows for a system of two first-order differential equations in two unknowns. To load the package, enter

```
<<VectorFieldPlots';
```

In order to plot the direction field for the predator-prey model in Example 3.12 with $a = b = c = r = 1$ and $K = 4$, we now type

```
VectorFieldPlot[{x*(1-x/4)-x*y, x*y-y}, {x,0,4.5}, {y,0,2.5}]
```

which produces Figure 39. The arrows indicate that solutions in the first quadrant flow counterclockwise and spiral in toward the asymptotically stable equilibrium point.

The command `NDSolve` is used to compute a numerical approximation of the solution of a system of ordinary differential equations. We can then display the approximation as a curve in phase space or as a graph of one of the dependent variables against the independent variable. As an example, consider Figure 22, which illustrates a solution converging to a limit cycle for the predator-prey model in Example 3.39. We first enter

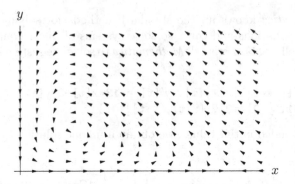

FIGURE 39.   A direction field for a predator-prey model.

```
NDSolve[{x'[t]==x[t]*(1-x[t])-9*x[t]*y[t]/(10*x[t]+1),
   y'[t]==0.3*y[t]*(1-10*y[t]/(10*x[t]+1)),
   x[0]==0.15, y[0]==0.25}, {x,y}, {t,0,150}]
```

and follow with

```
ParametricPlot[Evaluate[{x[t],y[t]} /.%], {t,0,150}]
```

to obtain that figure. If we now want a plot of x against t, then we can
enter

```
Plot[Evaluate[x[t]/.%%], {t,0,150}]
```

Note that % refers to the output immediately preceding the command, while
%% refers to the next-to-last output.

Finally, if we wish to perform a numerical experiment on a system
of equations, in which we will be changing the values of the parameters
and/or the initial conditions, then it is more efficient to define a function
that combines all the needed commands. Let's consider the Lorenz system
studied in section 3.7. The following user-defined function will make it easy
to change values of the parameters, initial conditions, and endpoints of the
time domain:

```
PhasePlot3D[{x0_,y0_,z0_}, {r_,b_,c_}, {t1_,t2_} ]:=
   Module[{t,n},
   n=NDSolve[{x'[t]==c*(y[t]-x[t]),
   y'[t]==r*x[t]-y[t]-x[t]*z[t],
   z'[t]==x[t]*y[t]-b*z[t],
   x[0]==x0, y[0]==y0, z[0]==z0},
   {x,y,z},{t,t1,t2}];
   ParametricPlot3D[Evaluate[{x[t],y[t],z[t]} /.n],
   {t,t1,t2}]]
```

Note that we changed the name of one parameter from σ to c for convenience. To produce Figure 27, we enter

```
PhasePlot3D[{3,4,5}, {28,8/3,10}, {0,35}]
```

It is staightforward to modify these commands to obtain other types of plots. For example, if we want a plot of x versus t (as in Figure 28), we can change the `ParametricPlot3D` in the definition of `PhasePlot3D` to `Plot` and change the function to be plotted to simply `x[t]`. Of course, we may also want to replace the name `PhasePlot3D` to something more suitable!

3.9 Exercises

3.1 Draw the phase plane diagram for the differential equation

$$x'' = -x + x^3.$$

What is the equation of the separatrix for this problem?

3.2 Draw the phase plane diagram for each of the following differential equations:

    (i) $x'' - 2xx' = 0$
    (ii) $x'' + |x| = 0$
    (iii) $x'' + e^x = 1$

3.3 Find all the equilibrium points for the system

$$\begin{aligned} x' &= x^2 y^3, \\ y' &= -x^3 y^2 \end{aligned}$$

and draw the phase plane diagram for this system.

3.4 Show that the differential equation $x'' + \lambda - e^x = 0$ has bifurcation at $\lambda = 0$ by drawing the phase plane diagram for $\lambda > 0$ and $\lambda < 0$ and noticing the drastic change in the phase plane diagram.

3.5 Show that if A is a 2×2 constant matrix with tr $A > 0$ and det $A \neq 0$, then the origin for the two dimensional system $x' = Ax$ is unstable.

3.6 Show that the system

$$\begin{aligned} x' &= x + 16y, \\ y' &= -8x - y \end{aligned}$$

is a Hamiltonian system, find the equations for the orbits, and sketch the phase plane diagram.

3.7 Work each of the following:

    (i) Show that

$$\begin{aligned} x' &= \frac{1 - x^2 + y^2}{y^2}, \\ y' &= -\frac{2x}{y} \end{aligned}$$

is a Hamiltonian system if $y \neq 0$.

(ii) Investigate how the orbits of the system in part (i) are related to the family of circles passing through the points $(1,0)$ and $(-1,0)$. [Note: The orbits in this problem are the curves of the electric field of a dipole located at $(1,0)$ and $(-1,0)$.]

3.8 For each of the following systems that are Hamiltonian systems, find a Hamiltonian function. If a system is not Hamiltonian system, say so.

(i)
$$
\begin{aligned}
x' &= e^y - 2x \\
y' &= 2y - 2x
\end{aligned}
$$

(ii)
$$
\begin{aligned}
x' &= x + y \\
y' &= -x + y
\end{aligned}
$$

(iii)
$$
\begin{aligned}
x' &= -x^3 - 4y^3 \\
y' &= 3x^2y + 2
\end{aligned}
$$

(iv)
$$
\begin{aligned}
x' &= x^2 \cos(xy) - x^2 \\
y' &= 2xy - \sin(xy) - xy \cos(xy)
\end{aligned}
$$

3.9 Find eigenpairs for each of the following and use your answers to draw the corresponding phase plane diagram.

(i)
$$
\begin{bmatrix} x \\ y \end{bmatrix}' = \begin{bmatrix} -5 & -1 \\ 6 & 0 \end{bmatrix} \begin{bmatrix} x \\ y \end{bmatrix}
$$

(ii)
$$
\begin{bmatrix} x \\ y \end{bmatrix}' = \begin{bmatrix} 1 & 0 \\ 2 & -2 \end{bmatrix} \begin{bmatrix} x \\ y \end{bmatrix}
$$

3.10 Find eigenpairs for each of the following and use your answers to draw the corresponding phase plane diagram. In (i) graph nullclines to draw a more accurate phase plane diagram.

(i)
$$
\begin{bmatrix} x \\ y \end{bmatrix}' = \begin{bmatrix} 1 & 4 \\ 1 & 1 \end{bmatrix} \begin{bmatrix} x \\ y \end{bmatrix}
$$

(ii)
$$
\begin{bmatrix} x \\ y \end{bmatrix}' = \begin{bmatrix} 0 & -2 \\ -1 & 1 \end{bmatrix} \begin{bmatrix} x \\ y \end{bmatrix}
$$

3.11 Draw the phase plane diagrams for each of the following linear systems using nullclines to help you graph a more accurate phase plane diagram:

(i)
$$
\begin{bmatrix} x \\ y \end{bmatrix}' = \begin{bmatrix} 2 & \frac{5}{2} \\ 0 & -3 \end{bmatrix} \begin{bmatrix} x \\ y \end{bmatrix}
$$

(ii)
$$
\begin{bmatrix} x \\ y \end{bmatrix}' = \begin{bmatrix} -\frac{9}{5} & \frac{2}{5} \\ \frac{2}{5} & -\frac{6}{5} \end{bmatrix} \begin{bmatrix} x \\ y \end{bmatrix}
$$

3.12 Draw the phase plane diagram for each of the following systems:

(i) $\begin{bmatrix} x \\ y \end{bmatrix}' = \begin{bmatrix} 0 & 4 \\ -4 & 0 \end{bmatrix} \begin{bmatrix} x \\ y \end{bmatrix}$

(ii) $\begin{bmatrix} x \\ y \end{bmatrix}' = \begin{bmatrix} 0 & 4 \\ -1 & 4 \end{bmatrix} \begin{bmatrix} x \\ y \end{bmatrix}$

(iii) $\begin{bmatrix} x \\ y \end{bmatrix}' = \begin{bmatrix} -8 & 10 \\ -4 & 4 \end{bmatrix} \begin{bmatrix} x \\ y \end{bmatrix}$

(iv) $\begin{bmatrix} x \\ y \end{bmatrix}' = \begin{bmatrix} -1 & -5 \\ 2 & 5 \end{bmatrix} \begin{bmatrix} x \\ y \end{bmatrix}$

3.13 The following is a model of competition between species having densities x and y:

$$\begin{aligned} x' &= x(2 - 2x - y), \\ y' &= y(2 - x - 2y). \end{aligned}$$

Draw the x and y nullclines and the direction of flow on each nullcline.

3.14 A certain epidemic is modeled by the equations

$$\begin{aligned} x' &= -axy + bz, \\ y' &= axy - cy, \\ z' &= cy - bz. \end{aligned}$$

Here x is the number of individuals who are susceptible to the disease, y is the number infected, z is the number recovered, and a, b, c are positive parameters. Note that recovered individuals immediately become susceptible again.

(i) Show that $x(t) + y(t) + z(t)$ is constant for all t, and use this fact to eliminate z and reduce the system to two equations in two unknowns.

(ii) Find and sketch the nullclines and the direction of flow.

3.15 A chemostat is a tank that holds a concentration $x(t)$ of bacteria at time t nourished by a concentration $y(t)$ of nutrient at time t. Fresh nutrient is continuously added to the tank, the tank is stirred, and the mix is drawn off at the same rate. A basic model is

$$\begin{aligned} x' &= a\left(\frac{y}{1+y}\right)x - x, \\ y' &= -\left(\frac{y}{1+y}\right)x - y + b, \end{aligned}$$

where $a > 1$, $b > 1/(a - 1)$. (See Edelstein-Keshet [14].) Sketch a phase plane diagram, showing nullclines and direction of flow.

3.16 Apply Liapunov's stability theorem (Theorem 3.18) to the simple scalar equation $x' = -kx$, where $k > 0$ is a constant. Same problem for $x' = -kx + f(x)$, where $k > 0$ is a constant and $f : \mathbb{R} \to \mathbb{R}$, is continuous and satisfies $xf(x) < 0$, $x \neq 0$.

3.17 If possible, use Theorem 3.26 to determine the stability of the equilibrium points for each of the following systems:

(i) $\begin{bmatrix} x \\ y \end{bmatrix}' = \begin{bmatrix} x + x^2 - 2xy \\ -y + xy \end{bmatrix}$

(ii) $\begin{bmatrix} x \\ y \end{bmatrix}' = \begin{bmatrix} -x + y + y^2 - 2xy \\ x \end{bmatrix}$

(iii) $\begin{bmatrix} x \\ y \end{bmatrix}' = \begin{bmatrix} x - y \\ x^2 - 1 \end{bmatrix}$

3.18 Show that for the pendulum problem with friction (Example 3.25) the kinetic energy plus the potential energy of the pendulum bob at time t is given by $\frac{1}{2}mL^2[\theta'(t)]^2 + mgL[1 - \cos\theta(t)]$ if the reference plane for computing the potential energy is where the pendulum bob is at when $\theta = 0$. Compare this energy expression to the Liapunov function in Example 3.25.

3.19 If possible, use Theorem 3.26 to determine the stability of the equilibrium points for each of the following systems:

(i) $\begin{bmatrix} x \\ y \end{bmatrix}' = \begin{bmatrix} x^5 - 2x + y \\ -x \end{bmatrix}$

(ii) $\begin{bmatrix} x \\ y \end{bmatrix}' = \begin{bmatrix} 4 - 4x^2 - y^2 \\ 3xy \end{bmatrix}$

3.20 Determine the stability of each equilibrium point of the system

$$\begin{aligned} x' &= y^2 - 3x + 2 \\ y' &= x^2 - y^2. \end{aligned}$$

3.21 Find constants α and β so that $V(x, y) := \alpha x^2 + \beta y^2$ defines a Liapunov function for the system

$$\begin{aligned} x' &= -x - 5y, \\ y' &= 3x - y^3 \end{aligned}$$

on $\mathbb{R}^2$.

3.22 Find constants α and β so that $V(x, y) := \alpha x^2 + \beta y^4$ is a Liapunov function for the system

$$\begin{aligned} x' &= -x - y^3, \\ y' &= x - y \end{aligned}$$

on $\mathbb{R}^2$.

3.23 Find a Liapunov function for the system

$$\begin{aligned} x' &= -x - y^2, \\ y' &= -\frac{1}{2}y + 2xy \end{aligned}$$

on $\mathbb{R}^2$. Identify a family of positively invariant sets.

3.24 Find a Liapunov function for the system

$$x' = y,$$
$$y' = -4x - cy$$

where $c > 0$.

3.25 Show that the system

$$x' = -4x^3(y-2)^2,$$
$$y' = 2x^4(2-y)$$

is a gradient system, and find any asymptotically stable equilibrium points.

3.26 Work each of the following:

(i) Show that

$$x' = 3y - 3x^2,$$
$$y' = 3x - 3y^2$$

is a gradient system.

(ii) Construct a strict Liapunov function for the system in part (i), and use it to find a positively invariant set in the first quadrant such that every orbit in the set converges to an equilibrium.

3.27 Apply the LaSalle invariance theorem to each of the following systems:

(i)

$$x' = -2x - 2xy,$$
$$y' = 2x^2 - y$$

(ii)

$$x' = y,$$
$$y' = -x + y^5 - 2y$$

3.28 Apply the LaSalle invariance theorem to each of the following systems:

(i)

$$x' = -x - y + xy^2,$$
$$y' = x - y - x^2y$$

(ii)

$$x' = -2x - 3xy^4,$$
$$y' = 2x^4y - y$$

(iii)

$$x' = -x + 6y^5,$$
$$y' = -x$$

3.29 What happens when you try to apply the LaSalle invariance theorem to the system

$$x' = xy^4,$$
$$y' = -x^2 y?$$

Draw the phase plane diagram for this system.

3.30 Apply the LaSalle invariance theorem to the system

$$x' = -xy^2 + 4y^3$$
$$y' = -x.$$

3.31 In the predator-prey Example 3.27, show that the equilibrium point $(K, 0)$ is asymptotically stable if $bK < c$.

3.32 For the competition model in Exercise 3.13, find and determine the stability of all the equilibrium points.

3.33 Find the equilibrium point in the first quadrant for the epidemic model in Exercise 3.14, and test its stability.

3.34 Show that the model of a chemostat in Exercise 3.15 has an equilibrium in the first quadrant, and test its stability.

3.35 Consider the system for modeling competition between species given by

$$x' = x(3 - x - 3y)$$
$$y' = y(3 - 3x - y).$$

Find all equilibria of this system and sketch the nullclines and the direction of flow on each nullcline. Determine the stability of each equilibrium point.

3.36 Work each of the following:

(i) Show that

$$x' = 3y^2 - 3x,$$
$$y' = 3y - 3x^2$$

is a Hamiltonian system.
(ii) Find and classify the equilibrium points for the system in part (i).

3.37 Show that the system

$$x' = 6xy^2 + 4x^3 y,$$
$$y' = -2y^3 - 6x^2 y^2 + 4x$$

is a Hamiltonian system and find all Hamiltonian functions for this system.

3.38 For the the chemostat model in Exercise 3.15, determine whether orbits that are near the asymptotically stable equilibrium spiral into it as $t \to \infty$.

3.39 Use the linearization theorem to show that the equilibrium point $(0,0)$ is an unstable spiral point for the system

$$\begin{aligned} x' &= x + y - 2xy \\ y' &= -2x + y + 3y^2. \end{aligned}$$

3.40 Show that the equilibrium point $(0,0)$ is an asymptotically stable node for the system

$$\begin{aligned} x' &= -x - y - 3x^2 y \\ y' &= -2x - 4y + 3y \sin x. \end{aligned}$$

3.41 In Example 3.35 we prove the existence of a nontrivial periodic solution of the system (3.46), (3.47). Show that for any constant α, $x(t) = \cos(t + \alpha)$, $y(t) = -\sin(t + \alpha)$ is a periodic solution of the system (3.46), (3.47).

3.42 Show that the system

$$\begin{aligned} x' &= -4x - y + x(x^2 + y^2), \\ y' &= x - 4y + y(x^2 + y^2) \end{aligned}$$

has a nontrivial periodic solution.

3.43 Show that the differential equation $x'' + [x^2 + (x')^2 - 1]x' + x = 0$ has a nontrivial periodic solution.

3.44 Show that the system

$$\begin{aligned} x' &= -x - y + x(3x^2 + y^2), \\ y' &= x - y + y(3x^2 + y^2) \end{aligned}$$

has a limit cycle.

3.45 Use the Poincaré-Bendixson theorem (Theorem 3.36) to show that the system

$$\begin{aligned} x' &= x - y - x^3, \\ y' &= x + y - y^3 \end{aligned}$$

has a periodic solution. (*Hint*: Show that the system has an invariant square.)

3.46 Find a planar system such that div $F(x, y) = 0$ and there are cycles.

3.47 Determine whether or not each of the following equations has a periodic solution:

(i) $x'' - (x')^2 - (1 + x^2) = 0$
(ii) $x'' - (x^2 + 1)x' + x^5 = 0$

3.48 Draw the phase plane diagram for the system

$$x' = -y + x(x^2 + y^2)\sin\left(\frac{\pi}{\sqrt{x^2 + y^2}}\right),$$

$$y' = x + y(x^2 + y^2)\sin\left(\frac{\pi}{\sqrt{x^2 + y^2}}\right).$$

Find the limit cycles for this system. Is the origin asymptotically stable?

3.49 Show that the system

$$x' = x - y - x\left(x^2 + \frac{3}{2}y^2\right),$$

$$y' = x + y - y\left(x^2 + \frac{3}{2}y^2\right)$$

has a limit cycle.

3.50 Consider the following alteration of Example 3.39

$$x' = x(1-x) - \frac{9xy}{10x+1},$$

$$y' = by\left(1 - \frac{20y}{10x+1}\right).$$

(i) Find the nullclines and a positively invariant rectangle.
(ii) Compute the equilibrium in the first quadrant and show that it
is asymptotically stable for all $b > 0$.

3.51 A special case of a model of glycolysis (a biochemical process in which cells obtain energy by breaking down sugar) due to *Sel'kov* is

$$x' = -x + .05y + x^2 y,$$

$$y' = .5 - .05y - x^2 y.$$

(i) Show that the five-sided polygonal region shown in Figure 40 is invariant. [*Note*: The curve $y = x/(.05 + x^2)$ is the y nullcline.]
(ii) Use the Poincaré-Bendixson theorem to deduce the existence of a limit cycle.

3.52 Apply the Poincaré-Bendixson theorem (Theorem 3.36) to the system

$$x' = 3x - y - xe^{x^2 + y^2},$$

$$y' = x + 3y - ye^{x^2 + y^2}.$$

3.53 Show that each of the following systems has no cycles:

(i)

$$x' = x - y + xy^2,$$

$$y' = x$$

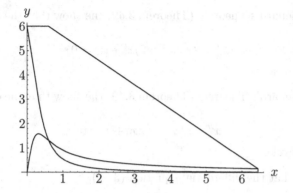

FIGURE 40. An invariant region for the model of glycolysis.

(ii)
$$x' = 1 + xy,$$
$$y' = 2x^2 + x^2y^2$$

(iii)
$$x' = -x - y + 2x^2 + y^2,$$
$$y' = x$$

3.54 Show that each of the following systems has no cycles:

(i)
$$x' = y + x^2y,$$
$$y' = xy + 2$$

(ii)
$$x' = y + x^3,$$
$$y' = x + y + 2y^3$$

(iii)
$$x' = -y,$$
$$y' = x + y - x^2 - y^2$$

3.55 Determine if the differential equation
$$x'' - (x')^2 - 1 - x^2 = 0$$
has a periodic solution.

3.56 Find the index of the mapping
$$F(x, y) = \begin{bmatrix} x^3 - 3xy^2 \\ 3x^2y - y^3 \end{bmatrix}$$
with respect to the simple closed curve C, $x^2 + y^2 = 1$, positively oriented.

3.57 Use Liénard's Theorem (Theorem 3.49) the show that the differential equation

$$x'' + 2x^2(x^2 - 2)x' + x^3 = 0$$

has a limit cycle.

3.58 Use Liénard's Theorem (Theorem 3.49) the show that the differential equation

$$x'' + x^6 x' - x^2 x' + x = 0$$

has a limit cycle.

3.59 Show that the differential equation (3.53)

$$ax'' + b(x^2 - 1)x' + cx = 0, \quad a, b, c > 0$$

can be transformed into the van der Pol equation by a suitable change of the independent variable.

3.60 Write the van der Pol equation 3.53 as a system in the standard way and determine the stability of the equilibrium point $(0,0)$ for the cases $\mu < 0$ and $\mu > 0$.

3.61 For what value of the parameter λ does the following system undergo a Hopf bifurcation?

$$
\begin{aligned}
x' &= x[x(1-x) - y], \\
y' &= (x - (1/\lambda))y
\end{aligned}
$$

3.62 Consider again the model of glycolysis in Exercise 3.51. Let's make the model somewhat more flexible with the addition of a positive parameter:

$$
\begin{aligned}
x' &= -x + .05y + x^2 y, \\
y' &= \lambda - .05y - x^2 y.
\end{aligned}
$$

Compute a value of the parameter λ at which a Hopf bifurcation occurs.

3.63 For the following linear system, compute the eigenvalues and eigenvectors, and discuss the nature of the orbits of the system.

$$
\begin{aligned}
x' &= y, \\
y' &= 2x + y, \\
z' &= 2y + z
\end{aligned}
$$

3.64 Compute the eigenvalues and eigenvectors for the following system, and sketch a few orbits for the system.

$$
\begin{aligned}
x' &= x + 2y, \\
y' &= -x - y, \\
z' &= 2x - z
\end{aligned}
$$

3.65 The following system is a simple model of a predator (density x) and two prey (densities y and z):

$$\begin{aligned} x' &= -rx + axy + xz, \\ y' &= by - xy, \\ z' &= cz(1-z) - dx\,z, \end{aligned}$$

where all parameters are positive. Compute the equilibrium in the first octant, and investigate its stability.

3.66 In the following system, the parameters a and b are positive:

$$\begin{aligned} x' &= x(1 - x - ay - bz), \\ y' &= y(1 - by - x - az), \\ z' &= z(1 - ax - by - z). \end{aligned}$$

(i) Show that this is a model of competition if (x, y, z) is in the first octant.
(ii) Show that the equilibrium in the first octant is $\frac{1}{1+a+b}(1, 1, 1)$.
(iii) Test the stability of the preceding equilibrium.

3.67 What happens to solutions of the Lorenz system that start on the z-axis?

3.68 Let $V(x, y, z) = \frac{1}{\sigma}x^2 + y^2 + z^2$.

(i) Along solutions of the Lorenz system, show that

$$\frac{d}{dt}V(x(t), y(t), z(t))$$

$$= -2\left(x(t) - \frac{r+1}{2}y(t)\right)^2$$

$$-2\left(1 - \left(\frac{r+1}{2}\right)^2\right)y^2(t) - 2bz^2(t).$$

(ii) Conclude that V is a strict Liapunov function if $0 < r < 1$ and that in this case all solutions of the Lorenz system go to the origin as $t \to \infty$.

3.69 Other period-doubling parameter regions exist for the Lorenz system. For example, fix $b = 8/3$ and $\sigma = 10$ and start r at 100 to obtain a cycle. Find a slightly smaller value of r that produces a doubling of the period. Then find a smaller value of r that looks chaotic.

3.70 A system that has properties similar to the Lorenz system is the Rössler system:

$$\begin{aligned} x' &= -y - z, \\ y' &= x + ay, \\ z' &= bx - cz + xz, \end{aligned}$$

where a, b, and c are positive.

   (i) Find the equilibrium points and test their stability.

   (ii) Let $a = .4$, $b = .3$, and increase c starting at $c = 1.4$ to find values of c at which the period doubles and then doubles again.

   (iii) Generate an image of the Rössler attractor, using $c = 4.449$.

3.71 Use a computer to investigate the forced "double well" oscillator

$$y'' + ry' + y^3 - y = a \cos t.$$

Fix $r = .25$ and sample values of a between .18 and .4 to find values of a and appropriate initial conditions that yield multiple limit cycles, transient chaos, and chaotic oscillations.

3.72 The system

$$\begin{aligned} x' &= yz, \\ y' &= -2xz, \\ z' &= xy \end{aligned}$$

describes the rotational motion of a book tossed into the air (see Bender and Orszag [5], p. 202).

   (i) Locate and classify all equilibria.

   (ii) Show that orbits lie on spheres centered at the origin.

   (iii) Use a computer to sketch some orbits on the unit sphere $x^2 + y^2 + z^2 = 1$.

Chapter 4

Perturbation Methods

4.1 Introduction

Even if a differential equation cannot be solved explicitly, we can often obtain useful information about the solutions by computing functions, called *analytic* approximations, that are close to the actual solutions. In this chapter, we give a brief account of some of the most successful methods of finding analytic approximations, namely *perturbation methods*. The idea is to identify a portion of the problem that is small in relation to the other parts and to take advantage of the size difference to reduce the original problem to one or more simpler problems. This fundamental idea has been a cornerstone of applied mathematics for over a century since it has proven remarkably successful for attacking an array of challenging and important problems. In order to introduce some of the basic terminology and to illustrate the idea of perturbation methods, let's begin with an elementary example.

Example 4.1 Consider the initial value problem

$$x' + x + \epsilon x^2 = 0, \quad x(0) = 1.$$

Here ϵ is a small, positive parameter, and the differential equation is said to be a small *perturbation* of the linear equation $x' + x = 0$, which has the general solution $x(t) = Ce^{-t}$. We expect that one of these solutions of the linear equation will serve as an analytic approximation of the solution of our initial value problem.

In order to standardize the procedure, let's try to compute the solution of the initial value problem as a *perturbation series*:

$$x(t, \epsilon) = x_0(t) + \epsilon x_1(t) + \epsilon^2 x_2(t) + \cdots.$$

Since ϵ is small, the $x_0(t)$ will represent the most significant approximation to $x(t, \epsilon)$, the term $\epsilon x_1(t)$ will correct that approximation by an amount directly related to the size of ϵ, the term $\epsilon^2 x_2(t)$ will further correct the approximation by a smaller amount (assuming that ϵ is much smaller than one), and so forth. The functions x_0, x_1, $\cdots$ will be computed by substituting the perturbation series directly into the initial value problem:

$$(x_0' + \epsilon x_1' + \cdots) + (x_0 + \epsilon x_1 + \cdots) + \epsilon(x_0 + \epsilon x_1 + \cdots)^2 = 0,$$
$$x_0(0) + \epsilon x_1(0) + \epsilon^2 x_2(0) + \cdots = 1.$$

W.G. Kelley and A.C. Peterson, *The Theory of Differential Equations:*
Classical and Qualitative, Universitext 278, DOI 10.1007/978-1-4419-5783-2_4,
© Springer Science+Business Media, LLC 2010

Now we use the familier algebraic procedure of setting the coefficient of each power of ϵ equal to zero in each equation. Beginning with the terms that do not involve ϵ, we have

$$
\begin{aligned}
x_0' + x_0 &= 0, \\
x_0(0) &= 1.
\end{aligned}
$$

The solution of this IVP is

$$
x_0(t) = e^{-t},
$$

which is the solution of the associated linear equation that satisfies the initial condition. Next, set the coefficient of ϵ equal to 0:

$$
\begin{aligned}
x_1' + x_1 + x_0^2 &= 0 \quad \text{or} \quad x_1' + x_1 + e^{-2t} = 0, \\
x_1(0) &= 0.
\end{aligned}
$$

This is a nonhomogeneous linear differential equation that can be solved by using the variation of constants formula in Chapter 1 or by finding an integrating factor. The solution is

$$
x_1(t) = e^{-2t} - e^{-t}.
$$

If we set the coefficient of ϵ^2 equal to 0, then we have

$$
\begin{aligned}
x_2' + x_2 + 2x_0 x_1 &= 0 \quad \text{or} \quad x_2' + x_2 + 2e^{-3t} - 2e^{-2t} = 0, \\
x_2(0) &= 0.
\end{aligned}
$$

The solution of this IVP is

$$
x_2(t) = -2e^{-2t} + e^{-3t} + e^{-t}.
$$

Consequently, our perturbation series is

$$
e^{-t} + \epsilon(e^{-2t} - e^{-t}) + \epsilon^2(e^{-3t} + e^{-t} - 2e^{-2t}) + \cdots .
$$

For this example, we can find out how good this approximation is by computing the exact solution. We leave it to the reader (Exercise 4.2) to show, using separation of variables, that the exact solution is

$$
x(t, \epsilon) = \frac{e^{-t}}{1 + \epsilon(1 - e^{-t})}.
$$

[We will denote both the exact solution and the perturbation series by $x(t, \epsilon)$.] We can see immediately that any finite portion of the perturbation series will not be a good approximation for all t because the exact solution goes to infinity as t approaches $\ln[\epsilon/(1 + \epsilon)]$, while the perturbation series does not have this property!

To investigate further, let's use the geometric series to write the exact solution in the form of an infinite series:

$$
x(t, \epsilon) = e^{-t}(1 - \epsilon(1 - e^{-t}) + \epsilon^2(1 - e^{-t})^2 + \cdots),
$$

which is valid if $\epsilon|1 - e^{-t}| < 1$. Note that this series is identical to the portion of the perturbation series that we computed earlier. However, as noted previously, the exact solution and the perturbation series are not the

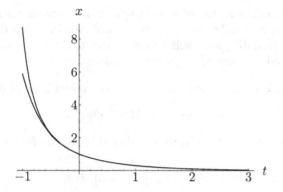

FIGURE 1. Approximations of the exact solution in Example 4.1.

same, and we can see that the problem is that the series converges only for a restricted range of t values.

If we restrict t to be nonnegative, then the series converges, assuming that ϵ is less than 1. Furthermore, the series is alternating and the terms are decreasing in magnitude, so we can estimate errors using the alternating series test. (We could, of course, compute the exact errors here, but that is not necessary for the conclusions that we wish to draw.) For example, suppose that we use just the first term e^{-t} in the perturbation series as our approximation. Then we have the absolute error estimate

$$|x(t, \epsilon) - e^{-t}| \le \epsilon(1 - e^{-t})e^{-t} \quad (t \ge 0)$$

as well as the estimate of the relative error

$$\frac{|x(t, \epsilon) - e^{-t}|}{e^{-t}} \le \epsilon(1 - e^{-t}) \quad (t \ge 0).$$

The relative error is important since it determines the number of significant digits in the approximation. If we want a better approximation, then we can use the first two terms in the perturbation series and obtain smaller error estimates:

$$|x(t, \epsilon) - e^{-t}(1 - \epsilon(1 - e^{-t}))| \le e^{-t}\epsilon^2(1 - e^{-t})^2,$$

and

$$\frac{|x(t, \epsilon) - e^{-t}(1 - \epsilon(1 - e^{-t}))|}{e^{-t}} \le \epsilon^2(1 - e^{-t})^2,$$

for $t \ge 0$. (Strictly speaking, this last estimate is not an estimate of the relative error, but an estimate of the ratio of the error to the dominant term in the approximation.) Figure 1 compares the first three terms in the perturbation series to the exact solution for $\epsilon = .4$. $\triangle$

In the last example, we were able to give rather precise error estimates. Such precision is usually not possible for problems in which the exact solution is not available, so it will be convenient to define a concept that is commonly used in *order of magnitude* estimates.

Definition 4.2 If there are constants M and ϵ_0 and a set S so that

$$|u(t, \epsilon)| \leq M|v(t, \epsilon)|$$

for $0 < \epsilon \leq \epsilon_0$ and $t \in S$, then we say that u is *big oh* of v on S as ϵ goes to 0 and write

$$u(t, \epsilon) = \mathcal{O}(v(t, \epsilon)) \quad (\epsilon \to 0) \quad (t \in S).$$

Using this definition, we have in the preceding example

$$x(t, \epsilon) - e^{-t} = \mathcal{O}(\epsilon) \quad (\epsilon \to 0) \quad (t \geq 0),$$

or, more precisely,

$$x(t, \epsilon) - e^{-t} = \mathcal{O}(\epsilon e^{-t}) \quad (\epsilon \to 0) \quad (t \geq 0).$$

This last expression is usually written in the equivalent form

$$x(t, \epsilon) = e^{-t}(1 + \mathcal{O}(\epsilon)) \quad (\epsilon \to 0) \quad (t \geq 0).$$

If the first two terms in the perturbation series are used, then we have

$$x(t, \epsilon) = e^{-t}(1 - \epsilon(1 - e^{-t}) + \mathcal{O}(\epsilon^2)) \quad (\epsilon \to 0) \quad (t \geq 0).$$

In order to use perturbation methods, a problem must be changed to non-dimensional form, so that any small parameters in the problem can be identified. The next example illustrates the procedure.

Example 4.3 Let y be the distance from the earth's surface to a rocket shot straight up into the air. Let R be the radius of the earth, g the acceleration of gravity at the earth's surface, and m the mass of the rocket. If we neglect all forces except the force of gravity, then by Newton's second law,

$$\begin{aligned} my'' &= -\frac{mgR^2}{(y + R)^2}, \\ y(0) &= 0, \quad y'(0) = v_0. \end{aligned}$$

We rescale the dependent and independent variables, using

$$y = \alpha x, \quad t = \beta \tau,$$

where x is the new dependent variable, τ is the new independent variable, and α and β are scaling constants. Note that we want α to have the same units as y and β to have the same units as t, so that the new variables will be dimensionless. By the chain rule,

$$\frac{d^2y}{dt^2} = \frac{d}{dt}\left(\frac{dy}{dt}\right)$$

$$= \frac{d}{d\tau}\left(\frac{d}{d\tau}\frac{\alpha x}{dt}\frac{d\tau}{dt}\right)\frac{d\tau}{dt}$$

$$= \frac{\alpha}{\beta^2}\frac{d^2x}{d\tau^2}.$$

Substituting into our initial value problem, we have

$$\frac{m\alpha}{\beta^2}\frac{d^2x}{d\tau^2} = -\frac{mgR^2}{(\alpha x + R)^2},$$

$$x(0) = 0, \quad \frac{\alpha}{\beta}\frac{dx}{d\tau}(0) = v_0,$$

or

$$\frac{d^2x}{d\tau^2} = -\frac{\beta^2 gR^2}{\alpha(\alpha x + R)^2},$$

$$x(0) = 0, \quad \frac{dx}{d\tau}(0) = \frac{\beta v_0}{\alpha}.$$

There are several ways to choose α and β to make the problem dimensionless. We take the following approach: First elimate the parameters in the initial conditions by making $\alpha = \beta v_0$. Our problem is now

$$\frac{d^2x}{d\tau^2} = -\frac{\beta g}{v_0\left(\frac{\beta v_0}{R}x + 1\right)^2},$$

$$x(0) = 0, \quad \frac{dx}{d\tau}(0) = 1.$$

The choice $\beta = v_0/g$ gives β the units of time, as desired, and results in the dimensionless equation

$$\frac{d^2x}{d\tau^2} = -\frac{1}{\left(\frac{v_0^2}{gR}x + 1\right)^2}.$$

Now if the square of the initial velocity of the rocket is small compared to the product of the radius of the earth and the acceleration of gravity, then the initial value problem contains a small parameter

$$\epsilon = \frac{v_0^2}{gR}.$$

The result of our calculation is a dimensionless problem with a small parameter:

$$\frac{d^2x}{d\tau^2} = -\frac{1}{(\epsilon x + 1)^2}, \tag{4.1}$$

$$x(0) = 0, \quad \frac{dx}{d\tau}(0) = 1. \tag{4.2}$$

Note that the process of making the problem dimensionless has reduced the number of parameters from four to one. △

Now that we have a problem with a small parameter, we can compute a perturbation series that will provide much useful information about the solution when ϵ is small.

Example 4.4 We substitute a perturbation series

$$x(\tau, \epsilon) = x_0(\tau) + \epsilon x_1(\tau) + \epsilon^2 x_2(\tau) + \cdots$$

into equations (4.1), (4.2) and get

$$\frac{d^2 x_0}{d\tau^2} + \epsilon \frac{d^2 x_1}{d\tau^2} + \epsilon^2 \frac{d^2 x_2}{d\tau^2} + \cdots = -\frac{1}{(1 + \epsilon x_0 + \epsilon^2 x_1 + \cdots)^2},$$

$$x_0(0) + \epsilon x_1(0) + \epsilon^2 x_2(0) + \cdots = 0,$$

$$\frac{dx_0}{d\tau}(0) + \epsilon \frac{dx_1}{d\tau}(0) + \epsilon^2 \frac{dx_2}{d\tau}(0) + \cdots = 1.$$

In order to equate coefficients of like powers of ϵ in the differential equation, we must first change the right-hand side into a series in powers of ϵ. We can do this quickly by recalling the binomial series:

$$(1 + z)^n = 1 + nz + \frac{n(n-1)}{2!} z^2 + \cdots,$$

which converges for $|z| < 1$. With $z = \epsilon x_0 + \epsilon^2 x_1 + \cdots$, the differential equation becomes

$$\frac{d^2 x_0}{d\tau^2} + \epsilon \frac{d^2 x_1}{d\tau^2} + \epsilon^2 \frac{d^2 x_2}{d\tau^2} + \cdots = -[1 + (-2)(\epsilon x_0 + \epsilon^2 x_1 + \cdots)$$

$$+ \frac{(-3)(-2)}{2!}(\epsilon x_0 + \epsilon^2 x_1 + \cdots)^2 + \cdots].$$

Equating the terms on each side of each equation that do not involve ϵ gives us the problem for x_0:

$$\frac{d^2 x_0}{d\tau^2} = -1,$$

$$x_0(0) = 0, \quad \frac{dx_0}{d\tau}(0) = 1.$$

The solution can be found by direct integration:

$$x_0 = \tau - \frac{\tau^2}{2}.$$

Next, we equate the coefficients of ϵ on each side of each equation to obtain

$$\frac{d^2 x_1}{d\tau^2} = 2x_0 = 2\left(\tau - \frac{\tau^2}{2}\right),$$

$$x_1(0) = 0, \quad \frac{dx_1}{d\tau}(0) = 0$$

and then integrate to find the solution

$$x_1 = \frac{\tau^3}{3} - \frac{\tau^4}{12}.$$

Similarly, the problem for x_2 is

$$\frac{d^2 x_2}{d\tau^2} = 2x_1 - 3x_0^2,$$

$$x_2(0) = 0, \quad \frac{dx_2}{d\tau}(0) = 0,$$

with solution given by

$$x_2(\tau) = -\frac{\tau^4}{4} + \frac{11\tau^5}{60} - \frac{11\tau^6}{360}.$$

For brevity, let's include only terms through ϵ in the perturbation series:

$$\tau - \frac{\tau^2}{2} + \epsilon \left(\frac{\tau^3}{3} - \frac{\tau^4}{12} \right) + \mathcal{O}(\epsilon^2) \quad (\epsilon \to 0) \quad (0 \leq \tau \leq T(\epsilon)),$$

where $T(\epsilon)$ is the time when the rocket returns to the earth's surface. Note that the next term in the series actually involves the expressions $\epsilon^2 \tau^4$, $\epsilon^2 \tau^5$, and $\epsilon^2 \tau^6$, but the dependence on τ can be supressed here since τ varies over a bounded interval.

In the original variables, the perturbation series is

$$v_0 t - \frac{1}{2} g t^2 + \frac{v_0^4}{g^2 R} \left[\frac{1}{3} \left(\frac{gt}{v_0} \right)^3 - \frac{1}{12} \left(\frac{gt}{v_0} \right)^4 \right] + \cdots.$$

The expression $v_0 t - gt^2/2$ represents the position of the rocket if we assume that the force of gravity is constant throughout the flight. The next segment of the perturbation series is positive and represents the *first-order* correction, revealing (as expected) that the rocket actually flies higher due to the fact that the force of gravity diminishes with increasing altitude. $\triangle$

We saw in Example 4.1 that a perturbation series does not necessarily yield a valid approximation of the solution, so our discussion for Example 4.4 is incomplete. The most commonly used method of checking an approximation of this type is to generate a numerical approximation for certain initial and parameter values and test it against the analytic approximation. Of course, this is only a partial check since we can't try all possible values. Here is a theorem that can be used to verify approximations for problems of the type discussed previously.

Theorem 4.5 *Let x be the unique solution of the initial value problem*

$$x' = f_0(t, x) + \epsilon f_1(t, x) + \cdots + \epsilon^{n-1} f_{n-1}(t, x) + \epsilon^n f_n(t, x, \epsilon),$$
$$x(t_0) = A,$$

where $f_0, f_1, \cdots, f_{n-1}$ are continuous in t and have n continuous derivatives with respect to x in the set $S = \{(t, x) : |t - t_0| \leq h, x \in D\}$, where

$A \in D$, D is a bounded subset of $\mathbb{R}^m$, and f_n is continuous on S for $0 \le \epsilon \le \epsilon_0$ and bounded on $S \times (0, \epsilon_0]$ for some $\epsilon_0 > 0$. If we substitute

$$x_0(t) + \epsilon x_1(t) + \cdots + \epsilon^{n-1} x_{n-1}(t)$$

into the initial value problem and compute the coefficient functions

$$x_0, x_1, \cdots, x_{n-1}$$

by the method outlined previously, then

$$x(t, \epsilon) - (x_0(t) + \epsilon x_1(t) + \cdots + \epsilon^{n-1} x_{n-1}(t)) = \mathcal{O}(\epsilon^n) \quad (\epsilon \to 0) \quad (t \in I),$$

where I is any interval in $[t_0 - h, t_0 + h]$ such that $x(t) \in D$ for $t \in I$.

Proof See Velhurst [51], Theorem 9.1.   □

In order to apply this theorem, we use Taylor's formula with the remainder in integral form:

$$f(z) = \sum_{i=0}^{n-1} \frac{f^{(i)}(0)}{i!} z^i + \frac{1}{(n-1)!} \int_0^z f^{(n)}(s)(z-s)^{n-1} \, ds,$$

where f is a function with n continuous derivatives on the interval between 0 and z. In Example 4.4, we use this formula with $f(z) = (1+z)^{-2}$ and $n = 2$ to write

$$-(1 + \epsilon x)^{-2} = -1 + 2\epsilon x - 6 \int_0^{\epsilon x} \frac{(\epsilon x - s)}{(1+s)^4} \, ds.$$

Note that the integral remainder is continuous in x for all positive values of x and ϵ. Furthermore,

$$6 \int_0^{\epsilon x} \frac{(\epsilon x - s)}{(1+s)^4} \, ds \le 6 \int_0^{\epsilon x} (\epsilon x - s) \, ds = 3\epsilon^2 x^2,$$

so the remainder is bounded for bounded values of x. Then Theorem 4.5 can be used to conclude that

$$x(t, \epsilon) - \left(\tau - \frac{\tau^2}{2} + \epsilon \left(\frac{\tau^3}{3} - \frac{\tau^4}{12} \right) \right) = \mathcal{O}(\epsilon^2) \quad (\epsilon \to 0) \quad (0 \le \tau \le T(\epsilon)).$$

The last example of this section examines a completely different type of perturbation problem.

Example 4.6 The following partial differential equation is often called Fisher's equation:

$$\frac{\partial y}{\partial t} = \frac{\partial^2 y}{\partial x^2} + y(1 - y).$$

It was proposed by R. A. Fisher [15] (see also Murray [37]) as a simple model of the propagation of an advantageous gene through a population. Here y is the relative frequency of the favored gene and x is a single space variable.

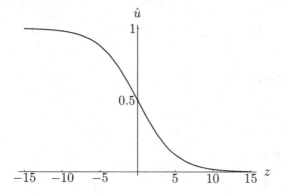

FIGURE 2. A traveling wave solution of Fisher's equation.

Let's assume the equation is in dimensionless form and look for traveling wave solutions:

$$y(x,t) = \hat{u}(x - ct) = \hat{u}(z),$$

where c is the wave speed. Then $\hat{u}(z)$ satisfies the boundary value problem

$$\frac{d^2\hat{u}}{dz^2} + c\frac{d\hat{u}}{dz} + \hat{u}(1 - \hat{u}) = 0,$$

$$\hat{u}(-\infty) = 1, \quad \hat{u}(0) = \frac{1}{2}, \quad \hat{u}(\infty) = 0.$$

(See Figure 2.) Two of the boundary conditions arise from the assumption that the frequency of the gene is initially small (ahead of the wave) and that it approaches 1 (behind the wave) as time increases. Since any translation of a traveling wave is also a solution, we have made the solution unique by setting $\hat{u}(0) = 1/2$.

Since it is known that this problem has solutions for large values of c, we convert it into a perturbation problem by setting $\epsilon = 1/c^2$, $w = \epsilon^{1/2}z$, $u(w) = \hat{u}(\frac{w}{\sqrt{\epsilon}})$:

$$\epsilon\frac{d^2u}{dw^2} + \frac{du}{dw} + u(1 - u) = 0,$$

$$u(-\infty) = 1, \quad u(0) = \frac{1}{2}, \quad u(\infty) = 0.$$

Substituting a perturbation series

$$u(w, \epsilon) = u_0(w) + \epsilon u_1(w) + \cdots$$

into the boundary value problem, we find that u_0 satisfies

$$\frac{du_0}{dw} + u_0(1 - u_0) = 0,$$

$$u_0(-\infty) = 1, \quad u_0(0) = \frac{1}{2}, \quad u_0(\infty) = 0.$$

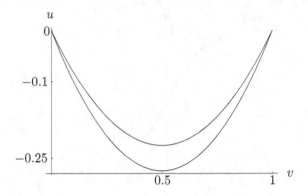

Figure 3. A positively invariant region.

The solution is found by separation of variables to be

$$u_0(w) = \frac{1}{1 + e^w}.$$

Similarly, we can show that

$$u_1(w) = \frac{e^w}{(1 + e^w)^2} \log\left(\frac{4e^w}{(1 + e^w)^2}\right).$$

How close is u_0 to an actual traveling wave solution? We can answer this question using phase plane analysis. Let $v = \frac{du}{dw}$. Then we have the system

$$\frac{du}{dw} = v,$$

$$\frac{dv}{dw} = \frac{1}{\epsilon}(u(u - 1) - v).$$

Note that in the phase plane the approximation u_0 corresponds to the parabola $v = u(u - 1)$ for $0 < u < 1$. To show that the desired solution $u(w, \epsilon)$ is near $u_0(w)$, we will construct a positively invariant set of the form

$$S = \{(u, v) : C(\epsilon)u(u - 1) \le v \le D(\epsilon)u(u - 1), 0 \le u \le 1\},$$

where $C(\epsilon) > 1$, $D(\epsilon) < 1$ are to be determined. (See Figure 3.)

Orbits will flow inward along the upper boundary of S, $v = D(\epsilon)u(u - 1)$, $0 < u < 1$, if

$$0 < \left[v, \frac{1}{\epsilon}(u(u - 1) - v)\right] \cdot \left[\begin{array}{c} D(\epsilon)(2u - 1) \\ -1 \end{array}\right],$$

for $0 < u < 1$. Simplifying and using $v = Du(u - 1)$, the inequality is equivalent to

$$0 < (1 - 2u)D^2 - \frac{D}{\epsilon} + \frac{1}{\epsilon}$$

for $0 < u < 1$. Thus we need

$$0 \leq -D^2 - \frac{D}{\epsilon} + \frac{1}{\epsilon}.$$

We can choose

$$D(\epsilon) = \frac{-1 + \sqrt{1 + 4\epsilon}}{2\epsilon} = 1 - \epsilon + \mathcal{O}(\epsilon^2) \quad (\epsilon \to 0).$$

Similarly, flow will be inward on the lower boundary, $v = Cu(u-1)$, if

$$0 < \left[v, \frac{1}{\epsilon}(u(u-1) - v) \right] \cdot \left[\begin{matrix} C(\epsilon)(1-2u) \\ 1 \end{matrix} \right],$$

for $0 < u < 1$. Equivalently,

$$0 < C^2(2u-1) + \frac{1}{\epsilon}(C-1) \quad (0 < u < 1),$$

so we require

$$0 \leq -C^2 + \frac{C}{\epsilon} - \frac{1}{\epsilon}.$$

Choose

$$C = \frac{1 - \sqrt{1 - 4\epsilon}}{2\epsilon} = 1 + \epsilon + \mathcal{O}(\epsilon^2) \quad (\epsilon \to 0),$$

provided that $\epsilon \leq .25$. Thus for $\epsilon \leq .25$ S is a positively invariant region. We leave it to the reader to show that $(1,0)$ is a saddle point and that $(0,0)$ is an asymptotically stable equilibrium (see Exercise 4.12). Since S is positively invariant, there is an orbit (part of the unstable manifold at $(1,0)$) that goes to $(1,0)$ as $w \to -\infty$ and for increasing w remains in S and goes to $(0,0)$ as $w \to \infty$. This is the desired solution $u(w, \epsilon)$.

Summing up, we have found constants $C(\epsilon)$ and $D(\epsilon)$ and a solution $u(w, \epsilon)$ of the boundary value problem such that

$$Cu(u-1) < u' < Du(u-1) \quad (0 < u < 1),$$

provided that $\epsilon \leq .25$. By integrating these inequalities, we have

$$\frac{1}{1 + e^{Cw}} \leq u(w, \epsilon) \leq \frac{1}{1 + e^{Dw}}$$

for $w \geq 0$. The inequalities are reversed for $w \leq 0$. Recall that the approximation $u_0(w)$ satisfies the same inequalities. We can use the mean value theorem (applied to $g(x) = 1/(1 + e^x)$) to obtain an error estimate for $w \geq 0$ (a similar estimate is true for $w \leq 0$):

$$
\begin{aligned}
|u(w, \epsilon) - u_0(w)| &\leq \frac{1}{1 + e^{Dw}} - \frac{1}{1 + e^{Cw}} \\
&= \frac{we^z}{(1 + e^z)^2}(C - D) \quad (Dw < z < Cw) \\
&< \frac{we^{Dw}}{(1 + e^{Dw})^2} \mathcal{O}(\epsilon) \quad (\epsilon \to 0).
\end{aligned}
$$

The estimate shows that the error is $\mathcal{O}(\epsilon)$ for all w and is exponentially small as $w \to \pm\infty$. In Exercise 4.13, the reader is asked to show that no traveling wave solution exists if $\epsilon > .25$. See Kelley [**29**], for extensions of this method to other equations. $\triangle$

4.2 Periodic Solutions

In Section 3.6, we discussed a number of ways by which systems of equations can be shown to have periodic solutions. Now we use perturbation methods to obtain approximations of periodic solutions for three different types of equations. In each case, we will check our approximations against numerical approximations of the cycles. For a rigorous justification of perturbation methods for periodic solutions, see Smith [**47**] or Grimshaw [**16**].

Example 4.7 The Hopf bifurcation theorem was used in Example 3.47 to show that the system

$$\begin{aligned} x' &= y, \\ y' &= -x + \lambda y - ay^3 \end{aligned}$$

has a periodic solution that encircles the origin if λ is a small positive number and a is positive. Let the intersection of the periodic solution with the positive x-axis be denoted $(A, 0)$. In order to approximate this cycle, we write the system as a single second order equation containing a small parameter:

$$u'' + u = \epsilon \left(u' - \frac{u'^3}{3} \right),$$

with initial conditions $u(0) = A$, $u'(0) = 0$. Now substitute $u(t, \epsilon) = u_0(t) + \epsilon u_1(t) + \cdots$ into the equation

$$(u_0'' + \epsilon u_1'' + \cdots) + (u_0 + \epsilon u_1 + \cdots) = \epsilon((u_0' + u_1' + \cdots) - (1/3)(u_0' + \epsilon u_1' + \cdots)^3).$$

Then

$$u_0'' + u_0 = 0.$$

Similarly, substituting the perturbation series into the initial conditions yields

$$u_0(0) = A, \quad u_0'(0) = 0,$$

so

$$u_0 = A \cos t.$$

Next,

$$\begin{aligned} u_1'' + u_1 &= u_0' - (1/3)u_0'^3 \\ &= -A \sin t + \frac{A^3}{3} \sin^3 t. \end{aligned}$$

The solution is

$$u_1 = B \cos t + C \sin t + \frac{1}{2} \left(1 - \frac{A^2}{4} \right) At \cos t + \frac{A^3}{96} \sin 3t.$$

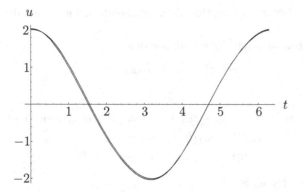

FIGURE 4. The periodic solution and perturbation approximation for Example 4.7 with $\epsilon = .4$.

Note that one of the terms making up u_1 is unbounded for large t, so it does not belong in the approximation of the periodic solution that we are seeking. Fortunately, it is possible to eliminate this term by choosing $A = 2$. Then we have

$$u_1(t) = B \cos t + C \sin t + \frac{1}{12} \sin 3t.$$

Since u_1 satisfies homogeneous initial conditions at $t = 0$, we find that $B = 0$ and $C = 1/4$. The approximation is then reduced to

$$u(t, \epsilon) = 2 \cos t - \frac{\epsilon}{4} \sin t + \frac{\epsilon}{12} \sin 3t + \cdots.$$

In Figure 4, a numerical approximation of the periodic solution and the perturbation approximation are plotted over one period for $\epsilon = .4$. Since the periods of the periodic solution and the approximation are slightly different, their graphs gradually separate if they are plotted over a longer time interval. Also, the number A where the periodic solution intersects the positive x-axis is actually a function of ϵ and has a value slightly greater than 2 for small values of ϵ.

$\triangle$

Example 4.8 Consider the IVP

$$\frac{d^2 u}{d\theta^2} + u = a + \epsilon u^2,$$

$$u(0) = b, \quad u'(0) = 0,$$

where a, b, and ϵ are positive parameters. This problem is a dimensionless form of the equation of motion (in polar coordinates with $u = 1/r$) of the orbit of a planet around the sun. The small nonlinear term is a correction, resulting from Einstein's theory of gravitation, to the classical Keplerian model. (See Smith [47].) In contrast to the preceding example, in which there was an isolated cycle, this differential equation has a family of periodic solutions.

We begin, as usual, by trying a series

$$u(\theta, \epsilon) = u_0(\theta) + \epsilon u_1(\theta) + \cdots :$$

$$u_0'' + \epsilon u_1'' + \cdots + u_0 + \epsilon u_1 + \cdots = a + \epsilon(u_0 + \epsilon u_1 + \cdots)^2,$$
$$u_0(0) + \epsilon u_1(0) + \cdots = b,$$
$$u_0'(0) + \epsilon u_1'(0) + \cdots = 0.$$

The problem for u_0 is

$$u_0'' + u_0 = a, \quad u_0(0) = b, \quad u_0'(0) = 0,$$

so

$$u_0(\theta) = a + (b - a)\cos\theta.$$

Next,

$$u_1'' + u_1 = u_0^2 = (a + (b - a)\cos\theta)^2,$$

with homogeneous initial conditions. The solution is

$$u_1(\theta) = a^2(1 - \cos\theta) + a(b - a)\theta\sin\theta + (1/3)(b - a)^2(2 - \cos^2\theta - \cos\theta).$$

Here again, we have among the terms in u_1 a term that is unbounded. In this case, we cannot eliminate this unbounded term, except by choosing $a = b$, which is not true for the application that motivates the problem. Consequently, we will use an alternate method, called the method of *renormalization*, to improve our approximation.

Numerical studies of the periodic solutions indicate that in fact the frequency is a function of ϵ. We can incorporate this fact into our approximation by defining

$$\omega = 1 + \omega_1\epsilon + \omega_2\epsilon^2 + \cdots$$

and a new independent variable $\phi = \omega\theta$, so that

$$\theta = \omega^{-1}\phi = (1 + \omega_1\epsilon + \cdots)^{-1}\phi = (1 - \epsilon\omega_1 + \cdots)\phi.$$

Substituting these expressions into the approximations u_0 and u_1, we have

$$u_0 = a + (b - a)\cos(\phi - \epsilon\omega_1\phi + \cdots),$$
$$u_1 = a^2(1 - \cos(\phi - \epsilon\omega_1\phi + \cdots))$$
$$+ a(b - a)(\phi - \epsilon\omega_1\phi + \cdots)\sin(\phi - \epsilon\omega_1\phi + \cdots)$$
$$+ (1/3)(b - a)^2(2 - \cos^2(\phi - \epsilon\omega_1\phi + \cdots) - \cos(\phi - \epsilon\omega_1\phi + \cdots)).$$

From Taylor's formula,

$$\cos(\phi - \epsilon\omega_1\phi + \cdots) = \cos\phi + \epsilon\omega_1\phi\sin\phi + \cdots,$$
$$\sin(\phi - \epsilon\omega_1\phi + \cdots) = \sin\phi - \epsilon\omega_1\phi\cos\phi + \cdots.$$

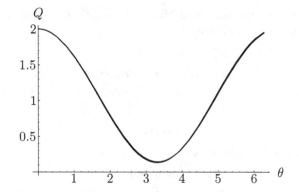

FIGURE 5. A periodic solution and perturbation approximation for Example 4.8 with $a = 1$, $b = 2$, and $\epsilon = .05$.

Using these formulas, our approximation becomes

$$u_0 + \epsilon u_1$$
$$= a + (b - a)(\cos\phi + \epsilon\omega_1\phi\sin\phi) + \epsilon a^2(1 - \cos\phi)$$
$$+ \epsilon(b - a)(\phi - \epsilon\omega_1\phi)\sin\phi + \epsilon(1/3)(b - a)^2(2 - \cos^2\phi - \cos\phi) + \cdots.$$

Note that unbounded terms in the approximation can be eliminated by choosing $\omega_1 = -a$! Neglecting terms of order ϵ^2, we now have the following approximation of the periodic solution of the initial value problem:

$$a + (b - a)\cos\phi + \epsilon a^2(1 - \cos\phi) + \epsilon(1/3)(b - a)^2(2 - \cos^2\phi - \cos\phi),$$

where $\phi = (1 - a\epsilon)\theta$. In Figure 5, the periodic solution is compared with this approximation for the case $a = 1$, $b = 2$, and $\epsilon = .05$. For the case of the orbit of the planet Mercury around the sun, the actual parameter values are approximately $a = .98$, $b = 1.01$, and ϵ is on the order of 10^{-7}.

$\triangle$

Example 4.9 Mathieu's equation was examined using Floquet theory in Exercise 2.61. Here we write it in a form suitable for applying perturbation methods:

$$u'' + (\alpha + 2\epsilon\cos t)u = 0,$$

where a and ϵ are positive and ϵ is small. We know that for certain special values of α and ϵ (these are curves in the parameter domain) there are periodic solutions of period 2π or 4π, as well as unbounded solutions. These curves serve as boundaries in the parameter domain for parameter regions in which the solutions are either unbounded or aperiodic bounded solutions. Let's see what information a straightforward perturbation approximation yields:

$$u(t, \epsilon) = u_0(t) + \epsilon u_1(t) + \epsilon^2 u_2(t) + \cdots.$$

Substituting this expression into Mathieu's equation, we find the following equations for the coefficients:

$$
\begin{aligned}
u_0'' + \alpha u_0 &= 0, \\
u_1'' + \alpha u_1 &= -2u_0 \cos t, \\
u_2'' + \alpha u_2 &= -2u_1 \cos t.
\end{aligned}
$$

Then

$$
u_0 = A \cos(\sqrt{\alpha} t + \beta),
$$

so the equation for u_1 becomes

$$
\begin{aligned}
u_1'' + \alpha u_1 &= -2A \cos t \cos(\sqrt{\alpha} t + \beta) \\
&= -A[\cos((1 + \sqrt{\alpha})t + \beta) + \cos((\sqrt{\alpha} - 1)t + \beta)].
\end{aligned}
$$

The solution (neglecting the homogeneous solution, which can be included in u_0) is

$$
u_1 = \frac{A \cos((\sqrt{\alpha} + 1)t + \beta)}{2\sqrt{\alpha} + 1} + \frac{A \cos((\sqrt{\alpha} - 1)t + \beta)}{1 - 2\sqrt{\alpha}},
$$

provided that α is not $1/4$. Similarly, for u_2 we obtain

$$
\begin{aligned}
u_2 &= \frac{A}{4(2\sqrt{\alpha} + 1)(\sqrt{\alpha} + 1)} \cos((\sqrt{\alpha} + 2)t + \beta) \\
&\quad - \frac{At}{\sqrt{\alpha}(1 - 4\alpha)} \sin(\sqrt{\alpha} t + \beta) \\
&\quad + \frac{A}{4(\sqrt{\alpha} - 1)(2\sqrt{\alpha} - 1)} \cos((\sqrt{\alpha} - 2)t + \beta),
\end{aligned}
$$

unless $\alpha = 1/4$ or $\alpha = 1$. Higher-order terms can be calculated in a similar way, and we find that the exceptional cases are in general $\alpha = n^2/4$, $n = 1, 2, 3, \cdots$.

The approximations generated so far do not appear to yield periodic solutions, so it is reasonable to expect that these solutions occur for small values of ϵ near the exceptional values $\alpha = n^2/4$. For example, let's consider a curve $\alpha(\epsilon)$ in the parameter domain that emanates from the point $(1/4, 0)$. First, write

$$
\alpha(\epsilon) = \frac{1}{4} + \epsilon\alpha_1 + \epsilon^2\alpha_2 + \cdots,
$$

and then again substitute

$$
u(t, \epsilon) = u_0(t) + \epsilon u_1(t) + \epsilon^2 u_2(t) + \cdots
$$

into Mathieu's equation. This time we have

$$
\begin{aligned}
u_0'' + \frac{1}{4}u_0 &= 0, \\
u_1'' + \frac{1}{4}u_1 &= -\alpha_1 u_0 - 2u_0 \cos t, \\
u_2'' + \frac{1}{4}u_2 &= -\alpha_1 u_1 - \alpha_2 u_0 - 2u_1 \cos t.
\end{aligned}
$$

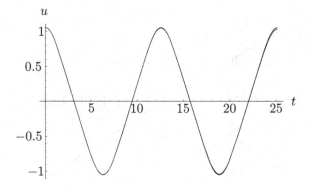

FIGURE 6. Comparison of a periodic solution and perturbation approximation for Mathieu's equation with $\epsilon = .1$ and $\alpha = .25 - \epsilon$.

Now

$$u_0(t) = A \cos\left(\frac{t}{2} + \beta\right),$$

which has period 4π, so

$$u_1'' + \frac{1}{4}u_1$$
$$= -\alpha_1 A \cos(.5t + \beta) - 2A \cos(.5t + \beta) \cos t$$
$$= -A(\alpha_1 + 1) \cos .5t \cos \beta + A(\alpha_1 - 1) \sin .5t \sin \beta - A \cos(1.5t + \beta).$$

There are exactly two ways we can avoid unbounded terms in u_1:

$$\alpha_1 = -1, \quad \beta = 0, \quad \text{or} \quad \alpha_1 = 1, \beta = \frac{\pi}{2}.$$

In the first case, we have

$$u_1'' + \frac{1}{4}u_1 = -A \cos 1.5t,$$

with periodic solutions $u_1 = .5A \cos 1.5t$, and the approximation is

$$u_0 + \epsilon u_1 = A(\cos .5t + .5\epsilon \cos 1.5t).$$

A comparison of this approximation with the actual periodic solution for $A = 1$, $\epsilon = .1$, and $\alpha = .25 - \epsilon$ is plotted in Figure 6 over two periods.

Likewise, in the second case we have the approximation

$$u_0 + \epsilon u_1 = -A(\sin .5t + .5\epsilon \sin 1.5t).$$

Consequently, there are two curves

$$\alpha(\epsilon) = .25 \pm \epsilon + \mathcal{O}(\epsilon^2) \quad (\epsilon \to 0)$$

emanating from $(.25, 0)$ (see Figure 7), along which there are solutions of period 4π. Our calculations also predict that the remaining solutions for Mathieu's equation with parameter values on these curves are unbounded for $t \geq 0$.

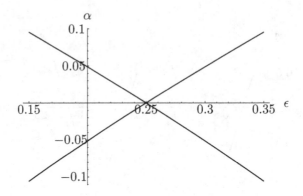

FIGURE 7. Curves in the parameter domain along which solutions of period 4π exist.

$\triangle$

The methods we have used in this section have been specialized to the problem at hand. There are more general methods, called the method of multiple scales and the method of averaging, that can be applied to large classes of these problems. See Smith [47], Verhulst [51], and Holmes [25] for descriptions and examples.

4.3 Singular Perturbations

There is an important class of problems called *singular* perturbation problems in which the regular perturbation method fails to provide a valid approximation on the whole domain. Such problems were first studied in the field of fluid dynamics in the early 1900s by Ludwig Prandtl [41]. (See O'Malley [38] for an interesting account of the development of the methods and applications of singular perturbations during the twentieth century.) We begin with a simple, but informative, example.

Example 4.10 Consider an object of mass M falling through a fluid, which resists the motion with a force proportional to the velocity of the object. At time 0 the object is at height H and at time T at height 0. We have the following boundary value problem for the height $y(t)$ of the object at time t:

$$My'' = -Mg - Ky',$$
$$y(0) = H, \quad y(T) = 0.$$

We convert the problem to dimensionless form by defining $y = Hx$ and $t = T\tau$. The resulting problem is (with a dot indicating differentiation with respect to τ)

$$\epsilon\ddot{x} = -\dot{x} - Q,$$

$$x(0) = 1, \quad x(1) = 0,$$

where $\epsilon = M/(KT)$ and $Q = MgT/(KH)$ are dimensionless parameters.

We assume ϵ is a small parameter and substitute a perturbation series

$$x = x_0(\tau) + \epsilon x_1(\tau) + \cdots$$

into the differential equation and boundary conditions:

$$
\begin{aligned}
\epsilon(\ddot{x}_0 + \epsilon \ddot{x}_1 + \cdots) &= -\dot{x}_0 - \epsilon \dot{x}_1 - \cdots - Q, \\
x_0(0) + \epsilon x_1(0) + \cdots &= 1, \\
x_0(1) + \epsilon x_1(1) + \cdots &= 0.
\end{aligned}
$$

Equating the ϵ-independent terms, we find that x_0 should satisfy the following conditions:

$$
\begin{aligned}
\dot{x}_0 &= -Q, \\
x_0(0) &= 1, \quad x_0(1) = 0.
\end{aligned}
$$

Unfortunately, it is impossible for x_0 to satisfy all these conditions, except in the unlikely event that $Q = 1$.

To determine how to compute a valid approximation, we can use a physical argument. The differential equation shows that the velocity of the object will approach a terminal velocity of $-Q$ as $\tau \to \infty$. See Figure 8 for a rough graph of the solution for the case $Q < 1$. This observation suggests that $x_0(\tau) = Q(1 - \tau)$ [this is the solution of the preceding problem for x_0 that satisfies the boundary condtion $x(1) = 0$] should serve as a good approximation of the solution as long as τ is not near 0. The region near $\tau = 0$, where the solution experiences a rapid rate of change is called a *boundary layer*, a term originating in the study of fluid velocity near a boundary. To approximate the solution in this region, we add to the initial approximation a *boundary layer correction* ϕ:

$$Q(1 - \tau) + \phi(\tau, \epsilon).$$

Substituting this trial approximation into our boundary value problem, we find that ϕ satisfies

$$\epsilon \ddot{\phi} = -\dot{\phi}.$$

In addition, we want our approximation to satisfy the boundary condition at $\tau = 0$, so

$$Q + \phi(0) = 1, \quad \text{or} \quad \phi(0) = 1 - Q.$$

Then we have

$$\phi(\tau, \epsilon) = (1 - Q)e^{-\tau/\epsilon}.$$

Finally, the desired approximation is

$$Q(1 - \tau) + (1 - Q)e^{-\tau/\epsilon}.$$

The reader will observe that this approximation does not satisfy the boundary condtion at $\tau = 1$, but it is very close since the boundary layer correction is extremely small at $\tau = 1$ if ϵ is small.

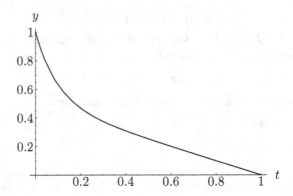

FIGURE 8.   A graph of the solution of the boundary value
problem in Example 4.10

For the sake of comparison, the exact solution of the boundary value
problem is

$$x(t, \epsilon) = \frac{(1-Q)e^{-\tau/\epsilon} + Q(1-\tau) - e^{-1/\epsilon}}{1 - e^{-1/\epsilon}}.$$

The difference between the approximation and the actual solution is
$\mathcal{O}(e^{-1/\epsilon})$ for $0 \le t \le 1$ as $\epsilon \to 0$.   △

In the following example, we study a nonlinear boundary value problem that has a number of different types of solutions, depending on the
boundary conditions.

Example 4.11 Consider the boundary value problem

$$\epsilon x'' - xx' - x^2 = 0,$$

$$x(0) = A, \quad x(1) = B,$$

where we again assume ϵ is a small, positive parameter.

Let's begin by choosing specific values of A and B, say $A = 1$ and
$B = 2$. If we substitute a trial perturbation series $x_0(t) + \epsilon x_1(t) + \cdots$ into
the differential equation, we find that

$$x_0 x_0' + x_0^2 = 0,$$

so

$$x_0(t) = Ce^{-t}$$

for some C. As in the preceding example, x_0 cannot satisfy both boundary
conditions, so we investigate the possibility of a boundary layer. If the
boundary layer occurs at $t = 0$, then we set $x_0(1) = 2$ to get $x_0 = 2e^{1-t}$.
Then the solution would have the basic shape of the graph in Figure 9.
Note that for certain values of t in the layer, x' and $-x''$ are both large
positive numbers while x is of moderate size. For such t, the left-hand side
of the differential equation

$$\epsilon x'' - xx' - x^2 = 0$$

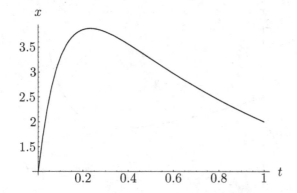

FIGURE 9. A boundary layer profile at $t = 0$.

contains at least one extremely negative term and no balancing positive term, so it cannot possibly yield 0. We conclude that there is no solution with boundary layer at $t = 0$ for these boundary values.

Now let's choose $x_0(0) = 1$ so that $x_0(t) = e^{-t}$ and consider a possible boundary layer at $t = 1$. Figure 10 indicates that in this case the positive term $\epsilon x''$ can balance the extremely negative term $-xx'$ in the layer and permit the differential equation to be satisfied. As in the previous example, we need a boundary layer correction ψ. Since it will be rapidly changing near $t = 1$, we assume it is a function of a fast variable $\sigma = (1 - t)/\epsilon^a$ for some a to be determined. In order to obtain the needed balance between the two terms in the differential equation, we require

$$\epsilon \frac{d^2\psi}{dt^2} = \frac{\epsilon}{\epsilon^{2a}} \frac{d^2\psi}{d\sigma^2}$$

to balance

$$-\psi \frac{d\psi}{dt} = \psi \frac{1}{\epsilon^a} \frac{d\psi}{d\sigma}.$$

Thus we want $\epsilon^{1-2a} = \epsilon^{-a}$, or $a = 1$.

Our trial approximation is

$$e^{-t} + \psi((1 - t)/\epsilon) = e^{\epsilon}\sigma - 1 + \psi(\sigma),$$

and, using a dot to indicate differentiation with respect to σ, the differential equation is

$$\ddot{x} + x\dot{x} - \epsilon x^2 = 0.$$

Subsituting, we have

$$\epsilon^2 e^{\epsilon\sigma-1} + \ddot{\psi} + \left(e^{\epsilon\sigma-1} + \psi\right)\left(\epsilon e^{\epsilon\sigma-1} + \dot{\psi}\right) - \epsilon\left(e^{\epsilon\sigma-1} + \psi\right)^2 = 0.$$

We can eliminate the $\mathcal{O}(\epsilon)$ terms by taking $\epsilon = 0$. Then ψ is chosen to satisfy

$$\ddot{\psi} + (e^{-1} + \psi)\dot{\psi} = 0,$$

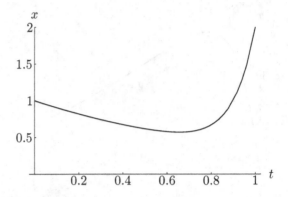

FIGURE 10.   A boundary layer profile at $t = 1$.

as well as $\psi(r)$, $\psi'(r) \to \infty$ as $r \to \infty$. In order that the approximation satisfy the boundary condition at $t = 1$, we also want

$$\psi(0) = 2 - e^{-1}.$$

In Exercise 4.23, the reader is asked to show that the solution in implicit form is

$$\sigma = \int_{\psi}^{2-e^{-1}} \frac{ds}{\int_0^s (e^{-1} + r)\, dr}.$$

The explicit solution is then easily obtained by computing the integrals and solving for ψ:

$$\psi(\sigma) = \frac{2(2 - e^{-1})e^{-\sigma/e}}{2e + 1 - (2e - 1)e^{-\sigma/e}}.$$

The resulting approximation,

$$e^{-t} + \frac{2(2 - e^{-1})e^{-\frac{1-t}{\epsilon e}}}{2e + 1 - (2e - 1)e^{-\frac{1-t}{\epsilon e}}},$$

is compared to the exact solution in Figure 11 for $\epsilon = .1$. Higher-order approximations can be computed by similar methods.

Now let's take A to be an arbitrary positive number and ask for what values of B can we find an approximation with boundary layer at $t = 1$? Following the same steps used above, we find that $x_0(t) = Ae^{-t}$ and that the boundary layer correction $\psi(\sigma)$ must solve the problem

$$\ddot{\psi} + (Ae^{-1} + \psi)\dot{\psi} = 0,$$
$$\psi(0) = B - A/e,$$

as well as the condition that $\psi(\sigma) \to 0$ as $\sigma \to \infty$. The solution in implicit form is

$$\sigma = \int_{\psi}^{B-A/e} \frac{ds}{\int_0^s (Ae^{-1} + r)\, dr} = \int_{\psi}^{B-A/e} \frac{ds}{Ae^{-1}s + s^2/2}.$$

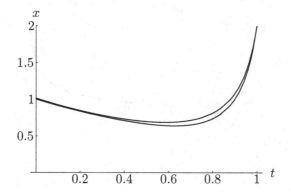

FIGURE 11. A comparison of the solution with the approximation for Example 4.11 with $\epsilon = .1$.

This integral can be computed to obtain ψ as a monotone function of σ, unless the denominator vanishes for some $0 < s \leq B - A/e$. Thus we must require $B > -A/e$. Consequently, this procedure will successfully produce an approximation of a solution with boundary layer at $t = 1$ for all $A > 0$ and $B > -A/e$.

In Exercise 4.25, we ask the reader to show that in case $B < -A/e$, it is possible to construct an approximation with boundary layer at $t = 0$ for all values of $A > 0$. This construction also works for the case $A < 0$, $B < 0$.

Looking over our discussion, we see there is one more major case to be considered: $A < 0$ and $B > 0$. In this case, it is impossible to construct approximations of the type computed previously, so we investigate whether there is an approximation with boundary layers at both $t = 0$ and $t = 1$ which follows the trivial solution $x_0 = 0$ outside of the layers.

Near $t = 1$, we again let $\sigma = (1-t)/\epsilon$, so the differential equation with independent variable σ is

$$\ddot{x} + x\dot{x} - \epsilon x^2 = 0.$$

If $\phi(\sigma)$ denotes the boundary layer correction, then the problem for ϕ is

$$\ddot{\phi} + \phi\dot{\phi} = 0,$$
$$\phi(0) = B, \quad \lim_{\sigma \to \infty} \phi(\sigma) = 0.$$

Thus ϕ is given implicity by

$$\sigma = \int_\phi^B \frac{ds}{\int_0^s r\,dr}$$

and explicity by

$$\phi = \frac{2B}{2 + B\sigma} = \frac{2B}{2 + B(1-t)/\epsilon}.$$

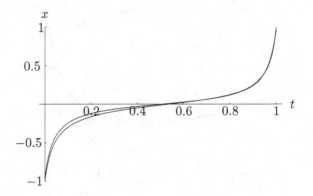

FIGURE 12.   A solution with twin boundary layers and the computed approximation for $\epsilon = .02$.

This is an algebraic boundary layer function, and it goes to zero (as $\sigma \to \infty$) much more slowly than the exponential boundary layer functions computed earlier.

The computation of the boundary layer correction at $t = 0$ is similar, and the final approximation (for the case $A < 0, B > 0$) is

$$\frac{2\epsilon A}{2\epsilon - At} + \frac{2\epsilon B}{2\epsilon + B(1 - t)}.$$

Figure 12 shows that this approximation is very accurate in case $\epsilon = .02$, $A = -1$, and $B = 1$.   △

In addition to the boundary layer phenomena described previously, singular perturbation problems can exhibit many other types of impulsive behavior. The following example illustrates the situation in which a solution undergoes a rapid change in the interior of the interval.

Example 4.12 Consider the boundary value problem

$$\epsilon x'' - xx' - x = 0,$$

$$x(0) = 2, \quad x(1) = -2.$$

The preceding differential equation is similar to the last one, but the behavior of the solutions is quite different. Straightforward analysis yields the approximations $x_0 = 0$ and $x_0 = C - t$, where C is arbitrary. Using arguments similar to those in the preceding example, we can show that it is not possible to construct an approximation for a solution with one or two boundary layers for these boundary values.

Let's look for a solution with an internal (shock) layer that connects $x_0 = 2-t$ (which satisfies the left boundary condition) to $x_0 = -1-t$ (which satisfies the right one). Figure 13 shows how the z-shaped shock layer correction connects the two lines. The point t_0 where the approximation

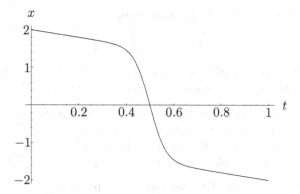

FIGURE 13. A shock layer profile for the solution of Example 4.12.

crosses the time axis is unknown and will have to be determined as part of the calculation.

Let $\tau = (t - t_0)/\epsilon$, and let $\gamma(\tau)$ denote the shock layer function. Our approximation will be of the form

$$2 - t - \gamma((t - t_0)/\epsilon) = 2 - t_0 - \epsilon\tau - \gamma(\tau).$$

Furthermore, we want

$$\gamma(\tau) \to 0, \quad (\tau \to -\infty)$$

so that the approximation will match up with $x_0 = 2 - t$ on the left side of the shock and

$$\gamma(\tau) \to 3, \quad (\tau \to \infty)$$

so that the approximation will match up with $x_0 = -1 - t$ on the right side. In terms of τ, the differential equation is (with dot indicating differentiation with respect to τ)

$$\ddot{x} - x\dot{x} - \epsilon x = 0.$$

Substituting the approximation, we get

$$-\ddot{\gamma} - (2 - t_0 - \epsilon\tau - \gamma)(-\epsilon - \dot{\gamma}) - \epsilon(2 - t_0 - \epsilon\tau - \gamma) = 0.$$

If we neglect all terms involving ϵ, then the equation for γ is

$$\ddot{\gamma} + (t_0 - 2 + \gamma)\dot{\gamma} = 0.$$

The implicit solution is

$$\tau = \int_{3/2}^{\gamma} \frac{ds}{(2 - t_0)x - s^2/2} = \frac{1}{2 - t_0} \ln \frac{\gamma(5 - 4t_0)}{3(4 - 2t_0 - \gamma)}.$$

Note that when $\tau \to -\infty$, $\gamma \to 0$, as expected. However, in order that $\gamma \to 3$ when $\tau \to \infty$, we must choose $t_0 = 1/2$. Thus the shock must occur at the middle of the interval! With this value of t_0, we have

$$\gamma = \frac{3}{1 + e^{-3\tau/2}} = \frac{3}{1 + e^{-3(t - .5)/(2\epsilon)}}.$$

Consequently, for $0 \le t \le .5$, our approximation is

$$2 - t - \frac{3}{1 + e^{-3(t-.5)/(2\epsilon)}}.$$

Actually, this approximation is correct for the whole interval $[0,1]$ since it converges to the approximation $-1 - t$ for $t > .5$ when $\epsilon \to 0$. $\quad \triangle$

These examples may give the reader the mistaken impression that the successful construction of an asymptotic approximation implies that there is an actual solution near the approximation. Unfortunately, this conclusion is not true in general. The following example is due to Eckhaus [13].

Example 4.13 Consider the boundary value problem

$$\epsilon x'' - x' = 2\epsilon - t,$$

$$x'(0) = 0, \quad x(1) = 0.$$

Substitution of a perturbation series $x_0 + \epsilon x_1 + \cdots$ yields $x_0' = t$, so $x_0 = t^2/2 + C$. If we choose $C = -1/2$, then

$$x_0(t) = \frac{t^2 - 1}{2}$$

satisfies both boundary conditions, so appears to be an asymptotic approximation of some solution. On the other hand, the actual solution of the boundary value problem is

$$x(t, \epsilon) = \frac{t^2 - 1}{2} + \epsilon[(1 - t) + \epsilon(e^{t/\epsilon} - e^{1/\epsilon})].$$

At $t = 0$,

$$x_0(0) - x(0, \epsilon) = -\epsilon + \epsilon(e^{1/\epsilon} - 1),$$

which goes to ∞ as $\epsilon \to 0$! Not only is x_0 a bad approximation, but it gets worse for smaller values of ϵ. $\quad \triangle$

Other interesting examples of spurious approximations are given by Eckhaus [13], Carrier and Pearson [6], and Bender and Orszag [5].

In many cases, asymptotic approximations of singular perturbation problems can be rigorously verified using the contraction mapping theorem (see Smith [47] or upper and lower solutions (see Howes [26], Chang and Howes [7] and Kelley [30]). When we study upper and lower solutions for boundary value problems in Chapter 7, we will give an example that illustrates how this method can be used to verify formal approximations.

4.4 Exercises

4.1 Find the first two terms in the perturbation series for

$$x' + 2x + \epsilon x^2 = 0, \quad x(0) = 3.$$

4.2 Solve the IVP in Example 4.1.

4.3 For the system

$$x' + x + \epsilon x^3 = 0, \quad (\epsilon > 0)$$
$$x(0) = 1,$$

(i) Substitute a perturbation series $x_0(t) + \epsilon x_1(t) + \cdots$ into the system and compute x_0 and x_1.

(ii) Expand the exact solution $x(t, \epsilon) = e^{-t}(1 + \epsilon(1 - e^{-2t}))^{-1/2}$ in a series and compare to your answer in part (i).

(iii) For $\epsilon = .1$, compare the exact solution to the computed approximation by graphing, and determine where the approximation breaks down.

4.4 Carry out the directions in the preceding exercise for the logistic equation with large carrying capacity:

$$x' = x(1 - \epsilon x), \quad (\epsilon > 0)$$
$$x(0) = 1.$$

In this case, the exact solution is $x(t, \epsilon) = e^t(1 - \epsilon + \epsilon e^t)^{-1}$.

4.5 Verify that the following equations are correct:

(i) $\frac{1}{t} - \frac{1}{t+\epsilon} = \mathcal{O}\left(\frac{\epsilon}{t^2}\right) \quad (\epsilon \to 0) \quad (t > 0)$

(ii) $e^{t+\epsilon} - e^t = \mathcal{O}(\epsilon e^t) \quad (\epsilon \to 0) \quad (t \in \mathbb{R})$

(iii) $\frac{\epsilon \sin(\epsilon t)}{1 - \epsilon t} = \mathcal{O}(\epsilon^2) \quad (\epsilon \to 0) \quad (1 \leq t \leq 100)$

(iv) $\frac{1 - \sqrt{1 - 4\epsilon}}{2\epsilon} = 1 + \epsilon + \mathcal{O}(\epsilon^2) \quad (\epsilon \to 0)$

4.6 Show that the logistic equation with initial condition:

$$y' = ry\left(1 - \frac{y}{K}\right),$$
$$y(0) = a$$

can be reduced to the dimensionless form in Exercise 4.4, where $\epsilon = a/K$.

4.7 Verify that the undamped oscillator problem

$$my'' + ky = 0,$$
$$y(0) = a, \quad y'(0) = b$$

can be transformed into the dimensionless form

$$\ddot{x} + x = 0,$$
$$x(0) = 1, \quad \dot{x}(0) = c,$$

where $c = (b/a)\sqrt{m/k}$.

4.8 Show how to change the damped oscillator

$$my'' + by' + ky = 0,$$
$$y(0) = 1, \quad y'(0) = 0$$

into the dimensionless form

$$\epsilon \ddot{x} + \dot{x} + x = 0,$$
$$x(0) = 1, \quad \dot{y}(0) = 0.$$

In terms of the original parameters, under what circumstances would ϵ be a small parameter?

4.9 In Example 4.4, $T(\epsilon)$ represents the time when the rocket returns to the earth's surface. Write $T(\epsilon) = T_0 + T_1 \epsilon + \cdots$, and compute T_0 and T_1 using the approximation $x_0(\tau) + \epsilon x_1(\tau)$ for $x(\tau, \epsilon)$.

4.10 Use Theorem 1.5 to verify that

$$x(\tau, \epsilon) - (x_0(\tau) + \epsilon x_1(\tau) + \epsilon^2 x_2(\tau)) = \mathcal{O}(\epsilon^3), \quad (\epsilon \to 0) \ \ (1 \le t \le T(\epsilon)),$$

where $x(\tau, \epsilon)$ is the solution in Example 4.4.

4.11 Work each of the following:

(i) Change the undamped pendulum problem

$$L\theta'' = -g \sin \theta,$$
$$\theta(0) = a, \quad \theta'(0) = 0,$$

to the dimensionless form

$$\epsilon \ddot{x} = -\sin(\epsilon x),$$
$$x(0) = 1, \quad \dot{x}(0) = 0,$$

where $\epsilon = a$ is assumed to be small.

(ii) Subsitute $x(t, \tau) = x_0(\tau) + \epsilon x_1(\tau) + \epsilon^2 x_2(\tau) + \cdots$ into the perturbation problem in part a, and show that $x_0(\tau) = \cos \tau$, $x_1(\tau) = 0$, and $x_2(\tau) = (\cos \tau - \cos 3\tau)/192 + (\tau \sin \tau)/16$.

(iii) Note that the approximation in part (ii) contains the unbounded term $(\tau \sin \tau)/16$, as $\tau \to \infty$. Consequently, the approximation will not be good for large values of τ. Let $\epsilon = .1$, and determine by computer how large τ must be to make the error of approximation more than .3.

4.12 In Example 4.6, verify that $(0,0)$ is asymptotically stable and that $(1,0)$ is a saddle point.

4.13 In Example 4.6, show that no traveling wave solution of the desired type exists if $\epsilon > .25$. [Hint: Consider the nature of the equilibrium $(0,0)$.]

4.14 Find infinitely many traveling wave solutions of

$$u_{xx}(x,t) + u_{tt}(x,t) + 2u(x,t) = 0$$

with speed 1.

4.15 Work each of the following:

(i) For the special case $\epsilon = 6/25$, show that the perturbation problem in Example 4.6 has the exact solution

$$u(w) = \left(1 + (\sqrt{2} - 1)e^{5w/6}\right)^{-2}.$$

(ii) Show that $u(w)$ in part a satisfies the estimates obtained in Example 4.6, namely

$$\frac{1}{1 + e^{5w/3}} \le u(w) \le \frac{1}{1 + e^{5w/6}} \quad (w \ge 0).$$

4.16 Carry out an analysis of the traveling wave solutions of Fisher's equation with convection

$$u_t + uu_x = u_{xx} + u(1 - u),$$

similar to that given in Example 4.6.

4.17 One of the classic nonlinear differential equations is van der Pol's equation

$$u'' + u = \epsilon u'(1 - u^2) \quad (\epsilon > 0).$$

It first appeared in the study of nonlinear electric circuits.

(i) Find an approximation of the limit cycle for van der Pol's equation by substituting a perturbation series $u = u_0(t) + \epsilon u_1(t) + \cdots$, and computing u_0 and u_1. Be sure that your approximation contains no unbounded terms.

(ii) Compare (by graphing) the approximation that you computed in part a to a numerical approximation of the limit cycle for $\epsilon = .5$.

4.18 The Duffing equation

$$u'' + u + \epsilon u^3 = 0 \quad (\epsilon > 0),$$

is fundamental in the study of nonlinear oscillations.

(i) Show that subsitution of the perturbation series $u = u_0(t) + \epsilon u_1(t) + \cdots$ results in the following approximation containing an unbounded term:

$$\alpha \cos(t + \beta) + \frac{\epsilon}{32}\alpha^3 \cos(3t + 3\beta) - \frac{3\epsilon}{8}\alpha^3 t \sin(t + \beta).$$

(ii) Use the method of renormalization to eliminate the unbounded term and to obtain a better approximation.

4.19 In Exercise 4.11, the undamped pendulum problem is reduced to the dimensionless form

$$\begin{aligned}
\epsilon \ddot{x} &= -\sin(\epsilon x), \\
x(0) &= 1, \\
\dot{x}(0) &= 0,
\end{aligned}$$

and a perturbation series $x_0 + \epsilon^2 x_2 + \cdots$ is computed in which x_2 is unbounded. Use the method of renormalization to improve the approximation.

4.20 Find a two-term perturbation series for

$$u'' + u = \epsilon u u'^2 \quad (\epsilon > 0)$$

that contains no unbounded terms.

4.21 For the Mathieu equation, let

$$\alpha(\epsilon) = 1 + \epsilon \alpha_1 + \epsilon^2 \alpha_2 + \cdots,$$

and compute α_1, α_2, so that there are solutions of period 2π along $\alpha(\epsilon)$. Also, find approximations of the periodic solutions.

4.22 Compute an approximation of the solution of the boundary value problem

$$\epsilon x'' = 2x' - 4, \quad x(0) = x(1) = 0$$

of the form $x_0(t) + \phi((1-t)/\epsilon)$, where ϕ is a boundary layer correction. Then compare your answer to the exact solution.

4.23 The purpose of this exercise is to solve

$$\frac{d^2\psi}{d\sigma^2} + (e^{-1} + \psi)\frac{d\psi}{d\sigma} = 0, \quad \psi(0) = 2 - e^{-1}.$$

We will also assume $\psi, \psi' \to 0$ as $\sigma \to \infty$.

  (i) Let $z = \frac{d\psi}{d\sigma}$, and show that the differential equation can be written in the form $\frac{dz}{d\psi} + e^{-1} + \psi = 0$.

  (ii) Using separation of variables and the conditions on ψ at infinity, show that $z = -\int_0^\psi (e^{-1} + r)\, dr$.

  (iii) Using separation of variables again and the initial condition, obtain

$$\sigma = \int_\psi^{2-e^{-1}} \frac{ds}{\int_0^s (e^{-1} + r)\, dr}.$$

  (iv) Solve the equation in part (iii) for ψ to obtain the explicit solution.

4.24 Compute an approximation (x_0 + boundary layer correction) for Example 4.11 in case $A = 4$, $B = -1$.

4.25 Show that in Example 4.11 it is possible to construct an approximation with boundary layer at $t = 0$ in case $A > 0$, $B < -A/e$.

4.26 For the case $A < 0$, $B > 0$ in Example 4.11, verify that the boundary layer correction at $t = 0$ is $2A\epsilon/(2\epsilon - At)$.

4.27 Construct an approximation of the form

$$x_0(t) + \phi(t/\sqrt{\epsilon}) + \psi((1-t)/\sqrt{\epsilon})$$

for the boundary value problem

$$\epsilon x'' - \epsilon x x' - x = -t, \quad x(0) = 2, \quad x(1) = 0.$$

4.28 Find an approximation of the type x_0 + boundary layer correction for the problem

$$\epsilon x'' - xx' - x = 0, \quad x(0) = 3, \quad x(1) = -1.$$

4.29 Find an approximation of the type x_0 + shock layer correction for the problem

$$\epsilon x'' - xx' - x = 0, \quad x(0) = 2, \quad x(1) = -3/2.$$

4.30 For what pairs of numbers A, B is it possible to construct an approximation to

$$\epsilon x'' - xx' - x = 0, \quad x(0) = A, \quad x(1) = B$$

that features a shock layer correction?

4.31 If in Example 4.10 we assume the retarding force is proportional to the square of the velocity, then the problem in dimensionless form is

$$
\begin{aligned}
\epsilon \ddot{x} &= \dot{x}^2 - Q^2, \\
x(0) &= 1, \quad x(1) = 0,
\end{aligned}
$$

If $0 < Q < 1$, compute an approximation of the form $x_0(\tau) + \phi(\tau, \epsilon)$. [Hint: ϕ should satisfy $\epsilon \ddot{\phi} = \dot{\phi}^2$, $\phi(0) = 1 - Q$, $\phi(1) = 0$.]

Chapter 5

The Self-Adjoint Second-Order Differential Equation

5.1 Basic Definitions

In this chapter we are concerned with the second-order (formally) self-adjoint linear differential equation

$$(p(t)x')' + q(t)x = h(t).$$

We assume throughout that p, q, and h are continuous on some given interval I and $p(t) > 0$ on I. Let

$$\mathbb{D} := \{x \colon x \text{ and } px' \text{ are continuously differentiable on } I\}.$$

We then define the linear operator L on $\mathbb{D}$ by

$$Lx(t) = (p(t)x'(t))' + q(t)x(t),$$

for $t \in I$. Then the self-adjoint equation can be written briefly as $Lx = h(t)$. If $h(t) \equiv 0$, we get the *homogeneous* self-adjoint differential equation $Lx = 0$; otherwise we say the differential equation $Lx = h(t)$ is *nonhomogeneous*.

Definition 5.1 If $x \in \mathbb{D}$ and $Lx(t) = h(t)$ for $t \in I$, we say that x is a *solution* of $Lx = h(t)$ on the interval I.

Example 5.2 In this example we show that any second-order linear differential equation of the form

$$p_2(t)x'' + p_1(t)x' + p_0(t)x = g(t), \tag{5.1}$$

where we assume that $p_2(t) \neq 0$ on I and p_i, $i = 0, 1, 2$, and g are continuous on an interval I can be written in self-adjoint form.

Assume x is a solution of (5.1); then

$$x''(t) + \frac{p_1(t)}{p_2(t)}x'(t) + \frac{p_0(t)}{p_2(t)}x(t) = \frac{g(t)}{p_2(t)},$$

for $t \in I$. Multiplying by the *integrating factor* $e^{\int \frac{p_1(t)}{p_2(t)}\,dt}$ for the first two terms, we obtain

$$e^{\int \frac{p_1(t)}{p_2(t)}\,dt}x''(t) + \frac{p_1(t)}{p_2(t)}e^{\int \frac{p_1(t)}{p_2(t)}\,dt}x'(t) + \frac{p_0(t)}{p_2(t)}e^{\int \frac{p_1(t)}{p_2(t)}\,dt}x(t) = \frac{g(t)}{p_2(t)}e^{\int \frac{p_1(t)}{p_2(t)}\,dt}$$

W.G. Kelley and A.C. Peterson, *The Theory of Differential Equations:*
Classical and Qualitative, Universitext 278, DOI 10.1007/978-1-4419-5783-2_5,
© Springer Science+Business Media, LLC 2010

$$\{e^{\int \frac{p_1(t)}{p_2(t)}dt}x'(t)\}' + \frac{p_0(t)}{p_2(t)}e^{\int \frac{p_1(t)}{p_2(t)}dt}x(t) = \frac{g(t)}{p_2(t)}e^{\int \frac{p_1(t)}{p_2(t)}dt}.$$

Hence we obtain that x is a solution of the self-adjoint equation

$$(p(t)x')' + q(t)x = h(t),$$

where

$$
\begin{aligned}
p(t) &= e^{\int \frac{p_1(t)}{p_2(t)}dt} > 0, \\
q(t) &= \frac{p_0(t)}{p_2(t)}e^{\int \frac{p_1(t)}{p_2(t)}dt}, \\
h(t) &= \frac{g(t)}{p_2(t)}e^{\int \frac{p_1(t)}{p_2(t)}dt},
\end{aligned}
$$

for $t \in I$. Note that p, q, and h are continuous on I. Actually p is continuously differentiable on I. In many studies of the self-adjoint differential equation it is assumed that p is continuously differentiable on I, but we will only assume p is continuous on I. In this case the self-adjoint equation $Lx = h(t)$ is more general than the linear differential equation (5.1). $\triangle$

Example 5.3 Write the differential equation

$$t^2x'' + 3tx' + 6x = t^4,$$

for $t \in I := (0,\infty)$ in self-adjoint form.

Dividing by t^2, we obtain

$$x'' + \frac{3}{t}x' + \frac{6}{t^2}x = t^2.$$

Hence

$$e^{\int \frac{3}{t}dt} = e^{3\log t} = t^3$$

is an integrating factor for the first two terms. Multiplying by the integrating factor t^3 and simplifying, we obtain the self-adjoint differential equation

$$\left(t^3x'\right)' + 6tx = t^5.$$

$\triangle$

We now state and prove the following existence-uniqueness theorem for the self-adjoint nonhomogeneous differential equation $Lx = h(t)$.

Theorem 5.4 (Existence-Uniqueness Theorem) *Assume that p, q, and h are continuous on I and $p(t) > 0$ on I. If $a \in I$, then the initial value problem (IVP)*

$$Lx = h(t),$$

$$x(a) = x_0, \quad x'(a) = x_1,$$

where x_0 and x_1 are given constants, has a unique solution and this solution exists on the whole interval I.

Proof We first write $Lx = h(t)$ as an equivalent vector equation. Let x be a solution of $Lx = h(t)$ and let

$$y(t) := p(t)x'(t).$$

Then

$$x'(t) = \frac{1}{p(t)}y(t).$$

Also, since x is a solution of $Lx = h(t)$,

$$\begin{aligned} y'(t) &= (p(t)x'(t))' \\ &= -q(t)x(t) + h(t). \end{aligned}$$

Hence if we let

$$z(t) := \left[\begin{array}{c} x(t) \\ y(t) \end{array} \right],$$

then z is a solution of the vector equation

$$z' = A(t)z + b(t),$$

where

$$A(t) := \left[\begin{array}{cc} 0 & \frac{1}{p(t)} \\ -q(t) & 0 \end{array} \right], \quad b(t) := \left[\begin{array}{c} 0 \\ h(t) \end{array} \right].$$

Note that the matrix function A and the vector function b are continuous on I. Conversely, it is easy to see that if

$$z(t) := \left[\begin{array}{c} x(t) \\ y(t) \end{array} \right]$$

defines a solution of the vector equation

$$z' = A(t)z + b(t),$$

then x is a solution of the scalar equation $Lx = h(t)$ and $y(t) = p(t)x'(t)$. By Theorem 2.3 there is a unique solution z of the IVP

$$z' = A(t)z + b(t),$$

$$z(a) = \left[\begin{array}{c} x_0 \\ y_1 \end{array} \right],$$

and this solution is a solution on the whole interval I. But this implies that there is a unique solution of $Lx = h(t)$ satisfying

$$x(a) = x_0, \quad p(a)x'(a) = y_1,$$

and this solution exists on the whole interval I. It follows that the initial value problem

$$Lx = (p(t)x')' + q(t)x = h(t),$$

$$x(a) = x_0, \quad x'(a) = x_1$$

has a unique solution on I. Some authors (for good reasons) prefer to call the conditions $x(a) = x_0$, $p(a)x'(a) = y_1$ initial conditions, but we will call the conditions $x(a) = x_0$, $x'(a) = x_1$ initial conditions. $\square$

Definition 5.5 Assume x, y are differentiable functions on an interval I; then we define the *Wronskian* of x and y by

$$w[x(t), y(t)] = \begin{vmatrix} x(t) & y(t) \\ x'(t) & y'(t) \end{vmatrix} := x(t)y'(t) - x'(t)y(t),$$

for $t \in I$.

Theorem 5.6 (Lagrange Identity) *Assume $x, y \in \mathbb{D}$; then*

$$y(t)Lx(t) - x(t)Ly(t) = \{y(t); x(t)\}',$$

for $t \in I$, where $\{y(t); x(t)\}$ is the Lagrange bracket of y and x, which is defined by

$$\{y(t); x(t)\} := p(t)w[y(t), x(t)], \quad t \in I,$$

where $w[y(t), x(t)]$ is the Wronskian of y and x.

Proof Let $x, y \in \mathbb{D}$ and consider

$$\{y(t); x(t)\}' = \{y(t)p(t)x'(t) - x(t)p(t)y'(t)\}',$$

$$= y(t)\left(p(t)x'(t)\right)' + y'(t)p(t)x'(t) - x(t)\left(p(t)y'(t)\right)' - x'(t)p(t)y'(t),$$

$$= y(t)\{(p(t)x'(t))' + q(t)x(t)\} - x(t)\{(p(t)y'(t))' + q(t)y(t)\},$$

$$= y(t)Lx(t) - x(t)Ly(t),$$

for $t \in I$, which is what we wanted to prove. □

Corollary 5.7 (Abel's Formula) *If x, y are solutions of $Lx = 0$ on I, then*

$$w[x(t), y(t)] = \frac{C}{p(t)},$$

for all $t \in I$, where C is a constant.

Proof Assume x, y are solutions of $Lx = 0$ on I. By the Lagrange identity,

$$x(t)Ly(t) - y(t)Lx(t) = \{x(t); y(t)\}',$$

for all $t \in I$. Since x and y are solutions on I,

$$\{x(t); y(t)\}' = 0,$$

for all $t \in I$. Hence

$$\{x(t); y(t)\} = C,$$

where C is a constant. It follows that

$$w[x(t), y(t)] = \frac{C}{p(t)},$$

for all $t \in I$, which is Abel's formula. □

Definition 5.8 If x and y are continuous on $[a, b]$, we define the *inner product* of x and y by

$$< x, y >= \int_a^b x(t)y(t)\, dt.$$

Corollary 5.9 (Green's Formula) *If $[a, b] \subset I$ and x, $y \in \mathbb{D}$, then*

$$< y, Lx > - < Ly, x >= \{y(t); x(t)\}_a^b,$$

where $\{F(t)\}_a^b := F(b) - F(a)$.

Proof Let $x, y \in \mathbb{D}$; then by the Lagrange identity,

$$y(t)Lx(t) - x(t)Ly(t) = \{y(t); x(t)\}',$$

for all $t \in I$. Integrating from a to b, we get the desired result

$$< y, Lx > - < Ly, x >= \{y(t); x(t)\}_a^b.$$

$\square$

Corollary 5.10 *If u, v are solutions of $Lx = 0$, then either*
(a) $w[u(t), v(t)] \neq 0$ for all $t \in I$
or
(b) $w[u(t), v(t)] = 0$ for all $t \in I$.
Case (a) occurs iff u, v are linearly independent on I and case (b) occurs iff u, v are linearly dependent on I.

Proof Assume u, v are solutions of $Lx = 0$. Then by Abel's formula (Corollary 5.7),

$$w[u(t), v(t)] = \frac{C}{p(t)},$$

for all $t \in I$, where C is a constant. If $C \neq 0$, then part (a) of this theorem holds, while if $C = 0$, part (b) of this theorem holds. The remainder of the proof of this corollary is left to the reader (see Exercise 5.5). $\square$

We now show that the differential equation $Lx = 0$ has two linearly independent solutions on I. To see this let $a \in I$ and let u be the solution of the IVP

$$Lu = 0, \quad u(a) = 1, \quad u'(a) = 0,$$

and let v be the solution of the IVP

$$Lv = 0, \quad v(a) = 0, \quad v'(a) = 1.$$

Since the Wronskian of these two solutions at a is different from zero, these two solutions are linearly independent on I by Corollary 5.10.

Theorem 5.11 *If x_1, x_2 are linearly independent solutions of $Lx = 0$ on I, then*

$$x = c_1 x_1 + c_2 x_2 \tag{5.2}$$

is a general solution of $Lx = 0$. By this we mean every function in this form is a solution of $Lx = 0$ and all solutions of $Lx = 0$ are in this form.

Proof Assume x_1, x_2 are linearly independent solutions on I of $Lx = 0$. Let x be of the form (5.2). Then

$$\begin{aligned} Lx &= L[c_1 x_1 + c_2 x_2] \\ &= c_1 Lx_1 + c_2 Lx_2 \\ &= 0, \end{aligned}$$

so x is a solution of $Lx = 0$.

Conversely, assume that x is a solution of $Lx = 0$ on I. Let $t_0 \in I$ and let

$$x_0 := x(t_0), \quad x_1 := x'(t_0).$$

Let

$$y(t) = c_1 x_1(t) + c_2 x_2(t).$$

We now show that we can pick constants c_1, c_2 such that

$$y(t_0) = x_0, \quad y'(t_0) = x_1.$$

These last two equations are equivalent to the equations

$$\begin{aligned} c_1 x_1(t_0) + c_2 x_2(t_0) &= x_0, \\ c_1 x_1'(t_0) + c_2 x_2'(t_0) &= x_1. \end{aligned}$$

The determinant of the coefficients for this system is

$$w[x_1, x_2](t_0) \neq 0$$

by Corollary 5.10. Let c_1, c_2 be the unique solution of this system. Then

$$y = c_1 x_1 + c_2 x_2$$

is a solution of $Lx = 0$ satisfying the same initial conditions as x at t_0. By the uniqueness of solutions of IVPs (Theorem 5.4), x and y are the same solution. Hence we get the desired result

$$x = c_1 x_1 + c_2 x_2.$$

$\square$

5.2 An Interesting Example

In this section we indicate how Theorem 5.4 can be used to define functions and how we can use this to derive properties of these functions.

Definition 5.12 We define s and c to be the solutions of the IVPs

$$s'' + s = 0, \quad s(0) = 0, \quad s'(0) = 1,$$
$$c'' + c = 0, \quad c(0) = 1, \quad c'(0) = 0,$$

respectively.

Theorem 5.13 (Properties of s and c)

    (i) $s'(t) = c(t), \quad c'(t) = -s(t)$
    (ii) $s^2(t) + c^2(t) = 1$

(iii) $s(-t) = -s(t), \quad c(-t) = c(t)$
(iv) $s(t+\alpha) = s(t)c(\alpha) + s(\alpha)c(t), \quad c(t+\alpha) = c(t)c(\alpha) - s(t)s(\alpha)$
(v) $s(t-\alpha) = s(t)c(\alpha) - s(\alpha)c(t), \quad c(t-\alpha) = c(t)c(\alpha) + s(t)s(\alpha)$

for $t, \alpha \in \mathbb{R}$.

Proof Since c and s' solve the same IVP

$$x'' + x = 0, \quad x(0) = 1, \quad x'(0) = 0,$$

we get by the uniqueness theorem that

$$s'(t) = c(t),$$

for $t \in \mathbb{R}$. Similarly, $-s$ and c' solve the same IVP

$$x'' + x = 0, \quad x(0) = 0, \quad x'(0) = -1,$$

and so by the uniqueness theorem

$$c'(t) = -s(t),$$

for $t \in \mathbb{R}$. By Abel's theorem the Wronskian of c and s is a constant, so

$$\begin{vmatrix} c(t) & s(t) \\ c'(t) & s'(t) \end{vmatrix} = \begin{vmatrix} c(0) & s(0) \\ c'(0) & s'(0) \end{vmatrix} = 1.$$

It follows that

$$\begin{vmatrix} c(t) & s(t) \\ -s(t) & c(t) \end{vmatrix} = 1,$$

and hence we obtain

$$s^2(t) + c^2(t) = 1.$$

Next note that since $s(-t)$ and $-s(t)$ both solve the IVP

$$x'' + x = 0, \quad x(0) = 0, \quad x'(0) = -1,$$

we have by the uniqueness theorem that

$$s(-t) = -s(t),$$

for $t \in \mathbb{R}$. By a similar argument we get that c is an even function. Since $s(t+\alpha)$ and $s(t)c(\alpha) + s(\alpha)c(t)$ both solve the IVP

$$x'' + x = 0, \quad x(0) = s(\alpha), \quad x'(0) = c(\alpha),$$

we have by the uniqueness theorem that

$$s(t+\alpha) = s(t)c(\alpha) + s(\alpha)c(t),$$

for $t \in \mathbb{R}$. By a similar argument

$$c(t+\alpha) = c(t)c(\alpha) - s(t)s(\alpha),$$

for $t \in \mathbb{R}$. Finally, using parts (iii) and (iv), we can easily prove part (v). $\qquad\square$

5.3 Cauchy Function and Variation of Constants Formula

In this section we will derive a variation of constants formula for the nonhomogeneous second-order self-adjoint differential equation

$$Lx = (p(t)x')' + q(t)x = h(t), \qquad (5.3)$$

where we assume h is a continuous function on I.

Theorem 5.14 *If u, v are linearly independent solutions on I of the homogeneous differential equation $Lx = 0$ and z is a solution on I of the nonhomogeneous differential equation $Lx = h(t)$, then*

$$x = c_1 u + c_2 v + z,$$

where c_1 and c_2 are constants, is a general solution of the nonhomogeneous differential equation $Lx = h(t)$.

Proof Let u, v be linearly independent solutions of the homogeneous differential equation $Lx = 0$, let z be a solution of the nonhomogeneous differential equation $Lx = h(t)$, and let

$$x := c_1 u + c_2 v + z,$$

where c_1 and c_2 are constants. Then

$$
\begin{aligned}
Lx &= c_1 Lu + c_2 Lv + Lz \\
&= h.
\end{aligned}
$$

Hence for any constants c_1 and c_2, $x := c_1 u + c_2 v + z$ is a solution of the nonhomogeneous differential equation $Lx = h(t)$.

Conversely, assume x_0 is a solution of the nonhomogeneous differential equation $Lx = h(t)$ and let

$$x := x_0 - z.$$

Then

$$Lx = Lx_0 - Lz = h - h = 0.$$

Hence x is a solution of the homogeneous differential equation $Lx = 0$ and so by Theorem 5.11 there are constants c_1 and c_2 such that

$$x = c_1 u + c_2 v$$

and this implies that

$$x_0 = c_1 u + c_2 v + z.$$

$\square$

Definition 5.15 Define the *Cauchy function* $x(\cdot, \cdot)$ for $Lx = 0$ to be the function $x : I \times I \to \mathbb{R}$ such that for each fixed $s \in I$, $x(\cdot, s)$ is the solution of the initial value problem

$$Lx = 0, \quad x(s) = 0, \quad x'(s) = \frac{1}{p(s)}.$$

Example 5.16 Find the Cauchy function for $(p(t)x')' = 0$.

For each fixed s,

$$(p(t)x'(t,s))' = 0, \quad t \in I.$$

Hence

$$p(t)x'(t,s) = \alpha(s).$$

The condition $x'(s,s) = \frac{1}{p(s)}$ implies that $\alpha(s) = 1$. It follows that

$$x'(t,s) = \frac{1}{p(t)}.$$

Integrating from s to t and using the condition $x(s,s) = 0$, we get that

$$x(t,s) = \int_s^t \frac{1}{p(\tau)} d\tau.$$

$\triangle$

Example 5.17 Find the Cauchy function for

$$\left(e^{-5t}x'\right)' + 6e^{-5t}x = 0.$$

Expanding this equation out and simplifying, we get the equivalent equation

$$x'' - 5x' + 6x = 0.$$

It follows that the Cauchy function is of the form

$$x(t,s) = \alpha(s)e^{2t} + \beta(s)e^{3t}.$$

The initial conditions

$$x(s,s) = 0, \quad x'(s,s) = \frac{1}{p(s)} = e^{5s}$$

lead to the equations

$$\alpha(s)e^{2s} + \beta(s)e^{3s} = 0,$$
$$2\alpha(s)e^{2s} + 3\beta(s)e^{3s} = e^{5s}.$$

Solving these simultaneous equations, we get

$$\alpha(s) = -e^{3s}, \quad \beta(s) = e^{2s},$$

and so

$$x(t,s) = e^{3t}e^{2s} - e^{2t}e^{3s}.$$

$\triangle$

Theorem 5.18 *If u and v are linearly independent solutions of $Lx = 0$, then the Cauchy function for $Lx = 0$ is given by*

$$x(t,s) = \frac{\begin{vmatrix} u(s) & v(s) \\ u(t) & v(t) \end{vmatrix}}{p(s) \begin{vmatrix} u(s) & v(s) \\ u'(s) & v'(s) \end{vmatrix}},$$

for $t, s \in I$.

Proof Since u and v are linearly independent solutions of $Lx = 0$, the Wronskian of these two solutions is never zero, by Corollary 5.10, so we can define

$$y(t, s) := \frac{\begin{vmatrix} u(s) & v(s) \\ u(t) & v(t) \end{vmatrix}}{p(s) \begin{vmatrix} u(s) & v(s) \\ u'(s) & v'(s) \end{vmatrix}},$$

for $t, s \in I$. Since for each fixed s in I, $y(\cdot, s)$ is a linear combination of u and v, $y(\cdot, s)$ is a solution of $Lx = 0$. Also, $y(s, s) = 0$ and $y'(s, s) = \frac{1}{p(s)}$ so we have by the uniqueness of solutions of initial value problems that $y(t, s) = x(t, s)$ for $t \in I$ for each fixed $s \in I$, which gives the desired result. $\qquad \square$

Example 5.19 Use Theorem 5.18 to find the Cauchy function for

$$(p(t)x')' = 0.$$

Let $a \in I$; then $u(t) := 1$, $v(t) := \int_a^t \frac{1}{p(\tau)} d\tau$ define linearly independent solutions of $(p(t)x')' = 0$. Hence by Theorem 5.18, the Cauchy function is given by

$$
\begin{aligned}
x(t, s) &= \frac{\begin{vmatrix} 1 & \int_a^s \frac{1}{p(\tau)} d\tau \\ 1 & \int_a^t \frac{1}{p(\tau)} d\tau \end{vmatrix}}{p(s) \begin{vmatrix} 1 & \int_a^s \frac{1}{p(\tau)} d\tau \\ 0 & \frac{1}{p(s)} \end{vmatrix}} \\
&= \int_s^t \frac{1}{p(\tau)} d\tau.
\end{aligned}
$$

$\triangle$

Example 5.20 Use Theorem 5.18 to find the Cauchy function for

$$\left(e^{-5t}x'\right)' + 6e^{-5t}x = 0.$$

From Example 5.17,

$$u(t) = e^{2t}, \quad v(t) = e^{3t}, \quad t \in \mathbb{R}$$

define (linearly independent) solutions of $\left(e^{-5t}x'\right)' + 6e^{-5t}x = 0$ on $\mathbb{R}$. Hence by Theorem 5.18, the Cauchy function is given by

$$
\begin{aligned}
x(t, s) &= \frac{\begin{vmatrix} e^{2s} & e^{3s} \\ e^{2t} & e^{3t} \end{vmatrix}}{e^{-5s} \begin{vmatrix} e^{2s} & 3s \\ 2e^{2s} & 3e^{3s} \end{vmatrix}} \\
&= e^{3t}e^{2s} - e^{2t}e^{3s}.
\end{aligned}
$$

$\triangle$

Theorem 5.21 *Assume $f(\cdot,\cdot)$ and the first-order partial derivative $f_t(\cdot,\cdot)$ are continuous real-valued functions on $I \times I$ and $a \in I$. Then*

$$\frac{d}{dt} \int_a^t f(t,s)\, ds = \int_a^t f_t(t,s)\, ds + f(t,t),$$

for $t \in I$.

Proof Letting $x = x(t)$ and $y = y(t)$ in the appropriate places and using the chain rule of differentiation,

$$
\begin{aligned}
\frac{d}{dt} \int_a^t f(t,s)\, ds &= \frac{d}{dt} \int_a^x f(y,s)\, ds \\
&= \frac{\partial}{\partial x} \left(\int_a^x f(y,s)\, ds \right) \frac{dx}{dt} + \frac{\partial}{\partial y} \left(\int_a^x f(y,s)\, ds \right) \frac{dy}{dt},
\end{aligned}
$$

then with $x(t) = t$ and $y(t) = t$ we get

$$f(y,x)\frac{dx}{dt} + \int_a^x f_y(y,s)\, ds \frac{dy}{dt} = \int_a^t f_t(t,s)\, ds + f(t,t),$$

for $t \in I$. $\qquad\square$

In the next theorem we derive the important variation of constants formula.

Theorem 5.22 (Variation of Constants Formula) *Assume h is continuous on I and assume $a \in I$. Then the solution of the initial value problem*

$$Lx = h(t), \quad x(a) = 0, \quad x'(a) = 0$$

is given by

$$x(t) = \int_a^t x(t,s)h(s)\, ds, \quad t \in I,$$

where $x(\cdot,\cdot)$ is the Cauchy function for $Lx = 0$.

Proof Let

$$x(t) := \int_a^t x(t,s)h(s)\, ds,$$

for $t \in I$ and note that $x(a) = 0$. We will let

$$x'(t,s) := x_t(t,s).$$

Then

$$
\begin{aligned}
x'(t) &= \int_a^t x'(t,s)h(s)\, ds + x(t,t)h(t) \\
&= \int_a^t x'(t,s)h(s)\, ds.
\end{aligned}
$$

Hence $x'(a) = 0$ and

$$p(t)x'(t) = \int_a^t p(t)x'(t,s)h(s)\, ds.$$

It follows that

$$(p(t)x'(t))' = \int_a^t (p(t)x'(t,s))' h(s) \, ds + p(t)x'(t,t)h(t)$$

$$= \int_a^t (p(t)x'(t,s))' h(s) \, ds + h(t).$$

Hence

$$Lx(t) = \int_a^t \left\{ (p(t)x'(t,s))' + q(t)x(t,s) \right\} h(s) \, ds + h(t)$$

$$= \int_a^t Lx(t,s)h(s) \, ds + h(t)$$

$$= h(t),$$

for $t \in I$. The uniqueness follows from Theorem 5.4. □

Corollary 5.23 *Assume h is continuous on I and assume $a \in I$. The solution of the initial value problem*

$$Lx = h(t),$$

$$x(a) = A, \quad x'(a) = B,$$

where A and B are constants, is given by

$$x(t) = u(t) + \int_a^t x(t,s)h(s) \, ds,$$

for $t \in I$, where u is the solution of the IVP $Lu = 0$, $u(a) = A$, $u'(a) = B$, and $x(\cdot, \cdot)$ is the Cauchy function for $Lx = 0$.

Proof Let

$$x(t) := u(t) + \int_a^t x(t,s)h(s) \, ds,$$

where u, $x(\cdot, \cdot)$, and h are as in the statement of this theorem. Then

$$x(t) = u(t) + v(t),$$

if we let

$$v(t) := \int_a^t x(t,s)h(s) \, ds.$$

Then by Theorem 5.22,

$$Lv(t) = h(t), \quad t \in I,$$

$$v(a) = 0, \quad v'(a) = 0.$$

It follows that

$$x(a) = u(a) + v(a) = A,$$

and

$$x'(a) = u'(a) + v'(a) = B.$$

Finally, note that

$$Lx(t) = Lu(t) + Lv(t) = h(t),$$

for $t \in I$.   □

We now give a simple example to illustrate the variation of constants formula.

Example 5.24 Use the variation of constants formula to solve the IVP

$$\left(e^{-5t}x'\right)' + 6e^{-5t}x = e^t,$$

$$x(0) = 0, \quad x'(0) = 0.$$

By Example 5.17 (or Example 5.20) we get that the Cauchy function for $\left(e^{-5t}x'\right)' + 6e^{-5t}x = 0$ is given by

$$x(t, s) = e^{3t}e^{2s} - e^{2t}e^{3s},$$

for $t, s \in I$. Hence the desired solution is given by

$$
\begin{aligned}
x(t) &= \int_0^t x(t, s)h(s)\, ds \\
&= \int_0^t \left(e^{3t}e^{2s} - e^{2t}e^{3s}\right) e^s ds \\
&= e^{3t}\int_0^t e^{3s} ds - e^{2t}\int_0^t e^{4s} ds \\
&= \frac{1}{4}e^{2t} - \frac{1}{3}e^{3t} + \frac{1}{12}e^{6t}.
\end{aligned}
$$

△

5.4 Sturm-Liouville Problems

In this section we will be concerned with the *Sturm-Liouville differential equation*

$$(p(t)x')' + (\lambda r(t) + q(t))\, x = 0. \tag{5.4}$$

In addition to the standard assumptions on the coefficient functions p and q we assume throughout that the coefficient function r is a real-valued continuous function on I and $r(t) \geq 0$, but is not identically zero, on I. Note that equation (5.4) can be written in the form

$$Lx = -\lambda r(t)x.$$

In this section we will be concerned with the Sturm-Liouville problem (SLP)

$$Lx = -\lambda r(t)x, \tag{5.5}$$

$$\alpha x(a) - \beta x'(a) = 0, \tag{5.6}$$

$$\gamma x(b) + \delta x'(b) = 0, \tag{5.7}$$

where $\alpha, \beta, \gamma, \delta$ are constants satisfying

$$\alpha^2 + \beta^2 > 0, \quad \gamma^2 + \delta^2 > 0.$$

Definition 5.25 We say λ_0 is an *eigenvalue* for the SLP (5.5)–(5.7) provided the SLP (5.5)–(5.7) with $\lambda = \lambda_0$ has a nontrivial solution x_0 (by a nontrivial solution we mean a solution that is not identically zero). We say that x_0 is an *eigenfunction* corresponding to λ_0 and we say that λ_0, x_0 is an *eigenpair* for the SLP (5.5)–(5.7).

Note that if λ_0, x_0 is an eigenpair for the SLP (5.5)–(5.7), then if k is any nonzero constant, λ_0, kx_0 is also an eigenpair of the SLP (5.5)–(5.7). We say λ_0 is a *simple eigenvalue* of the SLP (5.5)–(5.7) provided there is only one linearly independent eigenfunction corresponding to λ_0.

Example 5.26 Find eigenpairs for the Sturm-Liouville problem

$$x'' = -\lambda x, \tag{5.8}$$
$$x(0) = 0, \quad x(\pi) = 0. \tag{5.9}$$

The form for a general solution of (5.8) is different for the cases $\lambda < 0$, $\lambda = 0$, and $\lambda > 0$ so when we try to find eigenpairs for the SLP (5.8), (5.9) we will consider these three cases separately. First assume $\lambda < 0$. In this case let $\lambda = -\mu^2$, where $\mu > 0$. Then a general solution of (5.8) is defined by

$$x(t) = c_1 \cosh(\mu t) + c_2 \sinh(\mu t), \quad t \in [0, \pi].$$

To satisfy the first boundary condition $x(0) = 0$ we are forced to take $c_1 = 0$. The second boundary condition gives

$$x(\pi) = c_2 \sinh(\mu \pi) = 0,$$

which implies $c_2 = 0$ and hence when $\lambda < 0$ the SLP (5.8), (5.9) has only the trivial solution. Hence the SLP (5.8), (5.9) has no negative eigenvalues. Next we check to see if $\lambda = 0$ is an eigenvalue of the SLP (5.8), (5.9). In this case a general solution of (5.8) is

$$x(t) = c_1 t + c_2.$$

Since

$$x(0) = c_2 = 0,$$

we get $c_2 = 0$. Then the second boundary condition gives

$$x(\pi) = c_1 \pi = 0,$$

which implies $c_1 = 0$. Hence when $\lambda = 0$ the only solution of the boundary value problem (BVP) is the trivial solution and so $\lambda = 0$ is not an eigenvalue. Finally, assume $\lambda > 0$; then $\lambda = \mu^2 > 0$, where $\mu > 0$ and so a general solution of (5.8) in this case is given by

$$x(t) = c_1 \cos(\mu t) + c_2 \sin(\mu t), \quad t \in [0, \pi].$$

The boundary condition $x(0) = 0$ implies that $c_1 = 0$. The boundary condition $x(\pi) = 0$ leads to the equation

$$c_2 \sin(\mu \pi) = 0.$$

This last equation is true for $\mu = \mu_n := n$, $n = 1, 2, 3, \cdots$. It follows that the eigenvalues of the SLP (5.8), (5.9) are

$$\lambda_n = n^2, \quad n = 1, 2, 3, \cdots,$$

and corresponding eigenfunctions are defined by

$$x_n(t) = \sin(nt), \quad n = 1, 2, 3, \cdots,$$

for $t \in [0, \pi]$. Hence

$$\lambda_n = n^2, \quad x_n(t) = \sin(nt),$$

$n = 1, 2, 3, \cdots$ are eigenpairs for (5.8), (5.9). $\triangle$

Definition 5.27 Assume that $r : [a, b] \to \mathbb{R}$ is continuous and $r(t) \geq 0$, but not identically zero, on $[a, b]$. We define the *inner product with respect to the weight function* r of the continuous functions x and y on $[a, b]$ by

$$< x, y >_r = \int_a^b r(t) x(t) y(t) \, dt.$$

We say that x and y are *orthogonal with respect to the weight function* r *on the interval* $[a, b]$ provided

$$< x, y >_r = \int_a^b r(t) x(t) y(t) \, dt = 0.$$

Example 5.28 The functions defined by $x(t) = t^2$, $y(t) = 4 - 5t$ for $t \in [0, 1]$ are orthogonal with respect to the weight function defined by $r(t) = t$ on $[0, 1]$.

This is true because

$$\begin{aligned} < x, y >_r &= \int_a^b r(t) x(t) y(t) \, dt \\ &= \int_0^1 t \cdot t^2 \cdot (4 - 5t) \, dt \\ &= \int_0^1 (4t^3 - 5t^4) \, dt \\ &= 0. \end{aligned}$$

$\triangle$

Theorem 5.29 *All eigenvalues of the SLP (5.5)–(5.7) are real and simple. Corresponding to each eigenvalue there is a real-valued eigenfunction. Eigenfunctions corresponding to distinct eigenvalues of the SLP (5.5)–(5.7) are orthogonal with respect to the weight function* r *on* $[a, b]$.

Proof Assume λ_1, x_1 and λ_2, x_2 are eigenpairs for the SLP (5.5)–(5.7). By the Lagrange identity (Theorem 5.6),

$$x_1(t) L x_2(t) - x_2(t) L x_1(t) = \{p(t) w[x_1(t), x_2(t)]\}',$$

for $t \in [a, b]$. Hence

$$(\lambda_1 - \lambda_2)r(t)x_1(t)x_2(t) = \{p(t)w[x_1(t), x_2(t)]\}',$$

for $t \in [a, b]$. Integrating both sides from a to b, we get

$$(\lambda_1 - \lambda_2) < x_1, x_2 >_r = \{p(t)w[x_1(t), x_2(t)]\}_a^b.$$

Using the fact that x_1 and x_2 satisfy the boundary conditions (5.6) and (5.7), it can be shown (see Exercise 5.19) that

$$w[x_1(t), x_2(t)](a) = w[x_1(t), x_2(t)](b) = 0.$$

Hence we get that

$$(\lambda_1 - \lambda_2) < x_1, x_2 >_r = 0. \tag{5.10}$$

If $\lambda_1 \neq \lambda_2$, then we get that x_1, x_2 are orthogonal with respect to the weight function r on $[a, b]$. Assume λ_0, x_0 is an eigenpair for the SLP (5.5)–(5.7). From Exercise 5.18 we get that $\overline{\lambda_0}, \overline{x_0}$ is an eigenpair for the SLP (5.5)–(5.7). Hence from (5.10) for the eigenpairs λ_0, x_0 and $\overline{\lambda_0}, \overline{x_0}$ we get

$$(\lambda_0 - \overline{\lambda_0}) < x_0, \overline{x_0} >_r = 0.$$

It follows that $\lambda_0 = \overline{\lambda_0}$, that is, λ_0 is real.

Next assume that λ_0 is an eigenvalue and x_1, x_2 are corresponding eigenfunctions. Since x_1, x_2 satisfy the boundary condition (5.6), we have by Exercise 5.19 that

$$w[x_1(t), x_2(t)](a) = 0.$$

Since we also know that x_1, x_2 satisfy the same differential equation $Lx = \lambda_0 r(t)x$, we get x_1, x_2 are linearly dependent on $[a, b]$. Hence all eigenvalues of the SLP (5.5)–(5.7) are simple.

Finally, let $\lambda_0, x_0 = u + iv$, where u, v are real-valued functions on $[a, b]$, be an eigenpair for the SLP (5.5)–(5.7). Earlier we proved that all eigenvalues are real, so λ_0 is real. It is easy to see that $\lambda_0, x_0 = u + iv$ is an eigenpair implies that (since at least one of u, v is not identically zero) either u or v is a real-valued eigenfunction corresponding to λ_0. $\square$

In the next example we show how finding nontrivial solutions of a partial differential equation leads to solving a SLP.

Example 5.30 (Separation of Variables) In this example we use separation of variables to show how we can get solutions of the two-dimensional Laplace's equation

$$u_{xx} + u_{yy} = 0.$$

We look for solutions of the form

$$u(x, y) = X(x)Y(y).$$

From Laplace's equation we get

$$X''Y + XY'' = 0.$$

Separating variables and assuming $X(x) \neq 0$, $Y(y) \neq 0$, we get

$$\frac{X''}{-X} = \frac{Y''}{Y}.$$

Since the left-hand side of this equation depends only on x and the right-hand side depends only on y, we get that

$$\frac{X''(x)}{-X(x)} = \frac{Y''(y)}{Y(y)} = \lambda,$$

where λ is a constant. This leads to the Sturm-Liouville differential equations

$$X'' = -\lambda X, \quad Y'' = \lambda Y. \tag{5.11}$$

It follows that if X is a solution of the first differential equation in (5.11) and Y is a solution of the second equation in (5.11), then

$$u(x, y) = X(x)Y(y)$$

is a solution of Laplace's partial differential equation. Also note that if we want $u(x, y) = X(x)Y(y)$ to be nontrivial and satisfy the boundary conditions

$$\alpha u(a, y) - \beta u_x(a, y) = 0, \quad \gamma u(b, y) + \delta u_x(b, y) = 0,$$

then we would want X to satisfy the boundary conditions

$$\alpha X(a) - \beta X'(a) = 0, \quad \gamma X(b) + \delta X'(b) = 0.$$

$\triangle$

Theorem 5.31 *If $q(t) \leq 0$ on $[a, b]$, $\alpha\beta \geq 0$, and $\gamma\delta \geq 0$, then all eigenvalues of the SLP (5.5)–(5.7) are nonnegative.*

Proof Assume λ_0 is an eigenvalue of the SLP (5.5)–(5.7). By Theorem 5.29 there is a real-valued eigenfunction x_0 corresponding to λ_0. Then

$$(p(t)x_0'(t))' + (\lambda_0 r(t) + q(t)) x_0(t) = 0,$$

for $t \in [a, b]$. Multiplying both sides by $x_0(t)$ and integrating from a to b, we have

$$\int_a^b x_0(t) \left(p(t)x_0'(t)\right)' dt + \lambda_0 \int_a^b r(t)x_0^2(t) \, dt + \int_a^b q(t)x_0^2(t) \, dt = 0.$$

Since $q(t) \leq 0$ on $[a, b]$,

$$\lambda_0 \int_a^b r(t)x_0^2(t) \, dt \geq - \int_a^b x_0(t)[p(t)x_0'(t)]' \, dt.$$

Integrating by parts, we get

$$\lambda_0 \int_a^b r(t)x_0^2(t) \, dt \geq -\{p(t)x_0(t)x_0'(t)\}_a^b + \int_a^b p(t)[x_0'(t)]^2 \, dt$$

$$\geq -\{p(t)x_0(t)x_0'(t)\}_a^b. \tag{5.12}$$

Now we will use the fact that x_0 satisfies the boundary condition (5.6) to show that

$$p(a)x_0(a)x_0'(a) \geq 0.$$

If $\beta = 0$, then $x_0(a) = 0$ and consequently

$$p(a)x_0(a)x_0'(a) = 0.$$

On the other hand, if $\beta \neq 0$, then

$$p(a)x_0(a)x_0'(a) = \frac{\alpha}{\beta}[x_0(a)]^2 \geq 0.$$

Similarly, using x_0 satisfies the boundary condition (5.7) and the fact that $\gamma\delta \geq 0$, it follows that

$$p(b)x_0(b)x_0'(b) \leq 0.$$

It now follows from (5.12) that

$$\lambda_0 \geq 0.$$

$\square$

The following example (see page 179, [9]) is important in the study of the temperatures in an infinite slab, $0 \leq x \leq 1$, $-\infty < y < \infty$, where the left edge at $x = 0$ is insulated and surface heat transfer takes place at the the right edge $x = 1$ into a medium with temperature zero.

Example 5.32 Find eigenpairs for the SLP

$$X'' = -\lambda X, \tag{5.13}$$

$$X'(0) = 0, \quad hX(1) + X'(1) = 0, \tag{5.14}$$

where h is a positive constant.

Since $q(t) = 0 \leq 0$ on $[0, 1]$, $\alpha\beta = 0 \geq 0$, and $\gamma\delta = h \geq 0$, we have from Theorem 5.31 that all eigenvalues of the SLP (5.13), (5.14) are nonnegative. If $\lambda = 0$, then $X(x) = c_1 x + c_2$. Then $X'(0) = 0$ implies that $c_1 = 0$. Also,

$$hX(1) + X'(1) = c_2 h = 0$$

implies that $c_2 = 0$ and hence X is the trivial solution. Therefore, zero is not an eigenvalue of the SLP (5.13), (5.14). Next assume that $\lambda = \mu^2 > 0$, where $\mu > 0$; then a general solution of (5.13) is given by

$$X(x) = c_1 \cos(\mu x) + c_2 \sin(\mu x), \quad x \in [0, 1].$$

Then

$$X'(0) = c_2 \mu = 0$$

implies that $c_2 = 0$ and so

$$X(x) = c_1 \cos(\mu x), \quad x \in [0, 1].$$

It follows that

$$hX(1) + X'(1) = c_1 h \cos(\mu) - c_1 \mu \sin(\mu) = 0.$$

Hence we want to choose $\mu > 0$ so that

$$\tan(\mu) = \frac{h}{\mu}.$$

Let $0 < \mu_1 < \mu_2 < \mu_3 < \cdots$ be the positive numbers satisfying

$$\tan(\mu_n) = \frac{h}{\mu_n},$$

$n = 1, 2, 3, \cdots$; then

$$\lambda_n = \mu_n^2$$

are the eigenvalues and

$$X_n(x) = \cos(\mu_n x)$$

are corresponding eigenfunctions.

$\triangle$

We can also prove the following theorem. For this result and other results of this type, see Sagan [45].

Theorem 5.33 *The eigenvalues for the SLP* (5.5)–(5.7) *satisfy*

$$\lambda_1 < \lambda_2 < \lambda_3 < \cdots$$

and

$$\lim_{n \to \infty} \lambda_n = \infty.$$

Furthermore, if x_n is an eigenfunction corresponding to the eigenvalue λ_n, $n = 1, 2, 3, \cdots$, then x_{n+1} has exactly n zeros in (a, b).

We end this section by briefly considering the periodic Sturm-Liouville problem (PSLP)

$$Lx = -\lambda r(t)x, \tag{5.15}$$

$$x(a) = x(b), \quad x'(a) = x'(b). \tag{5.16}$$

The following example is important (see Churchill and Brown [9]) in the study of the heat flow in a circular plate that is insulated on its two faces and where the initial temperature distribution is such that along each radial ray eminating from the center of the disk and having polar angle θ, $-\pi \leq \theta \leq \pi$ the temperature is constant.

Example 5.34 Find eigenpairs for the PSLP

$$u'' = -\lambda u, \tag{5.17}$$

$$u(-\pi) = u(\pi), \quad u'(-\pi) = u'(\pi). \tag{5.18}$$

If $\lambda < 0$, let $\lambda = -\mu^2$, where $\mu > 0$. In this case a general solution of (5.17) is given by

$$u(\theta) = c_1 \cosh(\mu\theta) + c_2 \sinh(\mu\theta), \quad \theta \in [-\pi, \pi].$$

The first boundary condition implies

$$c_1 \cosh(\mu\pi) - c_2 \sinh(\mu\pi) = c_1 \cosh(\mu\pi) + c_2 \sinh(\mu\pi),$$

which is equivalent to the equation

$$2c_2 \sinh(\mu\pi) = 0.$$

This implies that $c_2 = 0$. Hence $u(\theta) = c_1 \cosh(\mu\theta)$. The second boundary condition gives

$$-c_1\mu \sinh(\mu\pi) = c_1\mu \sinh(\mu\pi),$$

which implies that $c_1 = 0$. Hence there are no negative eigenvalues. Next assume $\lambda = 0$; then a general solution of (5.17) is

$$u(\theta) = c_1\theta + c_2, \quad \theta \in [-\pi, \pi].$$

The first boundary condition gives us

$$-c_1\pi + c_2 = c_1\pi + c_2,$$

which implies that $c_1 = 0$. Hence $u(\theta) = c_2$, which satisfies the second boundary condition. Hence $\lambda_0 = 0$ is an eigenvalue and $u_0(\theta) = 1$ defines a corresponding eigenfunction. Finally, consider the case when $\lambda > 0$. In this case we let $\lambda = \mu^2$, where $\mu > 0$. A general solution of (5.17) in this case is

$$u(\theta) = c_1 \cos(\mu\theta) + c_2 \sin(\mu\theta), \quad \theta \in [-\pi, \pi].$$

The first boundary condition gives us

$$c_1 \cos(\mu\pi) - c_2 \sin(\mu\pi) = c_1 \cos(\mu\pi) + c_2 \sin(\mu\pi),$$

which is equivalent to the equation

$$c_2 \sin(\mu\pi) = 0.$$

Hence the first boundary condition is satisfied if we take

$$\mu = \mu_n := n,$$

$n = 1, 2, 3, \cdots$. It is then easy to check that the second boundary condition is also satisfied. Hence

$$\lambda_n = n^2,$$

$n = 1, 2, 3, \cdots$ are eigenvalues and corresponding to each of these eigenvalues are two linearly independent eigenfunctions given by $u_n(\theta) = \cos(n\theta)$, $v_n(\theta) = \sin(n\theta)$, $\theta \in [-\pi, \pi]$. $\triangle$

Theorem 5.35 *Assume $p(a) = p(b)$; then eigenfunctions corresponding to distinct eigenvalues of the PSLP (5.15), (5.16) are orthogonal with respect to the weight function r on $[a, b]$.*

Proof Assume λ_1, x_1 and λ_2, x_2 are eigenpairs for the PSLP (5.15), (5.16), with $\lambda_1 \neq \lambda_2$. By the Lagrange identity (Theorem 5.6),

$$x_1(t)Lx_2(t) - x_2(t)Lx_1(t) = \{p(t)w[x_1(t), x_2(t)]\}',$$

for $t \in [a, b]$. It follows that

$$(\lambda_1 - \lambda_2)r(t)x_1(t)x_2(t) = \{p(t)w[x_1(t), x_2(t)]\}',$$

for $t \in [a, b]$. After integrating from a to b, we get

$$(\lambda_1 - \lambda_2) < x_1, x_2 >_r = \{p(t)w[x_1(t), x_2(t)]\}_a^b.$$

Since

$$\begin{aligned}
\{p(t)w[x_1(t), x_2(t)]\}(b) &= p(b)[x_1(b)x_2'(b) - x_2(b)x_1'(b)] \\
&= p(a)[x_1(a)x_2'(a) - x_2(a)x_1'(a)] \\
&= \{p(t)w[x_1(t), x_2(t)]\}(a),
\end{aligned}$$

we get that

$$(\lambda_1 - \lambda_2) < x_1, x_2 >_r = 0.$$

Since $\lambda_1 \neq \lambda_2$, we get the desired result

$$< x_1, x_2 >_r = 0.$$

$\square$

5.5 Zeros of Solutions and Disconjugacy

The study of the zeros of nontrivial solutions of $Lx = 0$ is very important, as we will see later in this chapter. We now give some elementary facts concerning zeros of nontrivial solutions of $Lx = 0$. First note for any nonempty subinterval J of I there is a nontrivial solution with a zero in J. To see this, let $t_0 \in J$, and let x be the solution of the initial value problem $Lx = 0$, $x(t_0) = 0$, $x'(t_0) = 1$. If x is a differentiable function satisfying $x(t_0) = x'(t_0) = 0$, then we say x has a *double zero* at t_0, whereas if $x(t_0) = 0$, $x'(t_0) \neq 0$, we say x has a *simple zero* at t_0. Note that by the uniqueness theorem there is no nontrivial solution with a double zero at some point in I. Hence all nontrivial solutions of $Lx = 0$ have only simple zeros in I. If we consider the equation $x'' + x = 0$ and $J \subset I$ is an interval of length less than π, then no nontrivial solution has two zeros in I. This leads to the following definition.

Definition 5.36 We say that $Lx = 0$ is *disconjugate* on $J \subset I$ provided no nontrivial solution of $Lx = 0$ has two or more zeros in J.

Example 5.37 It is easy to see that the following equations are disconjugate on the corresponding intervals J.

    (i) $x'' + x = 0$, $J = [0, \pi)$
    (ii) $x'' = 0$, $J = \mathbb{R}$
    (iii) $x'' - 5x' + 6x = 0$, $J = \mathbb{R}$

$\triangle$

At the other extreme there are self-adjoint differential equations $Lx = 0$, where there are nontrivial solutions with infinitely many zeros in I, as shown in the following example.

Example 5.38

(i) $x(t) := \sin t$ defines a nontrivial solution of $x'' + x = 0$ with infinitely many zeros in $\mathbb{R}$;

(ii) $x(t) := \sin \frac{1}{t}$ defines a nontrivial solution of $(t^2 x')' + \frac{1}{t^2} x = 0$ with infinitely many zeros in the bounded interval $J := (0, 1]$.

$\triangle$

This leads to the following definition.

Definition 5.39 We say that a nontrivial solution x of $Lx = 0$ is *oscillatory* on $J \subset I$ provided x has infinitely many zeros in J. If $Lx = 0$ has a nontrivial oscillatory solution on J, then we say the differential equation $Lx = 0$ is *oscillatory* on J. If $Lx = 0$ has no nontrivial oscillatory solution on J, then we say the differential equation $Lx = 0$ is *nonoscillatory* on J. If x is a solution of $Lx = 0$ that does not have infinitely many zeros in J, we say x is *nonoscillatory* on J.

Hence $x'' + x = 0$ is oscillatory on $\mathbb{R}$ and $(t^2 x')' + \frac{1}{t^2} x = 0$ is oscillatory on $(0, 1]$.

The next example will motivate the next theorem.

Example 5.40 The self-adjoint differential equation $x'' + x = 0$, $t \in I := \mathbb{R}$ has the functions defined by $x(t) = \cos t$, $y(t) = \sin t$ as linearly independent solutions on $\mathbb{R}$. Note that the zeros of these solutions separate each other in $\mathbb{R}$. $\triangle$

Theorem 5.41 (Sturm Separation Theorem) *If x, y are linearly independent solutions on I of the self-adjoint differential equation $Lx = 0$, then their zeros separate each other in I. By this we mean that x and y have no common zeros and between any two consecutive zeros of one of these solutions there is exactly one zero of the other solution.*

Proof Assume that x, y are linearly independent solutions on I of the self-adjoint differential equation $Lx = 0$. Then by Corollary 5.10

$$w(t) := w[x(t), y(t)] \neq 0,$$

for $t \in I$. Assume x and y have a common zero in I; that is, there is a $t_0 \in I$ such that

$$x(t_0) = y(t_0) = 0.$$

But then

$$w(t_0) = \begin{vmatrix} 0 & 0 \\ x'(t_0) & y'(t_0) \end{vmatrix} = 0,$$

which is a contradiction. Next assume x has consecutive zeros at $t_1 < t_2$ in I. We claim that y has a zero in (t_1, t_2). Assume to the contrary that $y(t) \neq 0$ for $t \in (t_1, t_2)$. Then without loss of generality we can assume that $y(t) > 0$ on the closed interval $[t_1, t_2]$. Also without loss of generality we can assume that $x(t) > 0$ on (t_1, t_2). But then

$$w(t_1) = \begin{vmatrix} 0 & y(t_1) \\ x'(t_1) & y'(t_1) \end{vmatrix} = -y(t_1) x'(t_1) < 0,$$

and

$$w(t_2) = \begin{vmatrix} 0 & y(t_2) \\ x'(t_2) & y'(t_2) \end{vmatrix} = -y(t_2)x'(t_2) > 0.$$

Hence by the intermediate value theorem there is a point $t_3 \in (t_1, t_2)$ such that

$$w(t_3) = 0,$$

which is a contradiction. By interchanging x and y in the preceding argument we get that between any two consecutive zeros of y there has to be a zero of x. It follows that the zeros of x and y have to separate each other in I. □

Note that it follows from the Sturm separation theorem that either all nontrivial solutions of $Lx = 0$ are oscillatory on I or all nontrivial solutions are nonoscillatory on I.

Definition 5.42 An interval I is said to be a *compact interval* provided it is a closed and bounded interval.

Theorem 5.43 *If J is a compact subinterval of I, then $Lx = 0$ is nonoscillatory on J.*

Proof To the contrary, assume that $Lx = 0$ is oscillatory on a compact subinterval J of I. Then there is a nontrivial solution x of $Lx = 0$ with infinitely many zeros in J. It follows that there is a infinite sequence $\{t_n\}$ of distinct points contained in J such that

$$x(t_n) = 0,$$

for $n = 1, 2, 3, \cdots$ and

$$\lim_{n \to \infty} t_n = t_0,$$

where $t_0 \in I$. Without loss of generality we can assume that the sequence $\{t_n\}$ is either strictly increasing or strictly decreasing with

$$t_0 := \lim_{n \to \infty} t_n \in J.$$

We will only consider the case where the sequence $\{t_n\}$ is strictly increasing since the other case is similar. Since

$$x(t_n) = x(t_{n+1}) = 0,$$

we have by Rolle's theorem that there is a $t'_n \in (t_n, t_{n+1})$ such that

$$x'(t'_n) = 0.$$

It follows that

$$x(t_0) = \lim_{n \to \infty} x(t_n) = 0$$

and

$$x'(t_0) = \lim_{n \to \infty} x'(t'_n) = 0.$$

But by the uniqueness theorem (Theorem 5.4), $x(t_0) = 0$, $x'(t_0) = 0$, implies that x is the trivial solution which is a contradiction. □

The following example shows that boundary value problems are not as nice as initial value problems.

Example 5.44 Consider the conjugate BVP

$$x'' + x = 0,$$

$$x(0) = 0, \quad x(\pi) = B.$$

If $B = 0$, this BVP has infinitely solutions $x(t) = c_1 \sin t$, where c_1 is a constant. On the other hand, if $B \neq 0$, then this BVP has no solutions. △

Theorem 5.45 *If $Lx = 0$ is disconjugate on $[a, b] \subset I$, then the conjugate BVP*

$$Lx = h(t),$$

$$x(a) = A, \quad x(b) = B,$$

where A and B are given constants and h is a given continuous function on $[a, b]$, has a unique solution.

Proof Let u, v be linearly independent solutions of the homogeneous equation $Lx = 0$ and let z be a solution of the nonhomogeneous equation $Lx = h(t)$. Then, by Theorem 5.14,

$$x = c_1 u + c_2 v + z$$

is a general solution of $Lx = h(t)$. Hence there is a solution of the given BVP iff there are constants c_1, c_2 such that

$$c_1 u(a) + c_2 v(a) = A - z(a),$$
$$c_1 u(b) + c_2 v(b) = B - z(b).$$

Hence our BVP has a unique solution iff

$$\begin{vmatrix} u(a) & v(a) \\ u(b) & v(b) \end{vmatrix} \neq 0.$$

Assume

$$\begin{vmatrix} u(a) & v(a) \\ u(b) & v(b) \end{vmatrix} = 0.$$

Then there are constants a_1, a_2, not both zero, such that

$$a_1 u(a) + a_2 v(a) = 0,$$
$$a_1 u(b) + a_2 v(b) = 0.$$

Let $x := a_1 u + a_2 v$; then x is a nontrivial solution of $Lx = 0$ satisfying

$$x(a) = a_1 u(a) + a_2 v(a) = 0,$$
$$x(b) = a_1 u(b) + a_2 v(b) = 0,$$

which contradicts that $Lx = 0$ is disconjugate on $[a, b]$.   □

Theorem 5.46 *If $Lx = 0$ has a positive solution on $J \subset I$, then $Lx = 0$ is disconjugate on J. Conversely, if J is a compact subinterval of I and $Lx = 0$ is disconjugate on J, then $Lx = 0$ has a positive solution on J.*

Proof Assume that $Lx = 0$ has a positive solution on $J \subset I$. It follows from the Sturm separation theorem that no nontrivial solution can have two zeros in J. Hence $Lx = 0$ is disconjugate on J.

Conversely, assume $Lx = 0$ is disconjugate on a compact subinterval J of I. Let $a < b$ be the endpoints of J and let u, v be the solutions of $Lx = 0$ satisfying the initial conditions

$$u(a) = 0, \quad u'(a) = 1,$$

and

$$v(b) = 0, \quad v'(b) = -1,$$

respectively. Since $Lx = 0$ is disconjugate on J, we get that

$$u(t) > 0 \quad \text{on} \quad (a, b],$$

and

$$v(t) > 0 \quad \text{on} \quad [a, b).$$

It follows that

$$x := u + v$$

is a positive solution of $Lx = 0$ on $J = [a, b]$. $\qquad \square$

The following example shows that we cannot remove the word compact in the preceding theorem.

Example 5.47 The differential equation

$$x'' + x = 0$$

is disconjugate on the interval $J = [0, \pi)$ but has no positive solution on $[0, \pi)$. $\qquad \triangle$

The following example is a simple application of Theorem 5.46.

Example 5.48 Since $x(t) = e^{2t}$, $t \in \mathbb{R}$, defines a positive solution of

$$x'' - 5x' + 6x = 0,$$

on $\mathbb{R}$, this differential equation is disconjugate on $\mathbb{R}$. $\qquad \triangle$

We will now be concerned with the two self-adjoint equations

$$(p_1(t)x')' + q_1(t)x \; = \; 0, \tag{5.19}$$
$$(p_2(t)x')' + q_2(t)x \; = \; 0. \tag{5.20}$$

We always assume that the coefficient functions in these two equations are continuous on an interval I and $p_i(t) > 0$ on I for $i = 1, 2$.

Theorem 5.49 (Picone Identity) *Assume u and v are solutions of (5.19) and (5.20), respectively, with $v(t) \neq 0$ on $[a, b] \subset I$. Then*

$$\left\{ \left(\frac{u(t)}{v(t)} \right) [p_1(t)u'(t)v(t) - p_2(t)u(t)v'(t)] \right\}_a^b \tag{5.21}$$

$$= \int_a^b [q_2(t) - q_1(t)]u^2(t) \, dt + \int_a^b [p_1(t) - p_2(t)][u'(t)]^2 \, dt$$

$$+ \int_a^b p_2(t)v^2(t) \left[\left(\frac{u(t)}{v(t)} \right)' \right]^2 dt.$$

Proof Consider

$$\left\{ \left(\frac{u(t)}{v(t)} \right) [p_1(t)u'(t)v(t) - p_2(t)u(t)v'(t)] \right\}'$$

$$= u(t)[p_1(t)u'(t)]' + p_1(t)[u'(t)]^2$$

$$- \frac{u^2(t)}{v(t)}[p_2(t)v'(t)]' - [p_2(t)v'(t)] \left(\frac{u^2(t)}{v(t)} \right)'$$

$$= -q_1(t)u^2(t) + p_1(t)[u'(t)]^2 + q_2(t)u^2(t)$$

$$- [p_2(t)v'(t)] \left(\frac{2v(t)u(t)u'(t) - u^2(t)v'(t)}{v^2(t)} \right)$$

$$= [q_2(t) - q_1(t)]u^2(t) + [p_1(t) - p_2(t)][u'(t)]^2$$

$$+ p_2(t) \left(\frac{v^2(t)[u'(t)]^2 - 2u(t)u'(t)v(t)v'(t) + u^2(t)[v'(t)]^2}{v^2(t)} \right)$$

$$= [q_2(t) - q_1(t)]u^2(t) + [p_1(t) - p_2(t)][u'(t)]^2$$

$$+ p_2(t)v^2(t) \left(\frac{v(t)u'(t) - u(t)v'(t)}{v^2(t)} \right)^2$$

$$= [q_2(t) - q_1(t)]u^2(t) + [p_1(t) - p_2(t)][u'(t)]^2 + p_2(t)v^2(t) \left[\left(\frac{u(t)}{v(t)} \right)' \right]^2,$$

for $t \in [a, b]$. Integrating both sides from a to b, we get the Picone identity (5.21). $\square$

We next use the Picone identity to prove the Sturm comparison theorem, which is a generalization of the Sturm separation theorem.

Theorem 5.50 (Sturm Comparison Theorem) *Assume u is a solution of (5.19) with consecutive zeros at $a < b$ in I and assume that*

$$q_2(t) \geq q_1(t), \quad 0 < p_2(t) \leq p_1(t), \tag{5.22}$$

for $t \in [a, b]$. If v is a solution of (5.20) and if for some $t \in [a, b]$ one of the inequalities in (5.22) is strict or if u and v are linearly independent on $[a, b]$, then v has a zero in (a, b).

Proof Assume u is a solution of (5.19) with consecutive zeros at $a < b$ in I. Assume the conclusion of this theorem is not true. That is, assume v is a solution of (5.20) with

$$v(t) \neq 0, \quad t \in (a, b).$$

Assume $a < c < d < b$, then by the Picone identity (5.21) with a replaced by c and b replaced by d we have

$$\int_c^d [q_2(t) - q_1(t)]u^2(t) \, dt + \int_c^d [p_1(t) - p_2(t)][u'(t)]^2 \, dt \quad (5.23)$$

$$+ \int_c^d p_2(t)v^2(t) \left[\left(\frac{u(t)}{v(t)} \right)' \right]^2 dt$$

$$= \left\{ \left(\frac{u(t)}{v(t)} \right) [p_1(t)u'(t)v(t) - p_2(t)u(t)v'(t)] \right\}_c^d.$$

Using the inequalities in (5.22) and the fact that either one of the inequalities in (5.22) is strict at some point $t \in [a, b]$, or u and v are linearly independent on $[a, b]$, we get, letting $c \to a+$ and letting $d \to b-$,

$$\lim_{c \to a+, d \to b-} \left\{ \left(\frac{u(t)}{v(t)} \right) [p_1(t)u'(t)v(t) - p_2(t)u(t)v'(t)] \right\}_c^d > 0.$$

We would get a contradiction if the two limits in the preceding inequality are zero. We will only show this for the right-hand limit at a (see Exercise 5.28 for the other case). Note that if $v(a) \neq 0$, then clearly

$$\lim_{t \to a+} \left\{ \left(\frac{u(t)}{v(t)} \right) [p_1(t)u'(t)v(t) - p_2(t)u(t)v'(t)] \right\} = 0.$$

Now assume that $v(a) = 0$; then since v only has simple zeros, $v'(a) \neq 0$. Consider

$$\lim_{t \to a+} \left\{ \left(\frac{u(t)}{v(t)} \right) [p_1(t)u'(t)v(t) - p_2(t)u(t)v'(t)] \right\}$$

$$= - \lim_{t \to a+} \frac{u^2(t)p_2(t)v'(t)}{v(t)}$$

$$= - \lim_{t \to a+} \frac{u^2(t)[p_2(t)v'(t)]' + 2u(t)u'(t)p_2(t)v'(t)}{v'(t)}$$

$$= 0,$$

where we have used l'Hôpital's rule   □

Corollary 5.51 *Assume that*

$$q_2(t) \geq q_1(t), \quad 0 < p_2(t) \leq p_1(t), \quad (5.24)$$

for $t \in J \subset I$. If (5.20) is disconjugate on J, then (5.19) is disconjugate on J. If (5.19) is oscillatory on J, then (5.20) is oscillatory on J.

Proof Assume equation (5.19) is oscillatory on J. Then there is a nontrivial solution u of equation (5.19) with infinitely many zeros in J. Let v be a nontrivial solution of equation (5.20). If u and v are linearly dependent on J, then v has the same zeros as u and it would follow that the differential equation (5.20) is oscillatory on J. On the other hand, if u and v are linearly independent on J, then by Theorem 5.50, v has at least one zero between each pair of consecutive zeros of u and it follows that the differential equation (5.20) is oscillatory on J. The other part of this proof is similar (see Exercise 5.30). □

Example 5.52 Show that if

$$0 < p(t) \le t^2, \quad \text{and} \quad q(t) \ge b > \frac{1}{4},$$

for $t \in [1, \infty)$, then the self-adjoint equation $Lx = 0$ is oscillatory on $[1, \infty)$.

To see this we use Corollary 5.51. We compare $Lx = 0$ with the differential equation

$$\left(t^2 x'\right)' + bx = 0.$$

Expanding this equation out, we get the Euler–Cauchy differential equation

$$t^2 x'' + 2tx' + bx = 0.$$

By Exercise 5.29 this equation is oscillatory since $b > \frac{1}{4}$. Hence by Corollary 5.51 we get that $Lx = 0$ is oscillatory on $[1, \infty)$. △

5.6 Factorizations and Recessive and Dominant Solutions

Theorem 5.53 (Polya Factorization) *Assume $Lx = 0$ has a positive solution u on $J \subset I$. Then, for $x \in \mathbb{D}$,*

$$Lx(t) = \rho_1(t) \left\{ \rho_2(t) [\rho_1(t) x(t)]' \right\}',$$

for $t \in J$, where

$$\rho_1(t) := \frac{1}{u(t)} > 0, \quad \text{and} \quad \rho_2(t) := p(t) u^2(t) > 0,$$

for $t \in J$.

Proof Assume u is a positive solution of $Lx = 0$ and assume that $x \in \mathbb{D}$; then by the Lagrange identity,

$$
\begin{aligned}
u(t)Lx(t) &= \{p(t)w[u(t), x(t)]\}' \\
&= \{p(t)[u(t)x'(t) - u'(t)x(t)]\}' \\
&= \left\{ p(t)u^2(t) \frac{[u(t)x'(t) - u'(t)x(t)]}{u^2(t)} \right\}' \\
&= \left\{ p(t)u^2(t) \left[\frac{x(t)}{u(t)} \right]' \right\}' \\
&= \{\rho_2(t)[\rho_1(t)x(t)]'\}',
\end{aligned}
$$

for $t \in J$, which leads to the desired result.   □

We say that the differential equation

$$\rho_1(t) \{\rho_2(t)[\rho_1(t)x(t)]'\}' = 0$$

is the Polya factorization of the differential equation $Lx = 0$. Note that by the proof of Theorem 5.53 we only need to assume that $Lx = 0$ has a solution u without zeros on J to get the factorization, but if $u(t) < 0$ on J, then $\rho_1(t) < 0$ on J. When $\rho_i(t) > 0$ on J for $i = 1, 2$ we call our factorization a Polya factorization; otherwise we just call it a factorization.

Example 5.54 Find a Polya factorization of the differential equation $Lx = x'' + x = 0$ on $J = (0, \pi)$. Here $u(t) = \sin t$, $t \in J$, defines a positive solution on $J = (0, \pi)$. By Theorem 5.53 a Polya factorization is

$$Lx(t) = \frac{1}{\sin t} \left\{ \sin^2 t \left[\frac{x(t)}{\sin t} \right]' \right\}' = 0,$$

for $x \in \mathbb{D}$, $t \in (0, \pi)$.   △

The following example shows that Polya factorizations are not unique, and we will also indicate how this example motivates Theorem 5.58.

Example 5.55 Find two Polya factorizations for the differential equation

$$Lx = \left(e^{-6t}x' \right)' + 8e^{-6t}x = 0. \tag{5.25}$$

Expanding this differential equation out and then dividing by e^{-6t}, we get the equivalent equation

$$x'' - 6x' + 8x = 0.$$

Note that $u_1(t) := e^{4t}$ defines a positive solution of the differential equation (5.25) on $\mathbb{R}$ [which implies that the differential equation (5.25) is disconjugate on $\mathbb{R}$] and hence from Theorem 5.53 we get the Polya factorization

$$Lx(t) = \rho_1(t) \{\rho_2(t)[\rho_1(t)x(t)]'\}' = 0,$$

for $x \in \mathbb{D}$, $t \in \mathbb{R}$, where

$$\rho_1(t) = e^{-4t} \text{ and } \rho_2(t) = e^{2t},$$

for $t \in \mathbb{R}$. Note that $u_2(t) := e^{2t}$ also defines a positive solution of the differential equation (5.25) and hence from Theorem 5.53 we get the Polya factorization

$$Lx(t) = \gamma_1(t) \left\{ \gamma_2(t)[\gamma_1(t)x(t)]' \right\}' = 0,$$

for $t \in \mathbb{R}$, where

$$\gamma_1(t) = e^{-2t} \quad \text{and} \quad \gamma_2(t) = e^{-2t},$$

for $t \in \mathbb{R}$. Hence we have found two distinct factorizations of equation (5.25). Note in the first factorization we get

$$\int_0^\infty \frac{1}{\rho_2(t)} dt = \int_0^\infty e^{-2t} \, dt < \infty,$$

whereas in the second factorization we get

$$\int_0^\infty \frac{1}{\gamma_2(t)} dt = \int_0^\infty e^{2t} \, dt = \infty.$$

In the next theorem we show that under the hypothesis of Theorem 5.53 we can always get a factorization of $Lx = 0$ (called a Trench factorization), where the γ_2 in the Polya factorization satisfies

$$\int_a^b \frac{1}{\gamma_2(t)} dt = \infty$$

(usually we will want $b = \infty$ as in the preceding example). We will see in Theorem 5.59 why we want this integral to be infinite.

$$\triangle$$

We get the following result from Theorem 5.53.

Theorem 5.56 (Reduction of Order) *If u is a solution of $Lx = 0$ without zeros in $J \subset I$ and $t_0 \in J$, then*

$$v(t) := u(t) \int_{t_0}^t \frac{1}{p(\tau)u^2(\tau)} d\tau, \quad t \in J$$

defines a second linearly independent solution on J.

Proof Assume u is a solution of $Lx = 0$ without zeros in $J \subset I$ and $t_0 \in J$. Then by Theorem 5.53 we have the factorization

$$Lx(t) = \frac{1}{u(t)} \left(p(t)u^2(t) \left(\frac{x(t)}{u(t)} \right)' \right)' = 0,$$

for $t \in J$. It follows from this that the solution v of the IVP

$$p(t)u^2(t) \left(\frac{v}{u(t)} \right)' = 1, \quad v(t_0) = 0$$

is a solution of $Lx = 0$ on J. It follows that

$$\left(\frac{v(t)}{u(t)}\right)' = \frac{1}{p(t)u^2(t)},$$

for $t \in J$. Integrating both sides from t_0 to t and solving for $v(t)$, we get

$$v(t) = u(t) \int_{t_0}^t \frac{1}{p(\tau)u^2(\tau)} d\tau, \quad t \in J.$$

To see that u, v are linearly independent on J, note that

$$
\begin{aligned}
w[u(t), v(t)] &= \begin{vmatrix} u(t) & v(t) \\ u'(t) & v'(t) \end{vmatrix} \\
&= \begin{vmatrix} u(t) & u(t) \int_{t_0}^t \frac{1}{p(\tau)u^2(\tau)} d\tau \\ u'(t) & u'(t) \int_{t_0}^t \frac{1}{p(\tau)u^2(\tau)} d\tau + \frac{1}{p(t)u(t)} \end{vmatrix} \\
&= \begin{vmatrix} u(t) & 0 \\ u'(t) & \frac{1}{p(t)u(t)} \end{vmatrix} \\
&= \frac{1}{p(t)} \neq 0,
\end{aligned}
$$

for $t \in J$. Hence u, v are linearly independent solutions of $Lx = 0$ on J.   □

Example 5.57 Given that $u(t) = e^t$ defines a solution, solve the differential equation

$$\left(\frac{1}{te^{2t}} x'\right)' + \left(\frac{1+t}{t^2 e^{2t}}\right) x = 0,$$

$t \in I := (0, \infty)$.

By the reduction of order theorem (Theorem 5.56), a second linearly independent solution on $I = (0, \infty)$ is given by

$$
\begin{aligned}
v(t) &= u(t) \int_1^t \frac{1}{p(s)u^2(s)} ds \\
&= e^t \int_1^t \frac{se^{2s}}{e^{2s}} ds \\
&= e^t \int_1^t s \, ds \\
&= \frac{1}{2} t^2 e^t - \frac{1}{2} e^t.
\end{aligned}
$$

Hence a general solution is given by

$$
\begin{aligned}
x(t) &= c_1 e^t + c_2 \left(\frac{1}{2} t^2 e^t - \frac{1}{2} e^t\right) \\
&= \alpha e^t + \beta t^2 e^t.
\end{aligned}
$$

△

Theorem 5.58 (Trench Factorization) *Assume* $Lx = 0$ *has a positive solution on* $[a, b) \subset I$, *where* $-\infty < a < b \le \infty$. *Then there are positive functions* γ_i, $i = 1, 2$ *on* $[a, b)$ *such that for* $x \in \mathbb{D}$

$$Lx(t) = \gamma_1(t) \left\{ \gamma_2(t) [\gamma_1(t) x(t)]' \right\}',$$

for $t \in [a, b)$, *and*

$$\int_a^b \frac{1}{\gamma_2(t)} dt = \infty.$$

Proof Assume $Lx = 0$ has a positive solution on $[a, b) \subset I$, where $-\infty < a < b \le \infty$. Then by Theorem 5.53 the operator L has a Polya factorization on $[a, b)$. That is, if $x \in \mathbb{D}$, then

$$Lx(t) = \rho_1(t) \left\{ \rho_2(t) [\rho_1(t) x(t)]' \right\}',$$

for $t \in [a, b)$. Let

$$\alpha_i(t) = \frac{1}{\rho_i(t)},$$

$i = 1, 2$ for $t \in [a, b)$. Then

$$Lx(t) = \frac{1}{\alpha_1(t)} \left\{ \frac{1}{\alpha_2(t)} \left[\frac{x(t)}{\alpha_1(t)} \right]' \right\}',$$

for $x \in \mathbb{D}$, $t \in [a, b)$. If

$$\int_a^b \alpha_2(t) \, dt = \infty$$

we have a Trench factorization and the proof is complete in this case. Now assume that

$$\int_a^b \alpha_2(t) \, dt < \infty.$$

In this case we let

$$\beta_1(t) := \alpha_1(t) \int_t^b \alpha_2(s) \, ds, \quad \beta_2(t) := \frac{\alpha_2(t)}{\left[\int_t^b \alpha_2(s) \, ds \right]^2},$$

for $t \in [a, b)$. Then

$$\int_a^b \beta_2(t) \, dt = \lim_{c \to b-} \int_a^c \frac{\alpha_2(t)}{\left[\int_t^b \alpha_2(s) \, ds \right]^2} dt$$

$$= \lim_{c \to b-} \left\{ \left[\int_t^b \alpha_2(s) \, ds \right]^{-1} \right\}_a^c$$

$$= \infty.$$

Let $x \in \mathbb{D}$ and consider

$$
\left\{ \frac{x(t)}{\beta_1(t)} \right\}' = \left\{ \frac{\frac{x(t)}{\alpha_1(t)}}{\int_t^b \alpha_2(s)\, ds} \right\}'
$$

$$
= \frac{\int_t^b \alpha_2(s)\, ds \left[\frac{x(t)}{\alpha_1(t)} \right]' - \frac{x(t)}{\alpha_1(t)} [-\alpha_2(t)]}{\left[\int_t^b \alpha_2(s)\, ds \right]^2}
$$

$$
= \frac{\int_t^b \alpha_2(s)\, ds \left[\frac{x(t)}{\alpha_1(t)} \right]' + \frac{x(t)}{\alpha_1(t)} [\alpha_2(t)]}{\left[\int_t^b \alpha_2(s)\, ds \right]^2}.
$$

Hence

$$
\frac{1}{\beta_2(t)} \left[\frac{x(t)}{\beta_1(t)} \right]' = \left\{ \int_t^b \alpha_2(s)\, ds \right\} \left\{ \frac{1}{\alpha_2(t)} \left[\frac{x(t)}{\alpha_1(t)} \right]' \right\} + \frac{x(t)}{\alpha_1(t)}
$$

and so

$$
\left\{ \frac{1}{\beta_2(t)} \left[\frac{x(t)}{\beta_1(t)} \right]' \right\}' = \int_t^b \alpha_2(s)\, ds \left\{ \frac{1}{\alpha_2(t)} \left[\frac{x(t)}{\alpha_1(t)} \right]' \right\}'.
$$

Finally, we get that

$$
\frac{1}{\beta_1(t)} \left\{ \frac{1}{\beta_2(t)} \left[\frac{x(t)}{\beta_1(t)} \right]' \right\}' = \frac{1}{\alpha_1(t)} \left\{ \frac{1}{\alpha_2(t)} \left[\frac{x(t)}{\alpha_1(t)} \right]' \right\}'
$$

$$
= Lx(t),
$$

for $t \in [a, b)$.   □

We now can use the Trench factorization to prove the existence of recessive and dominant solutions at b.

Theorem 5.59 *Assume $Lx = 0$ is nonoscillatory on $[a, b) \subset I$, where $-\infty < a < b \le \infty$; then there is a solution u, called a recessive solution at b, such that if v is any second linearly independent solution, called a dominant solution at b, then*

$$
\lim_{t \to b-} \frac{u(t)}{v(t)} = 0,
$$

$$
\int_{t_0}^b \frac{1}{p(t)u^2(t)}\, dt = \infty,
$$

and

$$
\int_{t_0}^b \frac{1}{p(t)v^2(t)}\, dt < \infty,
$$

for some $t_0 < b$ sufficiently close. Furthermore,

$$\frac{p(t)v'(t)}{v(t)} > \frac{p(t)u'(t)}{u(t)},$$

for $t < b$ sufficiently close. Also, the recessive solution is unique up to multiplication by a nonzero constant.

Proof Since $Lx = 0$ is nonoscillatory on $[a, b)$, there is a $c \in [a, b)$ such that $Lx = 0$ has a positive solution on $[c, b)$. It follows that the operator L has a Trench factorization on $[c, b)$. That is, for any $x \in \mathbb{D}$,

$$Lx(t) = \gamma_1(t) \left\{ \gamma_2(t)[\gamma_1(t)x(t)]' \right\}',$$

for $t \in [c, b)$, where $\gamma_i(t) > 0$ on $[c, b)$, $i = 1, 2$ and

$$\int_c^b \frac{1}{\gamma_2(t)} dt = \infty.$$

Then from the factorization we get that

$$u(t) := \frac{1}{\gamma_1(t)},$$

defines a solution of $Lx = 0$ that is positive on $[c, b)$. Let z be the solution of the IVP

$$\gamma_2(t) \left(\gamma_1(t)z \right)' = 1,$$

$$z(c) = 0.$$

Then from the factorization we get that z is a solution of $Lx = 0$ that is given by

$$z(t) = \frac{1}{\gamma_1(t)} \int_c^t \frac{1}{\gamma_2(s)} ds,$$

for $t \in [c, d)$. Note that

$$\lim_{t \to b-} \frac{u(t)}{z(t)} = \lim_{t \to b-} \frac{\frac{1}{\gamma_1(t)}}{\frac{1}{\gamma_1(t)} \int_c^t \frac{1}{\gamma_2(s)} ds}$$

$$= \lim_{t \to b-} \frac{1}{\int_c^t \frac{1}{\gamma_2(s)} ds}$$

$$= 0.$$

Now let v be any solution such that u and v are linearly independent; then

$$v(t) = c_1 u(t) + c_2 z(t),$$

where $c_2 \neq 0$. Then

$$
\begin{aligned}
\lim_{t \to b-} \frac{u(t)}{v(t)} &= \lim_{t \to b-} \frac{u(t)}{c_1 u(t) + c_2 z(t)} \\
&= \lim_{t \to b-} \frac{\frac{u(t)}{z(t)}}{c_1 \frac{u(t)}{z(t)} + c_2} \\
&= 0.
\end{aligned}
$$

Next consider, for $t \in [c, d)$,

$$
\begin{aligned}
\left[\frac{z(t)}{u(t)}\right]' &= \frac{u(t)z'(t) - z(t)u'(t)}{u^2(t)} \\
&= \frac{w[u(t), z(t)]}{u^2(t)} \\
&= \frac{C}{p(t)u^2(t)},
\end{aligned}
$$

where C is a nonzero constant. Integrating from c to t, we get that

$$
\frac{z(t)}{u(t)} - \frac{z(c)}{u(c)} = C \int_c^t \frac{1}{p(t)u^2(t)}\,dt.
$$

Letting $t \to b-$, we get that

$$
\int_c^b \frac{1}{p(t)u^2(t)}\,dt = \infty.
$$

Next let v be a solution of $Lx = 0$ such that u and v are linearly independent. Pick $d \in [c, b)$ so that $v(t) \neq 0$ on $[d, b)$. Then consider

$$
\begin{aligned}
\left[\frac{u(t)}{v(t)}\right]' &= \frac{v(t)u'(t) - u(t)v'(t)}{v^2(t)} \\
&= \frac{w[v(t), u(t)]}{v^2(t)} \\
&= \frac{D}{p(t)v^2(t)},
\end{aligned}
$$

where D is a nonzero constant. Integrating from d to t, we get that

$$
\frac{u(t)}{v(t)} - \frac{u(d)}{v(d)} = D \int_d^t \frac{1}{p(s)v^2(s)}\,ds.
$$

Letting $t \to b-$, we get that

$$
\int_d^b \frac{1}{p(s)v^2(s)}\,ds < \infty.
$$

Finally, let u be as before and let v be a second linearly independent solution. Pick $t_0 \in [a, b)$ so that $v(t) \neq 0$ on $[t_0, b)$. The expression

$$
\frac{p(t)v'(t)}{v(t)}
$$

is the same if we replace $v(t)$ by $-v(t)$ so, without loss of generality, we can assume $v(t) > 0$ on $[t_0, b)$. For $t \in [t_0, b)$, consider

$$\frac{p(t)v'(t)}{v(t)} - \frac{p(t)u'(t)}{u(t)} = \frac{p(t)w[u(t), v(t)]}{u(t)v(t)}$$

$$= \frac{H}{u(t)v(t)},$$

where H is a constant. It remains to show that $H > 0$. To see this note that

$$\left[\frac{v(t)}{u(t)}\right]' = \frac{u(t)v'(t) - v(t)u'(t)}{u^2(t)}$$

$$= \frac{w[u(t), v(t)]}{u^2(t)}$$

$$= \frac{H}{p(t)u^2(t)},$$

for $t \in [t_0, b)$. Since

$$\lim_{t \to b-} \frac{v(t)}{u(t)} = \infty,$$

it follows that $H > 0$. The proof of the last statement in the statement of this Theorem is Exercise 5.36.   □

Example 5.60 Find a recessive and dominant solution of

$$x'' - 3x' + 2x = 0$$

at ∞ and show directly that the conclusions of Theorem 5.59 concerning these two solutions are true.

The self-adjoint form of this equation is

$$\left(e^{-3t}x'\right)' + 2e^{-3t}x = 0,$$

so $p(t) = e^{-3t}$ and $q(t) = 2e^{-3t}$. Two solutions of this differential equation are e^t and e^{2t}. It follows that our given differential equation is disconjugate on $\mathbb{R}$, but we want to show directly that if we take $u(t) = e^t$ and $v(t) = e^{2t}$ then the conclusions of Theorem 5.59 concerning these two solutions are true. First note that

$$\lim_{t \to \infty} \frac{u(t)}{v(t)} = \lim_{t \to \infty} \frac{e^t}{e^{2t}} = \lim_{t \to \infty} \frac{1}{e^t} = 0.$$

Also,

$$\int_0^\infty \frac{1}{p(t)u^2(t)} dt = \int_0^\infty e^t \, dt = \infty.$$

In addition,

$$\int_0^\infty \frac{1}{p(t)v^2(t)} dt = \int_0^\infty e^{-t} \, dt < \infty.$$

Finally, consider

$$\frac{p(t)v'(t)}{v(t)} = 2e^{-3t} > e^{-3t} = \frac{p(t)u'(t)}{u(t)},$$

for $t \in \mathbb{R}$.

$\triangle$

Example 5.61 Find a recessive and dominant solution of

$$x'' + x = 0$$

at π and show directly that the conclusions of Theorem 5.59 concerning these two solutions are true.

Let $u(t) := \sin t$ and $v(t) := \cos t$; then

$$\lim_{t \to \pi-} \frac{u(t)}{v(t)} = \lim_{t \to \pi-} \tan t = 0.$$

Also,

$$\int_{\frac{\pi}{2}}^{\pi} \frac{1}{p(t)u^2(t)} dt = \int_{\frac{\pi}{2}}^{\pi} \csc^2 t \, dt = \infty.$$

In addition,

$$\int_{\frac{3\pi}{4}}^{\pi} \frac{1}{p(t)v^2(t)} dt = \int_{\frac{3\pi}{4}}^{\pi} \sec^2 t \, dt < \infty.$$

Finally, consider

$$\frac{p(t)v'(t)}{v(t)} = -\tan t > \cot t = \frac{p(t)u'(t)}{u(t)},$$

for $t \in (\frac{\pi}{2}, \pi)$.

$\triangle$

Theorem 5.62 (Mammana Factorization) *Assume* $Lx = 0$ *has a solution* u *with* $u(t) \neq 0$ *on* $J \subset I$. *Then if* $x \in \mathbb{D}(J)$,

$$Lx(t) = \left[\frac{d}{dt} + \rho(t) \right] p(t) \left[\frac{d}{dt} - \rho(t) \right] x(t),$$

for $t \in J$, *where*

$$\rho(t) := \frac{u'(t)}{u(t)},$$

for $t \in J$.

Proof Let $x \in \mathbb{D}(J)$ and consider

$$\left[\frac{d}{dt} - \rho(t) \right] x(t) = \left[\frac{d}{dt} - \frac{u'(t)}{u(t)} \right] x(t)$$

$$= x'(t) - \frac{u'(t)x(t)}{u(t)},$$

for $t \in J$. Hence

$$\left[\frac{d}{dt} + \rho(t)\right] p(t) \left[\frac{d}{dt} - \rho(t)\right] x(t)$$

$$= \left[\frac{d}{dt} + \rho(t)\right] \left[p(t)x'(t) - p(t)u'(t)\frac{x(t)}{u(t)}\right]$$

$$= [p(t)x'(t)]' - [p(t)u'(t)]'\frac{x(t)}{u(t)} - p(t)u'(t)\frac{u(t)x'(t) - x(t)u'(t)}{u^2(t)}$$

$$+ \ p(t)x'(t)\frac{u'(t)}{u(t)} - p(t)u'(t)\frac{x(t)u'(t)}{u^2(t)}$$

$$= Lx(t),$$

for $t \in J$. □

Example 5.63 Find a Mammana factorization of the equation $Lx = x'' + x = 0$ on $I = (0, \pi)$.

Here $u(t) = \sin t$ is a positive solution on $I = (0, \pi)$. By Theorem 5.62 the Mammana factorization is

$$Lx(t) = \left[\frac{d}{dt} + \cot t\right] \left[\frac{d}{dt} - \cot t\right] x(t) = 0,$$

for $t \in (0, \pi)$. △

5.7 The Riccati Equation

Assume throughout this section that p, q are continuous on an interval I and that $p(t) > 0$ on I. We define the Riccati operator $R : C^1(I) \to C(I)$ by

$$Rz(t) = z'(t) + q(t) + \frac{1}{p(t)}z^2(t),$$

for $t \in I$. The nonlinear (see Exercise 5.38) first-order differential equation

$$Rz = z' + q(t) + \frac{1}{p(t)}z^2 = 0$$

is called the *Riccati differential equation*. This Riccati equation can be written in the form $z' = f(t, z)$, where $f(t, z) := -q(t) - \frac{z^2}{p(t)}$. Since $f(t, z)$ and $f_z(t, z) = -\frac{2z}{p(t)}$ are continuous on $I \times \mathbb{R}$, we have by Theorem 1.3 that solutions of IVPs in $I \times \mathbb{R}$ for the Riccati equation $Rz = 0$ exist, are unique, and have maximal intervals of existence. The following example shows that solutions of the Riccati equation do not always exist on the whole interval I. Recall that by Theorem 5.4 solutions of the self-adjoint differential equation $Lx = 0$ always exist on the whole interval I.

Example 5.64 The solution of the IVP

$$Rz = z' + 1 + z^2 = 0, \quad z(0) = 0$$

is given by

$$z(t) = -\tan t,$$

for $t \in (-\frac{\pi}{2}, \frac{\pi}{2})$. In this example $p(t) = q(t) = 1$ are continuous on $\mathbb{R}$ and $p(t) > 0$ on $\mathbb{R}$ so $I = \mathbb{R}$, but the solution of the preceding IVP only exists on $(-\frac{\pi}{2}, \frac{\pi}{2})$. $\triangle$

The following theorem gives a relationship between the self-adjoint operator L and the Riccati operator R.

Theorem 5.65 (Factorization Theorem) *Assume $x \in \mathbb{D}(J)$ and $x(t) \neq 0$ on $J \subset I$; then if*

$$z(t) := \frac{p(t)x'(t)}{x(t)}, \quad t \in J,$$

then

$$Lx(t) = x(t)Rz(t), \quad t \in J.$$

Proof Assume $x \in \mathbb{D}(J)$, $x(t) \neq 0$ on J, and make the *Riccati substitution*

$$z(t) := \frac{p(t)x'(t)}{x(t)}, \quad t \in J.$$

Then

$$
\begin{aligned}
x(t)Rz(t) &= x(t)\left[z'(t) + q(t) + \frac{z^2(t)}{p(t)}\right] \\
&= x(t)z'(t) + q(t)x(t) + \frac{z^2(t)x(t)}{p(t)} \\
&= x(t)\left(\frac{p(t)x'(t)}{x(t)}\right)' + q(t)x(t) + \frac{z^2(t)x(t)}{p(t)} \\
&= x(t)\frac{x(t)\left[p(t)x'(t)\right]' - p(t)x'(t)x'(t)}{x^2(t)} + q(t)x(t) + \frac{x(t)z^2(t)}{p(t)} \\
&= (p(t)x'(t))' - \frac{x(t)z^2(t)}{p(t)} + q(t)x(t) + \frac{x(t)z^2(t)}{p(t)} \\
&= (p(t)x'(t))' + q(t)x(t) \\
&= Lx(t),
\end{aligned}
$$

for $t \in J$. $\square$

Theorem 5.66 *Assume $J \subset I$. The self-adjoint differential equation $Lx = 0$ has a solution x without zeros on J iff the Riccati equation $Rz = 0$ has a solution z that exists on J. These two solutions satisfy*

$$z(t) = \frac{p(t)x'(t)}{x(t)}, \quad t \in J.$$

Proof Assume that x is a solution of the self-adjoint differential equation $Lx = 0$ such that $x(t) \neq 0$ on J. Make the Riccati substitution $z(t) := \frac{p(t)x'(t)}{x(t)}$, for $t \in J$; then by Theorem 5.65

$$x(t)Rz(t) = Lx(t) = 0, \quad t \in J.$$

It follows that z is a solution of the Riccati equation $Rz = 0$ on J.

Conversely, assume that the Riccati equation $Rz = 0$ has a solution z on the interval J. Let x be the solution of the IVP

$$x' = \frac{z(t)}{p(t)}x, \quad x(t_0) = 1,$$

where $t_0 \in J$. Then $x(t) > 0$ on J. Furthermore, since

$$z(t) = \frac{p(t)x'(t)}{x(t)},$$

we get from Theorem 5.65 that

$$Lx(t) = x(t)Rz(t) = 0, \quad t \in J,$$

and so x is a solution of $Lx = 0$ without zeros in J. $\qquad\square$

Example 5.67 Solve the Riccati differential equation (DE)

$$z' + \frac{2}{t^4} + t^2 z^2 = 0, \quad t > 0.$$

Here

$$p(t) = \frac{1}{t^2}, \quad q(t) = \frac{2}{t^4}.$$

Hence the corresponding self-adjoint DE is

$$\left(\frac{1}{t^2}x'\right)' + \frac{2}{t^4}x = 0.$$

Expanding this equation out we obtain the Euler–Cauchy DE

$$t^2 x'' - 2tx' + 2x = 0.$$

A general solution to this equation is

$$x(t) = At + Bt^2.$$

Therefore,

$$z(t) = \frac{p(t)x'(t)}{x(t)}$$

$$= \frac{\frac{A}{t^2} + \frac{2B}{t}}{At + Bt^2}.$$

When $B = 0$ we get the solution

$$z(t) = \frac{1}{t^3}.$$

When $B \neq 0$ we can divide the numerator and denominator by $\frac{B}{t^2}$ to get that

$$z(t) = \frac{C + 2t}{Ct^3 + t^4},$$

where C is a constant is a solution. $\triangle$

Next we state and prove the most famous oscillation result for the self-adjoint differential equation $Lx = 0$.

Theorem 5.68 (Fite-Wintner Theorem) *Assume $I = [a, \infty)$. If*

$$\int_a^\infty \frac{1}{p(t)} dt = \int_a^\infty q(t) \, dt = \infty,$$

then the self-adjoint differential equation $Lx = 0$ is oscillatory on $[a, \infty)$.

Proof Assume that the differential equation $Lx = 0$ is nonoscillatory on $[a, \infty)$. Then by Theorem 5.59 there is a dominant solution v of $Lx = 0$ and a number $T \in [a, \infty)$ such that $v(t) > 0$ on $[T, \infty)$ and

$$\int_T^\infty \frac{1}{p(t)v^2(t)} dt < \infty. \tag{5.26}$$

If we make the Riccati substitution

$$z(t) := \frac{p(t)v'(t)}{v(t)}, \quad t \in [T, \infty),$$

then by Theorem 5.66, z is a solution of the Riccati equation $Rz = 0$ on $[T, \infty)$. Hence

$$z'(t) = -q(t) - \frac{z^2(t)}{p(t)} \leq -q(t), \quad t \in [T, \infty).$$

Integrating from T to t, we get

$$z(t) - z(T) \leq -\int_T^t q(s) \, ds.$$

It follows that

$$\lim_{t \to \infty} z(t) = -\infty.$$

Pick $T_1 \geq T$ so that

$$z(t) = \frac{p(t)v'(t)}{v(t)} < 0, \quad t \in [T_1, \infty).$$

It follows that

$$v'(t) < 0, \quad t \in [T_1, \infty),$$

and therefore v is decreasing on $[T_1, \infty)$. Then we get that for $t \geq T_1$,

$$\int_{T_1}^t \frac{1}{p(s)v^2(s)} ds \geq \frac{1}{v^2(T_1)} \int_{T_1}^t \frac{1}{p(s)} ds,$$

which implies that

$$\int_{T_1}^\infty \frac{1}{p(t)v^2(t)} dt = \infty,$$

which contradicts (5.26). □

Example 5.69 Since $\int_1^\infty \frac{1}{t}\,dt = \infty$, we get from the Fite-Wintner theorem (Theorem 5.68) that for any positive constant α the differential equation

$$(tx')' + \frac{\alpha}{t}x = 0$$

is oscillatory on $[1, \infty)$. △

The following lemma is important in the calculus of variations, which we study briefly in Section 5.8.

Lemma 5.70 (Completing the Square Lemma) *Assume $[a, b] \subset I$ and that $\eta : [a, b] \to \mathbb{R}$ is continuous and assume η' is piecewise continuous on $[a, b]$. If z is a solution of the Riccati equation $Rz = 0$ on $[a, b]$, then*

$$
[z(t)\eta^2(t)]' = \{p(t)[\eta'(t)]^2 - q(t)\eta^2(t)\}
$$
$$
- \left[\sqrt{p(t)}\,\eta'(t) - \frac{1}{\sqrt{p(t)}}\eta(t)z(t) \right]^2,
$$

for those $t \in [a, b]$ where $\eta'(t)$ exists.

Proof Assume that η is as in the statement of this theorem and z is a solution of the Riccati equation $Rz = 0$ on $[a, b] \subset I$, and consider

$$
\begin{aligned}
[z(t)\eta^2(t)]' &= z'(t)\eta^2(t) + 2z(t)\eta(t)\eta'(t) \\
&= [-q(t) - \frac{1}{p(t)}z^2(t)]\eta^2(t) + 2z(t)\eta(t)\eta'(t) \\
&= \{p(t)[\eta'(t)]^2 - q(t)\eta^2(t)\} \\
&\quad - \left\{ p(t)[\eta'(t)]^2 - 2z(t)\eta(t)\eta'(t) + \frac{1}{p(t)}\eta^2(t)z^2(t) \right\} \\
&= \{p(t)[\eta'(t)]^2 - q(t)\eta^2(t)\} \\
&\quad - \left[\sqrt{p(t)}\,\eta'(t) - \frac{1}{\sqrt{p(t)}}\eta(t)z(t) \right]^2,
\end{aligned}
$$

for those $t \in [a, b]$ where $\eta'(t)$ exists. □

We next define a quadratic functional Q that is very important in the calculus of variations.

Definition 5.71 First we let A be the set of all functions $\eta : [a, b] \to \mathbb{R}$ such that η is continuous, η' is piecewise continuous on $[a, b]$, and $\eta(a) = \eta(b) = 0$. Then we define the quadratic functional $Q : A \to \mathbb{R}$ by

$$
Q[\eta] = \int_a^b \{p(t)[\eta'(t)]^2 - q(t)\eta^2(t)\}\,dt.
$$

We call A the set of *admissible functions*.

Definition 5.72 We say that Q is positive definite on A provided $Q\eta \geq 0$ for all $\eta \in A$ and $Q\eta = 0$ iff $\eta = 0$.

Theorem 5.73 *Let* $[a, b] \subset I$. *Then* $Lx = 0$ *is disconjugate on* $[a, b]$ *iff the quadratic functional* Q *is positive definite on* A.

Proof Assume $Lx = 0$ is disconjugate on $[a, b]$. Then, by Theorem 5.46, $Lx = 0$ has a positive solution on $[a, b]$. It then follows from Theorem 5.66 that the Riccati equation $Rz = 0$ has a solution z that exists on $[a, b]$. Let $\eta \in A$; then by the completing the square lemma (Lemma 5.70)

$$[z(t)\eta^2(t)]' = \{p(t)[\eta'(t)]^2 - q(t)\eta^2(t)\}$$
$$- \left[\sqrt{p(t)}\eta'(t) - \frac{1}{\sqrt{p(t)}}\eta(t)z(t)\right]^2,$$

for those $t \in [a, b]$ where $\eta'(t)$ exists. Integrating from a to b and simplifying, we get

$$Q\eta = [z(t)\eta^2(t)]_a^b + \int_a^b \left[\sqrt{p(t)}\eta'(t) - \frac{1}{\sqrt{p(t)}}\eta(t)z(t)\right]^2 dt$$
$$= \int_a^b \left[\sqrt{p(t)}\eta'(t) - \frac{1}{\sqrt{p(t)}}\eta(t)z(t)\right]^2 dt$$
$$\geq 0.$$

Also note that $Q\eta = 0$ only if

$$\eta'(t) = \frac{z(t)}{p(t)}\eta(t),$$

for $t \in [a, b]$. Since we also know that $\eta(a) = 0$, we get $Q\eta = 0$ only if $\eta = 0$. Hence we have that Q is positive definite on A.

Conversely, assume Q is positive definite on A. We will show that $Lx = 0$ is disconjugate on $[a, b]$. Assume not; then there is a nontrivial solution x of $Lx = 0$ with

$$x(c) = 0 = x(d),$$

where $a \leq c < d \leq b$. Define the function η by

$$\eta(t) = \begin{cases} 0, & \text{if } t \in [a, c] \\ x(t), & \text{if } t \in [c, d] \\ 0, & \text{if } t \in [d, b]. \end{cases}$$

Since $\eta \in A$ and $\eta \neq 0$ (the zero function), we get that

$$Q[\eta] > 0.$$

But, using integration by parts,

$$
\begin{aligned}
Q[\eta] &= \int_a^b \{p(t)[\eta'(t)]^2 - q(t)\eta^2(t)\} \, dt \\
&= \int_c^d \{p(t)[\eta'(t)]^2 - q(t)\eta^2(t)\} \, dt \\
&= [p(t)\eta(t)\eta'(t)]_c^d - \int_c^d [p(t)\eta'(t)]'\eta(t) \, dt - \int_c^d q(t)\eta^2(t) \, dt \\
&= -\int_c^d Lx(t)x(t) \, dt \\
&= 0,
\end{aligned}
$$

which is a contradiction. $\square$

Corollary 5.74 *Assume (5.19) is disconjugate on $J \subset I$ and*

$$q_1(t) \geq q_2(t) \quad and \quad 0 < p_1(t) \leq p_2(t),$$

for $t \in J$; then (5.20) is disconjugate on J.

Proof It suffices to show that (5.20) is disconjugate on any closed subinterval $[a, b] \subset J$. Define $Q_i : A \to \mathbb{R}$ by

$$
Q_i[\eta] := \int_a^b \{p_i(t)[\eta'(t)]^2 - q_i(t)\eta^2(t)\} \, dt,
$$

for $i = 1, 2$. Assume (5.19) is disconjugate on $J \subset I$. Then, by Theorem 5.73, Q_1 is positive definite on A. Note that

$$
\begin{aligned}
Q_2[\eta] &= \int_a^b \{p_2(t)[\eta'(t)]^2 - q_2(t)\eta^2(t)\} \, dt \\
&\geq \int_a^b \{p_1(t)[\eta'(t)]^2 - q_1(t)\eta^2(t)\} \, dt \\
&= Q_1[\eta],
\end{aligned}
$$

for all $\eta \in A$. It follows that Q_2 is positive definite on A and so by Theorem 5.73 the self-adjoint equation (5.20) is disconjugate on $[a, b]$. $\square$

Similar to the proof of Corollary 5.74, we can prove the following Corollary (see Exercise 5.40).

Corollary 5.75 *Assume (5.19) and (5.20) are disconjugate on $J \subset I$ and*

$$p(t) = \lambda_1 p_1(t) + \lambda_2 p_2(t) \quad and \quad q(t) = \lambda_1 q_1(t) + \lambda_2 q_2(t),$$

for $t \in J$, where $\lambda_1, \lambda_2 \geq 0$, not both zero; then $Lx = 0$ is disconjugate on J.

Definition 5.76 Assume $Lx = 0$ is nonoscillatory on $[a, b) \subset I$, where $-\infty < a < b \le \infty$. We say z_m is the *minimum solution* of the corresponding Riccati differential equation $Rz = 0$ for $t < b$, sufficiently close, provided z_m is a solution of the Riccati equation $Rz = 0$ for $t < b$, sufficiently close, and if z is any solution of the Riccati equation $Rz = 0$ for $t < b$, sufficiently close, then

$$z_m(t) \le z(t),$$

for all $t < b$, sufficiently close.

Theorem 5.77 *Assume $Lx = 0$ is nonoscillatory on $[a, b) \subset I$, where $-\infty < a < b \le \infty$; then the minimum solution z_m of the corresponding Riccati differential equation $Rz = 0$ for $t < b$, sufficiently close, is given by*

$$z_m(t) := \frac{p(t)u'(t)}{u(t)},$$

where u is a recessive solution of $Lx = 0$ at b. Assume that the solution z_m exists on $[t_0, b)$; then for any $t_1 \in [t_0, b)$ the solution z of the IVP

$$Rz = 0, \quad z(t_1) = z_1,$$

where $z_1 < z_m(t_1)$ has right maximal interval of existence $[t_1, \omega)$, where $\omega < b$ and

$$\lim_{t \to \omega-} z(t) = -\infty.$$

Proof Recall that by Theorem 5.66 if $J \subset I$, then the self-adjoint differential equation $Lx = 0$ has a solution x without zeros on J iff the Riccati equation $Rz = 0$ has a solution z that exists on J and these two solutions are related by

$$z(t) = \frac{p(t)x'(t)}{x(t)}, \quad t \in J.$$

By Theorem 5.59 the differential equation $Lx = 0$ has a recessive solution u at b and for any second linearly independent solution v it follows that

$$z_m(t) := \frac{p(t)u'(t)}{u(t)} < \frac{p(t)v'(t)}{v(t)},$$

for $t < b$ sufficiently close. Note that if x is a solution of $Lx = 0$ that is linearly dependent of u, then $x = ku$, where $k \ne 0$ and

$$\frac{p(t)x'(t)}{x(t)} = \frac{p(t)ku'(t)}{ku(t)} = \frac{p(t)u'(t)}{u(t)} = z_m(t),$$

for $t < b$ sufficiently close (this also shows that no matter what recessive solution you pick at b the z_m is the same and hence z_m is well defined).

Finally, assume z_m is a solution on $[t_0, b)$ and assume z is the solution of the IVP

$$Rz = 0, \quad z(t_1) = z_1,$$

where $z_1 < z_m(t_1)$. From the uniqueness of solutions of IVPs, $z(t) < z_m(t)$ on the right maximal interval of existence $[t_1, \omega)$ of the solution z. It follows that $\omega < b$ and, using Theorem 1.3, we get that

$$\lim_{t \to \omega-} z(t) = -\infty. \quad \square$$

Example 5.78 Find the minimum solution z_m of the Riccati differential equation

$$Rz = z' + \frac{2}{t^4} + t^2 z^2 = 0$$

that exists for all sufficiently large t.

From Example 5.67, $u(t) = t$ is a recessive solution of the corresponding self-adjoint differential equation $Lx = 0$ at ∞. It follows from Theorem 5.77 that the minimum solution of the Riccati equation for all sufficiently large t is

$$z_m(t) = \frac{p(t)u'(t)}{u(t)} = \frac{1}{t^3}.$$

$\triangle$

Example 5.79 Find the minimum solution z_m of the Riccati differential equation

$$Rz = z' + 1 + z^2 = 0$$

that exists for all $t < \pi$ sufficiently close.

The corresponding self-adjoint differential equation is $x'' + x = 0$. A recessive solution of this equation at π is $u(t) = \sin t$. It follows from Theorem 5.77 that the minimum solution at π is

$$z_m(t) = \frac{p(t)u'(t)}{u(t)} = \cot t,$$

for $t \in (0, \pi)$. It can be shown that a general solution of this Riccati equation is given by

$$z(t) = \cot(t - \alpha),$$

where α is an arbitrary constant. Let $t_1 \in (0, \pi)$ and let z be the solution of the IVP

$$Rz = 0, \quad z(t_1) = z_1.$$

Note that if the constant

$$z_1 < z_m(t_1) = \cot(t_1),$$

then there is an $\alpha_1 \in (-\frac{\pi}{2}, 0)$ such that

$$z(t) = \cot(t - \alpha_1),$$

which has right maximal interval of existence $[t_1, \alpha_1 + \pi)$. Note that $\alpha_1 + \pi < \pi$ and

$$\lim_{t \to (\alpha_1 + \pi)-} z(t) = -\infty.$$

On the other hand, if $z_1 \geq z_m(t_1) = \cot(t_1)$, then there is an $\alpha_2 \in [0, \frac{\pi}{2})$ such that

$$z(t) = \cot(t - \alpha_2) \geq z_m(t),$$

for $t < \pi$ sufficiently close. $\triangle$

In the next theorem we summarize some of the important results (with one improvement) in this chapter. Ahlbrandt and Hooker [**1**] called this theorem the *Reid roundabout theorem* to honor W. T. Reid's work [**43**], [**42**] in this area. There have been numerous research papers written concerning various versions of this theorem. See, for example, Ahlbrandt and Peterson [**2**] for a general discrete Reid roundabout theorem.

Theorem 5.80 (Reid Roundabout Theorem) *The following are equivalent:*

   (i) $Lx = 0$ *is disconjugate on* $[a, b]$.
  (ii) $Lx = 0$ *has a positive solution on* $[a, b]$.
 (iii) Q *is positive definite on* A.
 (iv) *The Riccati differential inequality* $Rw \leq 0$ *has a solution that exists on the whole interval* $[a, b]$.

Proof By Theorem 5.46, (i) and (ii) are equivalent. By Theorem 5.73, (i) and (iii) are equivalent. By Theorem 5.66, we get that (ii) implies (iv). It remains to prove that (iv) implies (i). To this end, assume that the Riccati inequality $Rw \leq 0$ has a solution w that exists on the whole interval $[a, b]$. Let

$$h(t) := Rw(t), \quad t \in [a, b];$$

then $h(t) \leq 0$, for $t \in [a, b]$ and

$$w'(t) + (q(t) - h(t)) + \frac{w^2(t)}{p(t)} = 0,$$

for $t \in [a, b]$, but then from Theorem 5.66 we have that

$$(p(t)x')' + (q(t) - h(t))\, x = 0$$

has a solution without zeros in $[a, b]$ and hence is disconjugate on $[a, b]$. But since

$$q(t) - h(t) \geq q(t),$$

for $t \in [a, b]$, we get from the Sturm comparison theorem (Theorem 5.50) that $Lx = 0$ is disconjugate on $[a, b]$. $\square$

We end this section with another oscillation theorem for $Lx = 0$.

Theorem 5.81 *Assume* $I = [a, \infty)$. *If* $\int_a^\infty \frac{1}{p(t)} dt = \infty$ *and there is a* $t_0 \geq a$ *and a* $u \in C^1[t_0, \infty)$ *such that* $u(t) > 0$ *on* $[t_0, \infty)$ *and*

$$\int_{t_0}^\infty \left[q(t)u^2(t) - p(t)\, (u'(t))^2 \right] dt = \infty;$$

then $Lx = 0$ *is oscillatory on* $[a, \infty)$.

Proof We prove this theorem by contradiction. So assume $Lx = 0$ is nonoscillatory on $[a, \infty)$. By Theorem 5.59, $Lx = 0$ has a dominant solution v at ∞ such that for $t_1 \geq a$, sufficiently large,

$$\int_{t_1}^{\infty} \frac{1}{p(t)v^2(t)} dt < \infty,$$

and we can assume that $v(t) > 0$ on $[t_1, \infty)$. Let t_0 and u be as in the statement of this theorem. Let $T = \max\{t_0, t_1\}$; then let

$$z(t) := \frac{p(t)v'(t)}{v(t)}, \quad t \geq T.$$

Then by Lemma 5.70, we have for $t \geq T$

$$\begin{aligned}
\left\{z(t)u^2(t)\right\}' &= p(t)\left(u'(t)\right)^2 - q(t)u^2(t) \\
&\quad - \left\{\sqrt{p(t)}u'(t) - \frac{u(t)z(t)}{\sqrt{p(t)}}\right\}^2 \\
&\leq p(t)\left(u'(t)\right)^2 - q(t)u^2(t),
\end{aligned}$$

for $t \geq T$. Integrating from T to t, we get

$$z(t)u^2(t) \leq z(T)u^2(T) - \int_T^t \left[q(t)u^2(t) - p(t)\left(u'(t)\right)^2\right] dt,$$

which implies that

$$\lim_{t \to \infty} z(t)u^2(t) = -\infty.$$

But then there is a $T_1 \geq T$ such that

$$z(t) = \frac{p(t)v'(t)}{v(t)} < 0,$$

for $t \geq T_1$. This implies that $v'(t) < 0$ for $t \geq T_1$ and hence v is decreasing for $t \geq T_1$. But then

$$\begin{aligned}
\int_{T_1}^{\infty} \frac{1}{p(s)} ds &= v^2(T_1) \int_{T_1}^{\infty} \frac{1}{p(s)v^2(T_1)} ds \\
&\leq v^2(T_1) \int_{T_1}^{\infty} \frac{1}{p(s)v^2(s)} ds < \infty,
\end{aligned}$$

which is a contradiction.   □

Example 5.82 Show that if $(a > 0)$

$$\int_a^{\infty} t^{\alpha} q(t) \, dt = \infty,$$

where $\alpha < 1$, then $x'' + q(t)x = 0$ is oscillatory on $[a, \infty)$.

We show that this follows from Theorem 5.81. First note that

$$\int_a^{\infty} \frac{1}{p(t)} dt = \int_a^{\infty} 1 dt = \infty.$$

Now let
$$u(t) := t^{\frac{\alpha}{2}},$$
and consider
$$\int_a^\infty \{q(t)u^2(t) - p(t)[u'(t)]^2\}\, dt = \int_a^\infty \{t^\alpha q(t) - \frac{\alpha^2}{4} t^{\alpha-2}\}\, dt = \infty,$$
since $\alpha < 1$ implies
$$\int_a^\infty t^{\alpha-2}\, dt < \infty.$$
Hence $x'' + q(t)x = 0$ is oscillatory on $[a, \infty)$ from Theorem 5.81. $\triangle$

5.8 Calculus of Variations

In this section we will introduce the calculus of variations and show how the previous material in this chapter is important in the calculus of variations.

Let's look at the following example for motivation.

Example 5.83 The problem is to find the curve $x = x(t)$ joining the given points (a, x_a) and (b, x_b) such that x is continuous and x' and x'' are piecewise continuous on $[a, b]$ and the length of this curve is a minimum.

Using the formula for the length of a curve, we see that we want to minimize
$$I[x] := \int_a^b \sqrt{1 + [x'(t)]^2}\, dt$$
subject to
$$x(a) = x_a, \quad x(b) = x_b,$$
where x_a, x_b are given constants and x is continuous and x' and x'' are piecewise continuous on $[a, b]$. $\triangle$

We assume throughout this section that $f = f(t, u, v)$ is a given continuous real-valued function on $[a, b] \times \mathbb{R}^2$ such that f has continuous partial derivatives up through the third order with respect to each of its variables on $[a, b] \times \mathbb{R}^2$. The *simplest problem of the calculus of variations* is to extremize
$$I[x] := \int_a^b f(t, x(t), x'(t))\, dt$$
subject to
$$x(a) = x_a, \quad x(b) = x_b,$$
where x_a and x_b are given constants and x is continuous and x' and x'' are piecewise continuous on $[a, b]$. In this section we will let D be the set of all continuous $x : [a, b] \to \mathbb{R}$ such that x' and x'' are piecewise continuous on $[a, b]$, with $x(a) = x_a$, $x(b) = x_b$. We define a norm on D by
$$\|x\| = \sup \{\max\{|x(t)|, |x'(t)|, |x''(t)|\}\},$$
where the sup is over those $t \in [a, b]$, where $x'(t)$ and $x''(t)$ exist. We then define the *set of admissible variations* $\mathcal{A}$ to be the set of all continuous

$\eta : [a, b] \rightarrow \mathbb{R}$ such that η' and η'' are piecewise continuous on $[a, b]$ and $\eta(a) = \eta(b) = 0$. Note that if $x \in D$ and $\eta \in \mathcal{A}$, then $x + \epsilon\eta \in D$, for any number ϵ. Also, $x, y \in D$ implies $\eta := x - y \in \mathcal{A}$.

Definition 5.84 We say that

$$I[x] := \int_a^b f(t, x(t), x'(t))\, dt,$$

subject to $x \in D$ has a local minimum at x_0 provided $x_0 \in D$ and there is a $\delta > 0$ such that if $x \in D$ with $\|x - x_0\| < \delta$, then

$$I[x] \geq I[x_0].$$

The terms *local maximum*, *global minimum*, and *global maximum* are defined in the obvious way. We say I has a *proper global minimum* at x_0 provided $x_0 \in D$ and

$$I[x] \geq I[x_0],$$

for all $x \in D$ and equality holds only if $x = x_0$.

Definition 5.85 Fix $x_0 \in D$ and let the function $h : \mathbb{R} \rightarrow \mathbb{R}$ be defined by

$$h(\epsilon) := I[x_0 + \epsilon\eta],$$

where $\eta \in \mathcal{A}$. Then we define the *first variation* J_1 *of* I *along* x_0 by

$$J_1[\eta] = J_1[\eta; x_0] := h'(0), \quad \eta \in \mathcal{A}.$$

The *second variation* J_2 *of* I *along* x_0 is defined by

$$J_2[\eta] = J_2[\eta; x_0] := h''(0), \quad \eta \in \mathcal{A}.$$

Theorem 5.86 *Let* $x_0 \in D$; *then the first and second variations along* x_0 *are given by*

$$J_1[\eta] = \int_a^b \left\{ f_u(t, x_0(t), x_0'(t)) - \frac{d}{dt}\, [f_v(t, x_0(t), x_0'(t))] \right\} \eta(t)\, dt,$$

and

$$J_2[\eta] = \int_a^b \{p(t)[\eta'(t)]^2 - q(t)\eta^2(t)\}\, dt,$$

respectively, for $\eta \in \mathcal{A}$, *where*

$$p(t) = R(t), \quad q(t) = Q'(t) - P(t), \tag{5.27}$$

where

$$P(t) = f_{uu}(t, x_0(t), x_0'(t)), \quad Q(t) = f_{uv}(t, x_0(t), x_0'(t))$$

and

$$R(t) = f_{vv}(t, x_0(t), x_0'(t)),$$

for those $t \in [a, b]$ *such that* x_0' *and* x_0'' *are continuous.*

Proof Fix $x_0 \in D$, let $\eta \in \mathcal{A}$, and consider the function $h : \mathbb{R} \to \mathbb{R}$ defined by

$$h(\epsilon) := I[x_0 + \epsilon \eta] = \int_a^b f(t, x_0(t) + \epsilon \eta(t), x_0'(t) + \epsilon \eta'(t)) \, dt.$$

Then

$$
\begin{aligned}
h'(\epsilon) &= \int_a^b [f_u(t, x_0(t) + \epsilon \eta(t), x_0'(t) + \epsilon \eta'(t)) \eta(t)] \, dt \\
&\quad + \int_a^b [f_v(t, x_0(t) + \epsilon \eta(t), x_0'(t) + \epsilon \eta'(t)) \eta'(t)] \, dt \\
&= \int_a^b [f_u(t, x_0(t) + \epsilon \eta(t), x_0'(t) + \epsilon \eta'(t)) \eta(t)] \, dt \\
&\quad + [f_v(t, x_0(t) + \epsilon \eta(t), x_0'(t) + \epsilon \eta'(t)) \eta(t)]_a^b \\
&\quad - \int_a^b \frac{d}{dt} [f_v(t, x_0(t) + \epsilon \eta(t), x_0'(t) + \epsilon \eta'(t))] \, \eta(t) \, dt \\
&= \int_a^b f_u(t, x_0(t) + \epsilon \eta(t), x_0'(t) + \epsilon \eta'(t)) \eta(t) \, dt \\
&\quad - \int_a^b \frac{d}{dt} [f_v(t, x_0(t) + \epsilon \eta(t), x_0'(t) + \epsilon \eta'(t))] \, \eta(t) \, dt,
\end{aligned}
$$

where we have integrated by parts. It follows that

$$J_1[\eta] = h'(0) = \int_a^b \left\{ f_u(t, x_0(t), x_0'(t)) - \frac{d}{dt} [f_v(t, x_0(t), x_0'(t))] \right\} \eta(t) \, dt,$$

for $\eta \in \mathcal{A}$. Differentiating $h'(\epsilon)$, we get

$$
\begin{aligned}
h''(\epsilon) &= \int_a^b f_{uu}(t, x_0(t) + \epsilon \eta(t), x_0'(t) + \epsilon \eta'(t)) \eta^2(t) \, dt \\
&\quad + 2 \int_a^b f_{uv}(t, x_0(t) + \epsilon \eta(t), x_0'(t) + \epsilon \eta'(t)) \eta(t) \eta'(t) \, dt \\
&\quad + \int_a^b [f_{vv}(t, x_0(t) + \epsilon \eta(t), x_0'(t) + \epsilon \eta'(t))] [\eta'(t)]^2 \, dt.
\end{aligned}
$$

Hence

$$J_2[\eta] = h''(0) = \int_a^b P(t) \eta^2(t) \, dt + 2 \int_a^b Q(t) \eta(t) \eta'(t) \, dt + \int_a^b R(t) [\eta'(t)]^2 \, dt,$$

where

$$P(t) = f_{uu}(t, x_0(t), x_0'(t)), \quad Q(t) = f_{uv}(t, x_0(t), x_0'(t)),$$

and

$$R(t) = f_{vv}(t, x_0(t), x_0'(t)).$$

But

$$2 \int_a^b Q(t)\eta(t)\eta'(t)\, dt \;=\; \int_a^b Q(t)[\eta^2(t)]'\, dt$$

$$=\; [Q(t)\eta^2(t)]_a^b - \int_a^b Q'(t)\eta^2(t)\, dt$$

$$=\; -\int_a^b Q'(t)\eta^2(t)\, dt,$$

where we have integrated by parts and used $\eta(a) = \eta(b) = 0$. It follows that

$$J_2[\eta] = \int_a^b \{p(t)(\eta'(t))^2 - q(t)\eta^2(t)\}\, dt,$$

where

$$p(t) = R(t), \quad q(t) = Q'(t) - P(t),$$

for those $t \in [a, b]$ where x_0' and x_0'' are continuous. $\qquad \square$

Lemma 5.87 (Fundamental Lemma of the Calculus of Variations) *Assume that $g : [a, b] \to \mathbb{R}$ is piecewise continuous and*

$$\int_a^b g(t)\eta(t)\, dt = 0$$

for all $\eta \in \mathcal{A}$; then $g(t) = 0$ for those $t \in [a, b]$ where g is continuous.

Proof Assume that the conclusion of this theorem does not hold. Then there is a $t_0 \in (a, b)$ such that g is continuous at t_0 and $g(t_0) \neq 0$. We will only consider the case where $g(t_0) > 0$ as the other case (see Exercise 5.46) is similar. Since g is continuous at t_0, there is a $0 < \delta < \min\{t_0 - a, b - t_0\}$ such that $g(t) > 0$ on $[t_0 - \delta, t_0 + \delta]$. Define η by

$$\eta(t) = \begin{cases} 0, & \text{if } t \in [a, t_0 - \delta) \cup (t_0 + \delta, b] \\ (t - t_0 + \delta)^2 (t - t_0 - \delta)^2, & \text{if } t \in [t_0 - \delta, t_0 + \delta]. \end{cases}$$

Then $\eta \in \mathcal{A}$ and

$$\int_a^b g(t)\eta(t)\, dt = \int_{t_0 - \delta}^{t_0 + \delta} (t - t_0 + \delta)^2 (t - t_0 - \delta)^2 g(t)\, dt > 0,$$

which is a contradiction. $\qquad \square$

Theorem 5.88 (Euler-Lagrange Equation) *Assume*

$$I[x] := \int_a^b f(t, x(t), x'(t))\, dt,$$

subject to $x \in D$ has a local extremum at x_0. Then x_0 satisfies the Euler-Lagrange equation

$$f_u(t, x(t), x'(t)) - \frac{d}{dt}[f_v(t, x(t), x'(t))] = 0, \tag{5.28}$$

for those $t \in [a, b]$ where x_0' and x_0'' are continuous.

Proof Assume

$$I[x] := \int_a^b f(t, x(t), x'(t)) \, dt,$$

subject to $x \in D$ has a local extremum at x_0. This implies that for each $\eta \in \mathcal{A}$, h has a local extremum at $\epsilon = 0$ and hence

$$h'(0) = 0,$$

for each $\eta \in \mathcal{A}$. Hence

$$J_1[\eta] = \int_a^b \left\{ f_u(t, x_0(t), x_0'(t)) - \frac{d}{dt} \left[f_v(t, x_0(t), x_0'(t)) \right] \right\} \eta(t) \, dt = 0,$$

for all $\eta \in \mathcal{A}$. It follows from Lemma 5.87 that

$$f_u(t, x_0(t), x_0'(t)) - \frac{d}{dt} \left[f_v(t, x_0(t), x_0'(t)) \right] = 0,$$

for those $t \in [a, b]$ where x_0' and x_0'' are continuous. $\square$

Example 5.89 Assume that we are given that

$$I[x] := \int_1^e \left\{ t \, [x'(t)]^2 - \frac{1}{t} x^2(t) \right\} dt,$$

subject to $x \in D$, where the boundary conditions are

$$x(1) = 0, \quad x(e) = 1,$$

has a global minimum at x_0 and $x_0 \in C^2[1, e]$. Find x_0.

In this example

$$f(t, u, v) = tv^2 - \frac{u^2}{t}$$

and so

$$f_u(t, u, v) = -2\frac{u}{t}, \quad f_v(t, u, v) = 2tv.$$

It follows that the Euler-Lagrange equation is the self-adjoint equation

$$(tx')' + \frac{1}{t} x = 0.$$

It follows that x_0 is the solution of the BVP

$$(tx')' + \frac{1}{t} x = 0,$$

$$x(1) = 0, \quad x(e) = 1.$$

Solving this BVP, we get

$$x_0(t) = \frac{\sin(\log t)}{\sin 1}.$$

$\triangle$

Definition 5.90 The Euler-Lagrange equation for the second variation J_2 along x_0 is called the *Jacobi equation* for the simplest problem of the calculus of variations.

The following theorem shows that the self-adjoint equation is very important in the calculus of variations.

Theorem 5.91 *The Jacobi equation for the simplest problem of the calculus of variations is a self-adjoint second-order differential equation $Lx = 0$.*

Proof The second variation is given by

$$J_2[\eta] = \int_a^b \{p(t)(\eta'(t))^2 - q(t)\eta^2(t)\}\, dt.$$

Let F be defined by

$$F(t, u, v) := p(t)v^2 - q(t)u^2.$$

Then

$$F_u(t, u, v) = -2q(t)u, \quad F_v(t, u, v) = 2p(t)v.$$

It follows that the Jacobi equation is the self-adjoint equation $Lx = 0$, where p and q are given by (5.27). $\qquad\square$

Theorem 5.92 (Legendre's Necessary Condition) *Assume the simplest problem of the calculus of variations has a local extremum at x_0. In the local minimum case*

$$f_{vv}(t, x_0(t), x_0'(t)) \geq 0,$$

for those $t \in [a, b]$ such that x_0' and x_0'' are continuous. In the local maximum case

$$f_{vv}(t, x_0(t), x_0'(t)) \leq 0,$$

for those $t \in [a, b]$ such that x_0' and x_0'' are continuous.

Proof Assume

$$I[x] := \int_a^b f(t, x(t), x'(t))\, dt,$$

subject to $x \in D$ has a local minimum at x_0. Then if $\eta \in \mathcal{A}$ and $h : \mathbb{R} \to \mathbb{R}$ is defined as usual by

$$h(\epsilon) := I[x_0 + \epsilon\eta],$$

it follows that h has a local minimum at $\epsilon = 0$. This implies that

$$h'(0) = 0, \quad \text{and} \quad h''(0) \geq 0,$$

for each $\eta \in \mathcal{A}$. Since $h''(0) = J_2[\eta]$, we have by Theorem 5.86 that

$$J_2[\eta] = \int_a^b \{p(t)[\eta'(t)]^2 - q(t)\eta^2(t)\}\, dt \geq 0,$$

for all $\eta \in \mathcal{A}$. Now fix an s in (a, b), where x_0' and x_0'' are continuous. Let $\epsilon > 0$ be a constant so that $a \le s - \epsilon < s + \epsilon \le b$, and p is continuous on $[s - \epsilon, s + \epsilon]$. Next define η_ϵ by

$$\eta_\epsilon(t) = \begin{cases} 0, & a \le t \le s - \epsilon \\ t - s + \epsilon, & s - \epsilon \le t \le s \\ -t + s + \epsilon, & s \le t \le s + \epsilon \\ 0, & s + \epsilon \le t \le b. \end{cases}$$

Since $\eta_\epsilon \in \mathcal{A}$,

$$J_2[\eta_\epsilon] = \int_a^b \{p(t)[\eta_\epsilon'(t)]^2 - q(t)\eta_\epsilon^2(t)\} \, dt \ge 0.$$

It follows that

$$\int_{s-\epsilon}^{s+\epsilon} p(t) \, dt \ge \int_{s-\epsilon}^{s+\epsilon} q(t)\eta_\epsilon^2(t) \, dt.$$

Let $M > 0$ be a constant such that $q(t) \ge -M$ for all those $t \in [a, b]$ such that x_0' and x_0'' are continuous, then, using the mean value theorem from calculus, we get that

$$p(\xi_\epsilon)(2\epsilon) \ge -M\epsilon^2(2\epsilon),$$

where $s - \epsilon < \xi_\epsilon < s + \epsilon$. It follows that

$$p(\xi_\epsilon) \ge -M\epsilon^2.$$

Letting $\epsilon \to 0+$, we get that

$$p(s) \ge 0.$$

It then follows that

$$p(t) = f_{vv}(t, x_0(t), x_0'(t)) \ge 0,$$

at all $t \in (a, b)$, where x_0' and x_0'' are continuous. A similar argument can be given for $t = a$ and $t = b$. The proof of the local maximum case is left to the reader (see Exercise 5.54). □

Example 5.93 Show that

$$I[x] := \int_0^2 [(\cos t)x^2(t) - 3(\sin t)x(t)x'(t) - (t - 1)(x'(t))^2] \, dt,$$

for $x \in D$ with boundary conditions $x(0) = 1$, $x(2) = 2$ has no local maximums or local minimums.

Here

$$f(t, u, v) = (\cos t)u^2 - 3(\sin t)uv - (t - 1)v^2,$$

and hence

$$f_{vv}(t, u, v) = -2(t - 1),$$

which changes sign on $[0, 2]$ and hence by Theorem 5.92, I has no local extrema in D. △

Theorem 5.94 (Weierstrass Integral Formula) *Assume that x is a solution of $Lx = 0$ on $[a, b]$, $\eta \in \mathcal{A}$, and $z = x + \eta$; then*

$$Q[z] = Q[x] + Q[\eta].$$

Proof Assume that x is a solution of $Lx = 0$ on $[a, b]$, $\eta \in \mathcal{A}$, and $z = x + \eta$. Let $h : \mathbb{R} \to \mathbb{R}$ be defined by

$$h(\epsilon) := Q[x + \epsilon\eta].$$

By Taylor's formula,

$$h(1) = h(0) + \frac{1}{1!}h'(0) + \frac{1}{2!}h''(\xi), \tag{5.29}$$

where $\xi \in (0, 1)$. We next find $h'(0)$ and $h''(\xi)$. First note that

$$h'(\epsilon) = 2\int_a^b \{p(t)[x'(t) + \epsilon\eta'(t)]\eta'(t) - q(t)[x(t) + \epsilon\eta(t)]\eta(t)\} \, dt.$$

It follows that

$$\begin{aligned}
h'(0) &= 2\int_a^b \{p(t)x'(t)\eta'(t) - q(t)x(t)\eta(t)\} \, dt \\
&= 2p(t)x'(t)\eta(t)]_a^b - \int_a^b Lx(t)\eta(t) \, dt \\
&= 0,
\end{aligned}$$

where we have integrated by parts. Next note that

$$\begin{aligned}
h''(\epsilon) &= 2\int_a^b \{p(t)[\eta'(t)]^2 - q(t)\eta^2(t)\} \, dt \\
&= 2Q[\eta].
\end{aligned}$$

Since $h(1) = Q[z]$, $h(0) = Q[x]$, $h'(0) = 0$, and $h''(\xi) = 2Q[\eta]$, we get from (5.29) the Weierstrass integral formula

$$Q[z] = Q[x] + Q[\eta]. \qquad \square$$

Theorem 5.95 *Assume $Lx = 0$ is disconjugate on $[a, b]$ and let x_0 be the solution of the BVP*

$$Lx = 0,$$

$$x(a) = x_a, \quad x(b) = x_b.$$

Then

$$Q[x] = \int_a^b \{p(t)[x'(t)]^2 - q(t)x^2(t)\} \, dt,$$

subject to $x \in D$ has a proper global minimum at x_0.

Proof Since $Lx = 0$ is disconjugate on $[a, b]$, the BVP

$$Lx = 0,$$

$$x(a) = x_a, \quad x(b) = x_b,$$

has a unique solution x_0 by Theorem 5.45. Let $x \in D$ and let

$$\eta := x - x_0,$$

then $\eta \in \mathcal{A}$ and $x = x_0 + \eta$. Hence, by the Weierstrass integral formula (Theorem 5.94),

$$Q[x] = Q[x_0] + Q[\eta].$$

But $Lx = 0$ is disconjugate on $[a, b]$ implies (see Theorem 5.73) that Q is positive definite on A and hence is positive definite on $\mathcal{A}$. Therefore,

$$Q[x] \geq Q[x_0],$$

for all $x \in D$ and equality holds only if $x = x_0$. Hence Q has a proper global minimum at x_0. $\qquad\square$

Example 5.96 Find the minimum value of

$$Q[x] = \int_0^1 \{e^{-4t}[x'(t)]^2 - 4e^{-4t}x^2(t)\} \, dt$$

subject to

$$x(0) = 1, \quad x(1) = 0.$$

Here

$$p(t) = e^{-4t}, \quad q(t) = 4e^{-4t}$$

and so the corresponding self-adjoint equation is

$$(e^{-4t}x')' + 4e^{-4t}x = 0.$$

Expanding this equation out, we get the equivalent differential equation

$$x'' - 4x' + 4x = 0.$$

Since $x(t) = e^{2t}$ defines a positive solution of this differential equation on $\mathbb{R}$, we have by Theorem 5.46 that this differential equation is disconjugate on $\mathbb{R}$ and hence on $[0, 1]$. Hence, by Theorem 5.95, Q has a proper global minimum at x_0, where x_0 is the solution of the BVP

$$x'' - 4x' + 4x = 0,$$

$$x(0) = 1, \quad x(1) = 0.$$

Solving this BVP, we get

$$x_0(t) = e^{2t} - te^{2t}.$$

Hence, after some routine calculations, we get that the global minimum for Q on D is

$$Q[x_0] = -1.$$

$\triangle$

Many calculus of variation problems come from Hamilton's principle, which we now briefly discuss. Assume an object has some forces acting on it. Let T be the object's kinetic energy and V its potential energy. The *Lagrangian* is then defined by

$$L = T - V.$$

In general,

$$L = L(t, x, x').$$

The integral

$$I[x] = \int_a^b L(t, x(t), x'(t)) \, dt$$

is called the *action integral*. We now state Hamilton's principle in the simple case of rectilinear motion.

Hamilton's Principle: Let $x_0(t)$ be the position of an object under the influence of various forces on an x-axis at time t. If $x_0(a) = x_a$ and $x_0(b) = x_b$, where x_a and x_b are given, then the action integral assumes a stationary value at x_0.

In the next example we apply Hamilton's principle to a very simple example.

Example 5.97 Consider the spring problem, where the kinetic energy function T and the potential energy function V are given by

$$T = \frac{1}{2}m[x']^2 \quad \text{and} \quad V = \frac{1}{2}kx^2,$$

respectively. Assume that we know that $x(0) = 0$, $x(\frac{\pi}{2}\sqrt{\frac{m}{k}}) = 1$. Then the action integral is given by

$$I[x] = \int_0^{\frac{\pi}{2}\sqrt{\frac{m}{k}}} \left\{ \frac{1}{2}m[x'(t)]^2 - \frac{1}{2}kx^2(t) \right\} dt.$$

Applying Theorem 5.95, we get that the displacement x_0 of the mass is the solution of the BVP

$$mx'' + kx = 0,$$

$$x(0) = 0, \quad x\left(\frac{\pi}{2}\sqrt{\frac{m}{k}}\right) = 1.$$

Solving this BVP, we get

$$x_0(t) = \sin\left(\sqrt{\frac{m}{k}}t\right).$$

$\triangle$

We end this section by considering the special case of the simplest problem of the calculus of variations where $f(t, u, v) = f(u, v)$ is independent of t. In this case $x(t)$ solves the Euler–Lagrange equation on $[a, b]$ provided

$$f_u(x(t), x'(t)) - \frac{d}{dt}[f_v(x(t), x'(t))], \quad t \in [a, b]. \tag{5.30}$$

We now show that if $x(t)$ is a solution of (5.30) on $[a, b]$, then $x(t)$ solves the integrated form of the Euler–Lagrange equation

$$f_v(x, x')x' - f(x, x') = C \tag{5.31}$$

where C is a constant. To see this, assume $x(t)$ is a solution of (5.30) on $[a, b]$ and consider

$$\frac{d}{dt}[f_v(x(t), x'(t))x'(t) - f(x(t), x'(t))]$$

$$= \frac{d}{dt}[f_v(x(t), x'(t))]\, x'(t) + f_v(x(t), x'(t))x''(t)$$

$$- f_u(x(t), x'(t))x'(t) - f_v(x(t), x'(t))x''(t)$$

$$= x'(t)\left[\frac{d}{dt}f_v(x(t), x'(t)) - f_u(x(t), x'(t))\right]$$

$$= 0$$

for $t \in [a, b]$.

We now use the integrated form of the Euler–Lagrange form (5.31) to solve the following example.

Example 5.98 Find the C^1 curve $x = x(t)$ joining the given points (a, x_a) and (b, x_b), $a < b$, $x_a > 0$, $x_b > 0$ such that the surface area of the surface of revolution obtained by revolving the curve $x = x(t)$, $a \le t \le b$ about the x-axis is a minimum. $\triangle$

Here we want to minimize

$$I[x] = \int_a^b 2\pi\sqrt{1 + (x'(t))^2}\, dt$$

subject to

$$x(a) = x_a, \quad x(b) = x_b.$$

We now solve the integrated form of the Euler–Lagrange equation (5.31). In this case $f(u, v) = 2\pi u\sqrt{1 + v^2}$ so

$$f_u(u, v) = 2\pi\sqrt{1 + v^2}, \quad f_v(u, v) = 2\pi\frac{uv}{\sqrt{1 + v^2}}.$$

Hence from (5.31) we get the differential equation

$$\frac{x(x')^2}{\sqrt{1 + (x')^2}} - x\sqrt{1 + (x')^2} = A.$$

Simplifying this equation we get

$$-x = A\sqrt{1 + (x')^2}.$$

This can be rewritten in the form

$$\frac{Ax'}{\sqrt{x^2 - A^2}} = 1.$$

Integrating we get

$$t = A \ln \left[\frac{x + \sqrt{x^2 - A^2}}{A} \right] + B.$$

Solving this equation we get the catenaries

$$x(t) = A \cosh \frac{t - B}{A}.$$

We then want to find A and B so that the boundary conditions $x(a) = x_a$, $x(b) = x_b$ are satisfied. It can be shown (not routine) that if we fix a, $x_a > 0$, and $x_b > 0$, then as we vary b there is a value b_0 such that if $0 < b < b_0$, then there are two solutions one of which renders a minimum. For $b = b_0$ there is a single extremum and it renders a minimum. Finally if $b > b_0$, then there is no extremum.

5.9 Green's Functions

At the outset of this section we will be concerned with the Green's functions for a general two-point boundary value problem for the self-adjoint differential equation $Lx = 0$. At the end of this section we will consider the Green's function for the periodic BVP.

First we consider the homogeneous boundary value problem

$$
\begin{align}
Lx &= 0, & (5.32) \\
\alpha x(a) - \beta x'(a) &= 0, & (5.33) \\
\gamma x(b) + \delta x'(b) &= 0, & (5.34)
\end{align}
$$

where we always assume that $[a, b] \subset I$, $\alpha^2 + \beta^2 > 0$, and $\gamma^2 + \delta^2 > 0$.

Theorem 5.99 *Assume the homogeneous BVP* (5.32)–(5.34) *has only the trivial solution. Then the nonhomogeneous BVP*

$$
\begin{align}
Lu &= h(t), & (5.35) \\
\alpha u(a) - \beta u'(a) &= A, & (5.36) \\
\gamma u(b) + \delta u'(b) &= B, & (5.37)
\end{align}
$$

where A and B are constants and h is continuous on $[a, b]$, has a unique solution.

Proof Assume the BVP (5.32)–(5.34) has only the trivial solution. Let x_1, x_2 be linearly independent solutions of $Lx = 0$; then

$$x(t) = c_1 x_1(t) + c_2 x_2(t)$$

is a general solution of $Lx = 0$. Note that x satisfies boundary conditions (5.33) and (5.34) iff c_1, c_2 are constants such that the two equations

$$
\begin{align}
c_1[\alpha x_1(a) - \beta x_1'(a)] + c_2[\alpha x_2(a) - \beta x_2'(a)] &= 0, & (5.38) \\
c_1[\gamma x_1(b) + \delta x_1'(b)] + c_2[\gamma x_2(b) + \delta x_2'(b)] &= 0 & (5.39)
\end{align}
$$

hold. Since we are assuming that the BVP (5.32)–(5.34) has only the trivial solution, it follows that the only solution of the system (5.38), (5.39) is

$$c_1 = c_2 = 0.$$

Therefore, the determinant of the coefficients in the system (5.38), (5.39) is different than zero; that is,

$$\begin{vmatrix} \alpha x_1(a) - \beta x_1'(a) & \alpha x_2(a) - \beta x_2'(a) \\ \gamma x_1(b) + \delta x_1'(b) & \gamma x_2(b) + \delta x_2'(b) \end{vmatrix} \neq 0. \tag{5.40}$$

Now we will show that the BVP (5.35)–(5.37) has a unique solution. Let u_0 be a fixed solution of $Lu = h(t)$; then a general solution of $Lu = h(t)$ is given by

$$u(t) = a_1 x_1(t) + a_2 x_2(t) + u_0(t).$$

It follows that u satisfies the boundary conditions (BCs) (5.36), (5.37) iff a_1, a_2 are constants satisfying the system of equations

$$a_1[\alpha x_1(a) - \beta x_1'(a)] + a_2[\alpha x_2(a) - \beta x_2'(a)] \tag{5.41}$$
$$= A - \alpha u_0(a) + \beta u_0'(a),$$
$$a_1[\gamma x_1(b) + \delta x_1'(b)] + a_2[\gamma x_2(b) + \delta x_2'(b)] \tag{5.42}$$
$$= B - \gamma u_0(a) - \delta u_0'(a).$$

Since (5.40) holds, the system (5.41), (5.42) has a unique solution a_1, a_2. This implies that the BVP (5.35)–(5.37) has a unique solution. $\quad\square$

In the next example we give a BVP of the type (5.32)–(5.34) that does not have just the trivial solution.

Example 5.100 Find all solutions of the BVP

$$(p(t)x')' = 0,$$
$$x'(a) = 0, \quad x'(b) = 0.$$

This BVP is equivalent to a BVP of the form (5.32)–(5.34), where $q(t) \equiv 0$, $\alpha = \gamma = 0$, $\beta \neq 0$, and $\delta \neq 0$. A general solution of this differential equation is

$$x(t) = c_1 + c_2 \int_a^t \frac{1}{p(s)} ds.$$

The boundary conditions lead to the equations

$$x'(a) = c_2 \frac{1}{p(a)} = 0, \quad x'(b) = c_2 \frac{1}{p(b)} = 0.$$

Thus $c_2 = 0$ and there is no restriction on c_1. Hence for any constant c_1, $x(t) = c_1$ is a solution of our BVP. In particular, our given BVP has nontrivial solutions. $\quad\triangle$

In the next theorem we give a neccessary and sufficient condition for some boundary value problems of the form (5.32)–(5.34) to have only the trivial solution. The proof of this theorem is Exercise 5.57. Note that Theorem 5.101 gives the last statement in Example 5.100 as a special case.

Theorem 5.101 *Let*

$$\rho := \alpha\gamma \int_a^b \frac{1}{p(s)} ds + \frac{\beta\gamma}{p(a)} + \frac{\alpha\delta}{p(b)}.$$

Then the BVP

$$(p(t)x')' = 0,$$

$$\alpha x(a) - \beta x'(a) = 0, \quad \gamma x(b) + \delta x'(b) = 0,$$

has only the trivial solution iff $\rho \neq 0$.

The function $G(\cdot, \cdot)$ in the following theorem is called the Green's function for the BVP (5.32)–(5.34).

Theorem 5.102 (Green's Function for General Two-Point BVP) *Assume the homogeneous BVP (5.32)–(5.34) has only the trivial solution. For each fixed $s \in [a, b]$, let $u(\cdot, s)$ be the solution of the BVP*

$$
\begin{align}
Lu &= 0, & (5.43) \\
\alpha u(a, s) - \beta u'(a, s) &= 0, & (5.44) \\
\gamma u(b, s) + \delta u'(b, s) &= -\gamma x(b, s) - \delta x'(b, s), & (5.45)
\end{align}
$$

where $x(\cdot, \cdot)$ is the Cauchy function for $Lx = 0$. Define

$$G(t, s) := \begin{cases} u(t, s), & \text{if } a \le t \le s \le b \\ v(t, s), & \text{if } a \le s \le t \le b, \end{cases} \quad (5.46)$$

where $v(t, s) := u(t, s) + x(t, s)$, for $t, s \in [a, b]$. Assume h is continuous on $[a, b]$; then

$$x(t) := \int_a^b G(t, s)h(s) \, ds,$$

for $t \in [a, b]$, defines the unique solution of the nonhomogeneous BVP $Lx = h(t)$, (5.33), (5.34). Furthermore, for each fixed $s \in [a, b]$, $v(\cdot, s)$ is a solution of $Lx = 0$ satisfying the boundary condition (5.34).

Proof The existence and uniqueness of $u(t, s)$ is guaranteed by Theorem 5.99. Since $v(t, s) := u(t, s) + x(t, s)$, we have for each fixed s that $v(\cdot, s)$ is a solution of $Lx = 0$. Using the boundary condition (5.45), it is easy to see that for each fixed s, $v(\cdot, s)$ satisfies the boundary condition (5.34). Let

$G(t, s)$ be as in the statement of this theorem and consider

$$
\begin{aligned}
x(t) \;&=\; \int_a^b G(t, s)h(s)\, ds \\
&=\; \int_a^t G(t, s)h(s)\, ds + \int_t^b G(t, s)h(s)\, ds \\
&=\; \int_a^t v(t, s)h(s)\, ds + \int_t^b u(t, s)h(s)\, ds \\
&=\; \int_a^t [u(t, s) + x(t, s)]h(s)\, ds + \int_t^b u(t, s)h(s)\, ds \\
&=\; \int_a^b u(t, s)h(s)\, ds + \int_a^t x(t, s)h(s)\, ds \\
&=\; \int_a^b u(t, s)h(s)\, ds + z(t),
\end{aligned}
$$

where, by the variation of constants formula (Theorem 5.22), $z(t) :=$
$\int_a^t x(t, s)h(s)\, ds$ defines the solution of the IVP

$$
Lz = h(t), \quad z(a) = 0, \quad z'(a) = 0.
$$

Hence

$$
\begin{aligned}
Lx(t) \;&=\; \int_a^b Lu(t, s)h(s)\, ds + Lz(t) \\
&=\; \int_a^b Lu(t, s)h(s)\, ds + h(t) \\
&=\; h(t).
\end{aligned}
$$

Thus x is a solution of $Lx = h(t)$. Note that

$$
\begin{aligned}
\alpha x(a) - \beta x'(a) \;&=\; \int_a^b [\alpha u(a, s) - \beta u'(a, s)]\, h(s)\, ds + \alpha z(a) - \beta z'(a) \\
&=\; \int_a^b [\alpha u(a, s) - \beta u'(a, s)]\, h(s)\, ds \\
&=\; 0,
\end{aligned}
$$

since for each fixed $s \in [a, b]$, $u(\cdot, s)$ satisfies the boundary condition (5.44). Therefore, x satisfies the boundary condition (5.33). It remains to show that x satisfies the boundary condition (5.34). Earlier in this proof we had that

$$
x(t) = \int_a^b u(t, s)h(s)\, ds + \int_a^t x(t, s)h(s)\, ds.
$$

It follows that

$$
\begin{aligned}
x(t) &= \int_a^b [v(t,s) - x(t,s)]h(s)\, ds + \int_a^t x(t,s)h(s)\, ds \\
&= \int_a^b v(t,s)h(s)\, ds - \int_t^b x(t,s)h(s)\, ds \\
&= \int_a^b v(t,s)h(s)\, ds + \int_b^t x(t,s)h(s)\, ds \\
&= \int_a^b v(t,s)h(s)\, ds + w(t),
\end{aligned}
$$

where, by the variation of constants formula (Theorem 5.22), $w(t) := \int_b^t x(t,s)h(s)\, ds$ defines the solution of the IVP

$$Lw = h(t), \quad w(b) = 0, \quad w'(b) = 0.$$

Hence

$$
\begin{aligned}
\gamma x(b) + \delta x'(b) &= \int_a^b [\gamma v(b,s) + \delta v'(b,s)]\, h(s)\, ds + \gamma w(b) + \delta w'(b) \\
&= \int_a^b [\gamma v(b,s) + \delta v'(b,s)]\, h(s)\, ds \\
&= 0,
\end{aligned}
$$

since for each fixed $s \in [a,b]$, $v(\cdot, s)$ satisfies the boundary condition (5.34). Hence x satisfies the boundary condition (5.34). □

We then get the following corollary, whose proof is left as an exercise (Exercise 5.63).

Corollary 5.103 *Assume the BVP (5.32)–(5.34) has only the trivial solution. If h is continuous on $[a,b]$, then the solution of the BVP*

$$
\begin{aligned}
Lx(t) &= h(t), \\
\alpha x(a) - \beta x'(a) &= A, \\
\gamma x(b) + \delta x'(b) &= B,
\end{aligned}
$$

where A and B are constants, is given by

$$x(t) = w(t) + \int_a^b G(t,s)h(s)\, ds,$$

where G is the Green's function for the BVP (5.32)–(5.34) and w is the solution of the BVP

$$Lw = 0,$$

$$\alpha w(a) - \beta w'(a) = A, \quad \gamma w(b) + \delta w'(b) = B.$$

In the next theorem we give another form of the Green's function for the BVP (5.32)–(5.34) and note that the Green's function is symmetric on the square $[a,b]^2$.

Theorem 5.104 *Assume that the BVP* (5.32)–(5.34) *has only the trivial solution. Let ϕ be the solution of the IVP*

$$L\phi = 0, \quad \phi(a) = \beta, \quad \phi'(a) = \alpha,$$

and let ψ be the solution of the IVP

$$L\psi = 0, \quad \psi(b) = \delta, \quad \psi'(b) = -\gamma.$$

Then the Green's function for the BVP (5.32)–(5.34) *is given by*

$$G(t, s) := \begin{cases} \frac{1}{c}\phi(t)\psi(s), & \text{if } a \le t \le s \le b \\ \frac{1}{c}\phi(s)\psi(t), & \text{if } a \le s \le t \le b, \end{cases} \tag{5.47}$$

where $c := p(t)w[\phi(t), \psi(t)]$ is a constant. Furthermore, the Green's function is symmetric on the square $[a, b]^2$; that is,

$$G(t, s) = G(s, t),$$

for $t, s \in [a, b]$.

Proof Let ϕ, ψ, and c be as in the statement of this theorem. By Abel's formula (Corollary 5.7), $c = p(t)w[\phi(t), \psi(t)]$ is a constant. We will use Theorem 5.102 to prove that G defined by (5.47) is the Green's function for the BVP (5.32)–(5.34). Note that

$$\alpha\phi(a) - \beta\phi'(a) = \alpha\beta - \beta\alpha = 0,$$

and

$$\gamma\psi(b) + \delta\psi'(b) = \gamma\delta - \delta\gamma = 0.$$

Hence ϕ satisfies the boundary condition (5.33) and ψ satisfies the boundary condition (5.34).

Let

$$u(t, s) := \frac{1}{c}\phi(t)\psi(s), \quad v(t, s) := \frac{1}{c}\phi(s)\psi(t),$$

for $t, s \in [a, b]$. Note that for each fixed $s \in [a, b]$, $u(\cdot, s)$ and $v(\cdot, s)$ are solutions of $Lx = 0$ on $[a, b]$. Also, for each fixed $s \in [a, b]$,

$$\alpha u(a, s) - \beta u'(a, s) = \frac{1}{c}\psi(s)[\alpha\phi(a) - \beta\phi'(a)] = 0,$$

and

$$\gamma v(b, s) + \delta v'(b, s) = \frac{1}{c}\phi(s)[\gamma\psi(b) + \delta\psi'(b)] = 0.$$

Hence for each fixed $s \in [a, b]$, $u(\cdot, s)$ satisfies the boundary condition (5.33) and $v(\cdot, s)$ satisfies the boundary condition (5.34). Let

$$k(t, s) := v(t, s) - u(t, s) = \frac{1}{c}[\phi(s)\psi(t) - \phi(t)\psi(s)].$$

It follows that for each fixed s, $k(\cdot, s)$ is a solution of $Lx = 0$. Also, $k(s, s) = 0$ and

$$
\begin{aligned}
k'(s, s) &= \frac{1}{c}[\phi(s)\psi'(s) - \phi'(s)\psi(s)] \\
&= \frac{\phi(s)\psi'(s) - \phi'(s)\psi(s)}{p(s)w[\phi(s), \psi(s)]} \\
&= \frac{1}{p(s)}.
\end{aligned}
$$

Therefore, $k(\cdot, \cdot) = x(\cdot, \cdot)$ is the Cauchy function for $Lx = 0$ and we have

$$
v(t, s) = u(t, s) + x(t, s).
$$

It remains to prove that for each fixed s, $u(\cdot, s)$ satisfies the boundary condition (5.45). To see this, consider

$$
\begin{aligned}
\gamma u(b, s) + \delta u'(b, s) &= [\gamma v(b, s) + \delta v'(b, s)] - [\gamma x(b, s) + \delta x'(b, s)] \\
&= -[\gamma x(b, s) + \delta x'(b, s)].
\end{aligned}
$$

Hence by Theorem 5.102, G defined by equation (5.47) is the Green's function for the BVP (5.32)–(5.34). It follows from (5.47) that the Green's function G is symmetric on the square $[a, b]^2$. □

A special case of Theorem 5.102 is when the boundary conditions (5.33), (5.34) are the conjugate (Dirichlet) boundary conditions

$$
x(a) = 0, \quad x(b) = 0. \tag{5.48}
$$

In this case we get the following result.

Corollary 5.105 *Assume that the homogeneous conjugate BVP $Lx = 0$, (5.48), has only the trivial solution. Then the Green's function for the conjugate BVP $Lx = 0$, (5.48), is given by (5.46), where for each $s \in [a, b]$, $u(\cdot, s)$ is the solution of the BVP $Lx = 0$,*

$$
u(a, s) = 0, \quad u(b, s) = -x(b, s), \tag{5.49}
$$

and $v(t, s) = u(t, s) + x(t, s)$, where $x(\cdot, \cdot)$ is the Cauchy function for $Lx = 0$. For each fixed $s \in [a, b]$, $v(\cdot, s)$ is a solution of $Lx = 0$ satisfying the boundary condition $v(b, s) = 0$. Furthermore, the Green's function is symmetric on the square $[a, b]^2$.

Example 5.106 Using an appropriate Green's function, we will solve the BVP

$$
x'' + x = t, \tag{5.50}
$$

$$
x(0) = 0, \quad x\left(\frac{\pi}{2}\right) = 0. \tag{5.51}
$$

It is easy to check that the homogeneous BVP $x'' + x = 0$, (5.51) has only the trivial solution, and hence we can use Corollary 5.105 to find the solution of the nonhomogeneous BVP (5.50), (5.51). The Cauchy function for $x'' + x = 0$ is given by

$$x(t, s) = \sin(t - s).$$

Since for each fixed $s \in [0, \frac{\pi}{2}]$, $u(\cdot, s)$ is a solution of $x'' + x = 0$,

$$u(t, s) = A(s) \cos t + B(s) \sin t.$$

Using the boundary conditions

$$u(0, s) = 0, \quad u\left(\frac{\pi}{2}, s\right) = -x\left(\frac{\pi}{2}, s\right) = -\sin\left(\frac{\pi}{2} - s\right) = -\cos s,$$

we get

$$A(s) = 0, \quad B(s) = -\cos(s).$$

Therefore,

$$u(t, s) = -\sin t \cos s.$$

Next

$$
\begin{aligned}
v(t, s) &= u(t, s) + x(t, s) \\
&= -\sin t \cos s + \sin(t - s) \\
&= -\sin s \cos t.
\end{aligned}
$$

Therefore, by Corollary 5.105, the Green's function for the BVP $x'' + x = 0$, (5.51) is given by

$$G(t, s) = \begin{cases} -\sin t \cos s, & \text{if} \quad 0 \le t \le s \le \frac{\pi}{2} \\ -\sin s \cos t, & \text{if} \quad 0 \le s \le t \le \frac{\pi}{2}. \end{cases}$$

Hence, by Theorem 5.102, the solution of the BVP (5.50), (5.51) is given by

$$
\begin{aligned}
x(t) &= \int_a^b G(t, s) h(s) \, ds \\
&= \int_0^{\frac{\pi}{2}} G(t, s) s \, ds \\
&= \int_0^t G(t, s) s \, ds + \int_t^{\frac{\pi}{2}} G(t, s) s \, ds \\
&= -\cos t \int_0^t s \sin s \, ds - \sin t \int_t^{\frac{\pi}{2}} s \cos s \, ds \\
&= -\cos t \left(-t \cos t + \sin t\right) - \sin t \left(\frac{\pi}{2} - t \sin t - \cos t\right) \\
&= t - \frac{\pi}{2} \sin t.
\end{aligned}
$$

$\triangle$

Example 5.107 Find the Green's function for the conjugate BVP

$$(p(t)x')' = 0,$$

$$x(a) = 0, \quad x(b) = 0.$$

By Example 5.16 (or Example 5.19), we get that the Cauchy function for $(p(t)x')' = 0$ is

$$x(t, s) = \int_s^t \frac{1}{p(\tau)} d\tau,$$

for $t, s \in [a, b]$. By Corollary 5.105, for each fixed s, $u(\cdot, s)$ solves the BVP

$$Lu(t, s) = 0, \quad u(a, s) = 0, \quad u(b, s) = -x(b, s).$$

Since $u(\cdot, s)$ is a solution for each fixed s,

$$u(t, s) = c_1(s) \cdot 1 + c_2(s) \int_a^t \frac{1}{p(\tau)} d\tau.$$

But $u(a, s) = 0$ implies that $c_1(s) = 0$, so

$$u(t, s) = c_2(s) \int_a^t \frac{1}{p(\tau)} d\tau.$$

But $u(b, s) = -x(b, s) = -\int_s^b \frac{1}{p(\tau)} d\tau$ implies that

$$c_2(s) = -\frac{\int_s^b \frac{1}{p(\tau)} d\tau}{\int_a^b \frac{1}{p(\tau)} d\tau}.$$

Hence

$$G(t, s) = u(t, s) = -\frac{\int_a^t \frac{1}{p(\tau)} d\tau \int_s^b \frac{1}{p(\tau)} d\tau}{\int_a^b \frac{1}{p(\tau)} d\tau},$$

for $a \le t \le s \le b$.

One could use the symmetry (by Theorem 5.104) of the Green's function to get the form of the Green's function for $a \le s \le t \le b$, but we will find G directly now. For $a \le s \le t \le b$,

$$
\begin{aligned}
G(t,s) &= v(t,s) \\
&= u(t,s) + x(t,s) \\
&= -\frac{\int_a^t \frac{1}{p(\tau)} d\tau \int_s^b \frac{1}{p(\tau)} d\tau}{\int_a^b \frac{1}{p(\tau)} d\tau} + \int_s^t \frac{1}{p(\tau)} d\tau \\
&= -\frac{\int_a^t \frac{1}{p(\tau)} d\tau \int_s^b \frac{1}{p(\tau)} d\tau - \int_a^b \frac{1}{p(\tau)} d\tau \int_s^t \frac{1}{p(\tau)} d\tau}{\int_a^b \frac{1}{p(\tau)} d\tau} \\
&= -\frac{\left[\int_a^b \frac{1}{p(\tau)} d\tau - \int_t^b \frac{1}{p(\tau)} d\tau\right] \int_s^b \frac{1}{p(\tau)} d\tau - \int_a^b \frac{1}{p(\tau)} d\tau \int_s^t \frac{1}{p(\tau)} d\tau}{\int_a^b \frac{1}{p(\tau)} d\tau} \\
&= -\frac{\int_a^b \frac{1}{p(\tau)} d\tau \int_t^b \frac{1}{p(\tau)} d\tau - \int_t^b \frac{1}{p(\tau)} d\tau \int_s^b \frac{1}{p(\tau)} d\tau}{\int_a^b \frac{1}{p(\tau)} d\tau} \\
&= -\frac{\int_t^b \frac{1}{p(\tau)} d\tau \int_a^s \frac{1}{p(\tau)} d\tau}{\int_a^b \frac{1}{p(\tau)} d\tau}.
\end{aligned}
$$

In summary, we get that the Green's function for the BVP

$$(p(t)x')' = 0,$$

$$x(a) = 0, \quad x(b) = 0$$

is given by

$$
G(t,s) := \begin{cases}
-\dfrac{\int_a^t \frac{1}{p(\tau)} d\tau \int_s^b \frac{1}{p(\tau)} d\tau}{\int_a^b \frac{1}{p(\tau)} d\tau}, & \text{if } a \le t \le s \le b \\[4ex]
-\dfrac{\int_t^b \frac{1}{p(\tau)} d\tau \int_a^s \frac{1}{p(\tau)} d\tau}{\int_a^b \frac{1}{p(\tau)} d\tau}, & \text{if } a \le s \le t \le b.
\end{cases}
$$

$\triangle$

Letting $p(t) = 1$ in Example 5.107, we get the following example.

Example 5.108 The Green's function for the conjugate BVP

$$x'' = 0,$$

$$x(a) = 0, \quad x(b) = 0$$

is given by

$$
G(t,s) := \begin{cases}
-\frac{(t-a)(b-s)}{b-a}, & \text{if } a \le t \le s \le b \\[2ex]
-\frac{(s-a)(b-t)}{b-a}, & \text{if } a \le s \le t \le b.
\end{cases}
$$

$\triangle$

We will use the following properties of the conjugate Green's function later in this chapter and in Chapter 7.

Theorem 5.109 *Let G be the Green's function for the BVP*

$$x'' = 0,$$

$$x(a) = 0, \quad x(b) = 0.$$

Then

$$-\frac{(b-a)}{4} \le G(t,s) \le 0,$$

for $t, s \in [a, b]$,

$$\int_a^b |G(t,s)|\, ds \le \frac{(b-a)^2}{8},$$

for $t \in [a, b]$, and

$$\int_a^b |G'(t,s)|\, ds \le \frac{(b-a)}{2},$$

for $t \in [a, b]$.

Proof It is easy to see that $G(t,s) \le 0$ for $t, s \in [a, b]$. For $a \le t \le s \le b$,

$$
\begin{aligned}
G(t,s) \;=\; u(t,s) &= -\frac{(t-a)(b-s)}{b-a} \\
&\ge -\frac{(s-a)(b-s)}{b-a} \\
&\ge -\frac{(b-a)}{4}.
\end{aligned}
$$

Similarly, for $a \le s \le t \le b$,

$$
\begin{aligned}
G(t,s) \;=\; v(t,s) &= -\frac{(s-a)(b-t)}{b-a} \\
&\ge -\frac{(s-a)(b-s)}{b-a} \\
&\ge -\frac{(b-a)}{4}.
\end{aligned}
$$

Next, for $t \in [a, b]$, consider

$$
\begin{aligned}
\int_a^b |G(t, s)|\, ds &= \int_a^t |G(t, s)|\, ds + \int_t^b |G(t, s)|\, ds \\
&= \int_a^t |v(t, s)|\, ds + \int_t^b |u(t, s)|\, ds \\
&= \int_a^t \frac{(s - a)(b - t)}{b - a}\, ds + \int_t^b \frac{(t - a)(b - s)}{b - a}\, ds \\
&= \frac{(b - t)}{b - a} \int_a^t (s - a)\, ds + \frac{(t - a)}{b - a} \int_t^b (b - s)\, ds \\
&= \frac{(b - t)(t - a)^2}{2(b - a)} + \frac{(t - a)(b - t)^2}{2(b - a)} \\
&= \frac{(b - t)(t - a)}{2} \\
&\leq \frac{(b - a)^2}{8}.
\end{aligned}
$$

Finally, for $t \in [a, b]$, consider

$$
\begin{aligned}
\int_a^b |G'(t, s)|\, ds &= \int_a^t |G'(t, s)|\, ds + \int_t^b |G'(t, s)|\, ds \\
&= \int_a^t |v'(t, s)|\, ds + \int_t^b |u'(t, s)|\, ds \\
&= \int_a^t \frac{(s - a)}{b - a}\, ds + \int_t^b \frac{(b - s)}{b - a}\, ds \\
&= \frac{(t - a)^2}{2(b - a)} + \frac{(b - t)^2}{2(b - a)} \\
&\leq \frac{(b - a)}{2}. \quad \square
\end{aligned}
$$

Example 5.110 Find the Green's function for the right focal BVP

$$(p(t)x')' = 0,$$

$$x(a) = 0, \quad x'(b) = 0.$$

Note that from Theorem 5.101, this BVP has only the trivial solution. By Example 5.16 (or Example 5.19) we get that the Cauchy function for $(p(t)x')' = 0$ is

$$x(t, s) = \int_s^t \frac{1}{p(\tau)}\, d\tau,$$

for $t, s \in I$. By Theorem 5.102, for each fixed s, $u(\cdot, s)$ solves the BVP

$$Lu(t, s) = 0, \ u(a, s) = 0, \ u'(b, s) = -x'(b, s).$$

Since for each fixed s, $u(\cdot, s)$ is a solution,

$$u(t, s) = c_1(s) \cdot 1 + c_2(s) \int_a^t \frac{1}{p(\tau)} d\tau.$$

But $u(a, s) = 0$ implies that $c_1(s) = 0$, so

$$u(t, s) = c_2(s) \int_a^t \frac{1}{p(\tau)} d\tau.$$

But $u'(b, s) = -x'(b, s) = -\frac{1}{p(b)}$ implies that $c_2(s) = -1$. Hence

$$G(t, s) = u(t, s) = -\int_a^t \frac{1}{p(\tau)} d\tau,$$

for $a \le t \le s \le b$. For $a \le s \le t \le b$,

$$\begin{aligned} G(t, s) &= v(t, s) \\ &= u(t, s) + x(t, s) \\ &= -\int_a^t \frac{1}{p(\tau)} d\tau + \int_s^t \frac{1}{p(\tau)} d\tau \\ &= -\int_a^s \frac{1}{p(\tau)} d\tau. \end{aligned}$$

In summary, we have that the Green's function for the right focal BVP

$$(p(t)x')' = 0,$$
$$x(a) = 0, \quad x'(b) = 0$$

is given by

$$G(t, s) := \begin{cases} -\int_a^t \frac{1}{p(\tau)} d\tau, & \text{if } a \le t \le s \le b \\ -\int_a^s \frac{1}{p(\tau)} d\tau, & \text{if } a \le s \le t \le b. \end{cases}$$

$\triangle$

Letting $p(t) = 1$ in Example 5.110, we get the following example.

Example 5.111 The Green's function for the right focal BVP

$$x'' = 0,$$
$$x(a) = 0, \quad x'(b) = 0$$

is given by

$$G(t, s) := \begin{cases} -(t - a), & \text{if } a \le t \le s \le b \\ -(s - a), & \text{if } a \le s \le t \le b. \end{cases}$$

$\triangle$

Theorem 5.112 *If $Lx = 0$ is disconjugate on $[a, b]$, then the Green's function for the conjugate BVP*

$$Lx = 0, \quad x(a) = 0, \quad x(b) = 0 \tag{5.52}$$

exists and satisfies

$$G(t, s) < 0,$$

for $t, s \in (a, b)$.

Proof Since $Lx = 0$ is disconjugate on $[a, b]$, the BVP (5.52) has only the trivial solution. Hence Corollary 5.105 gives the existence of the Green's function,

$$G(t, s) := \begin{cases} u(t, s), & \text{if } a \leq t \leq s \leq b \\ v(t, s), & \text{if } a \leq s \leq t \leq b, \end{cases}$$

for the conjugate BVP (5.52). For each fixed $s \in (a, b)$, $u(\cdot, s)$ is the solution of $Lx = 0$ satisfying

$$u(a, s) = 0, \quad u(b, s) = -x(b, s) < 0,$$

where the last inequality is true since $x(s, s) = 0$, $x'(s, s) > 0$, and $Lx = 0$ is disconjugate on $[a, b]$. But then it follows that

$$u(t, s) < 0,$$

for $t \in (a, b]$. Also, for each fixed $s \in (a, b)$, $v(\cdot, s)$ is the solution of $Lx = 0$ satisfying

$$v(b, s) = 0, \quad v(s, s) = u(s, s) + x(s, s) = u(s, s) < 0.$$

But then it follows that

$$v(t, s) < 0,$$

for $t \in [a, b)$. Therefore,

$$G(t, s) < 0,$$

for $t, s \in (a, b)$.   □

Theorem 5.113 (Comparison Theorem for BVPs) *Assume that $Lx = 0$ is disconjugate on $[a, b]$. If $u, v \in \mathbb{D}$ satisfy $u(a) \geq v(a)$, $u(b) \geq v(b)$ and $Lu(t) \leq Lv(t)$ for $t \in [a, b]$, then*

$$u(t) \geq v(t),$$

for $t \in [a, b]$.

Proof Let $z(t) := u(t) - v(t)$ and let $h(t) := Lz(t)$, for $t \in [a, b]$. Then

$$h(t) = Lz(t) = Lu(t) - Lv(t) \leq 0,$$

for $t \in [a, b]$. Also, let

$$A := z(a) = u(a) - v(a) \geq 0,$$

and

$$B := z(b) = u(b) - v(b) \geq 0.$$

Then z solves the BVP

$$Lz = h(t),$$
$$z(a) = A, \quad z(b) = B.$$

By Corollary 5.103,

$$z(t) = w(t) + \int_a^b G(t, s)h(s)\, ds, \tag{5.53}$$

where w is a solution of $Lx = 0$ satisfying $w(a) = A \geq 0$, $w(b) = B \geq 0$. Since $Lx = 0$ is disconjugate on $[a, b]$, $w(t) \geq 0$ and since by Theorem 5.112, $G(t, s) \leq 0$ for $t, s \in [a, b]$, it follows from (5.53) that

$$z(t) = u(t) - v(t) \geq 0,$$

on $[a, b]$. □

Theorem 5.114 (Liapunov Inequality) *If*

$$\int_a^b q^+(t)\, dt \leq \frac{4}{b - a}, \tag{5.54}$$

where $q^+(t) := \max\{q(t), 0\}$, *then* $x'' + q(t)x = 0$ *is disconjugate on* $[a, b]$.

Proof Assume the inequality (5.54) holds but $x'' + q(t)x = 0$ is not disconjugate on $[a, b]$. Then there is a nontrivial solution x such that x has consecutive zeros satisfying $a \leq t_1 < t_2 \leq b$. Without loss of generality, we can assume that

$$x(t) > 0,$$

on (t_1, t_2). Since x satisfies the BVP

$$x'' = -q(t)x(t),$$
$$x(t_1) = 0, \quad x(t_2) = 0,$$

we get that

$$x(t) = \int_{t_1}^{t_2} G(t, s)\, [-q(s)x(s)]\, ds,$$

where by Example 5.108 the Green's function G is given by

$$G(t, s) := \begin{cases} -\dfrac{(t - t_1)(t_2 - s)}{t_2 - t_1}, & \text{if } t_1 \leq t \leq s \leq t_2 \\[2mm] -\dfrac{(s - t_1)(t_2 - t)}{t_2 - t_1}, & \text{if } t_1 \leq s \leq t \leq t_2. \end{cases}$$

Pick $t_0 \in (t_1, t_2)$ so that

$$x(t_0) = \max\{x(t) : t_1 \leq t \leq t_2\}.$$

Consider

$$\begin{aligned} x(t_0) &= \int_{t_1}^{t_2} G(t_0, s)\, [-q(s)x(s)]\, ds \\ &\leq \int_{t_1}^{t_2} |G(t_0, s)| q^+(s)x(s)\, ds. \end{aligned}$$

If $q^+(t) = 0$ for $t \in [t_1, t_2]$, then $q(t) \leq 0$ for $t \in [t_1, t_2]$ and by Exercise 5.26 we have that $x'' + q(t)x = 0$ is disconjugate on $[t_1, t_2]$, but this is a contradiction. Hence it is not true that $q^+(t) = 0$ for $t \in [t_1, t_2]$ and it follows that

$$x(t_0) < x(t_0) \int_{t_1}^{t_2} |G(t_0, s)| q^+(s) \, ds.$$

Dividing both sides by $x(t_0)$, we get the inequality

$$1 < \int_{t_1}^{t_2} |G(t_0, s)| q^+(s) \, ds.$$

Using Theorem 5.109, we get the inequality

$$1 < \frac{t_2 - t_1}{4} \int_{t_1}^{t_2} q^+(t) \, dt.$$

But this implies that

$$\int_a^b q^+(t) \, dt \geq \int_{t_1}^{t_2} q^+(t) \, dt > \frac{4}{t_2 - t_1} \geq \frac{4}{b - a},$$

which contradicts the inequality (5.54).   □

Example 5.115 Use Theorem 5.114 to find T so that the differential equation

$$x'' + tx = 0$$

is disconjugate on $[0, T]$.

By Theorem 5.114 we want to pick T so that

$$\int_0^T t \, dt \leq \frac{4}{T}.$$

That is, we want

$$\frac{T^2}{2} \leq \frac{4}{T}.$$

It follows that $x'' + tx = 0$ is disconjugate on $[0, 2]$.   △

Example 5.116 Use Theorem 5.114 to show that the differential equation

$$x'' + k \sin t \, x = 0, \quad k > 0$$

is disconjugate on $[a, a + 2\pi]$ if $k \leq \frac{1}{\pi}$.

Since

$$\int_a^{a+2\pi} q^+(t) \, dt = \int_0^\pi k \sin t \, dt = 2k,$$

it follows from Theorem 5.114 that if $k \leq \frac{1}{\pi}$, then $x'' + k \sin t \, x = 0$ is disconjugate on $[a, a + 2\pi]$. It is interesting to note that it is known [11] that every solution of $x'' + k \sin t \, x = 0$, $k > 0$, is oscillatory (this is not easy to prove).   △

Theorem 5.114 says that if

$$\int_a^b q^+(t)\, dt \le \frac{C}{b-a},$$

where $C = 4$, then $x'' + q(t)x = 0$ is disconjugate on $[a, b]$. The next example shows that this result is sharp in the sense that $C = 4$ is the largest constant such that this result is true in general.

Example 5.117 ($C = 4$ is Sharp) Let $0 < \delta < \frac{1}{2}$ and choose $x : [0, 1] \to \mathbb{R}$ so that x has a continuous second derivative on $[0, 1]$ with

$$x(t) := \begin{cases} t, & \text{if} \quad 0 \le t \le \frac{1}{2} - \delta \\ 1 - t, & \text{if} \quad \frac{1}{2} + \delta \le t \le 1, \end{cases}$$

and x satisfies

$$x''(t) \le 0$$

on $[0, 1]$. Next let

$$q(t) := \begin{cases} -\dfrac{x''(t)}{x(t)}, & \text{if} \quad 0 < t < 1 \\ 0, & \text{if} \quad t = 0, 1. \end{cases}$$

Note that q is nonnegative and continuous on $[0, 1]$, and x is a nontrivial solution of $x'' + q(t)x = 0$ with $x(0) = 0$, $x(1) = 0$. Hence $x'' + q(t)x = 0$ is not disconjugate on $[0, 1]$. Also, note that

$$
\begin{aligned}
\int_0^1 q^+(t)\, dt &= \int_{\frac{1}{2}-\delta}^{\frac{1}{2}+\delta} \frac{-x''(t)}{x(t)}\, dt \\
&\le \frac{1}{\frac{1}{2}-\delta} \int_{\frac{1}{2}-\delta}^{\frac{1}{2}+\delta} [-x''(t)]\, dt \\
&= \frac{2}{1-2\delta}\left[x'(\frac{1}{2}-\delta) - x'(\frac{1}{2}+\delta) \right] \\
&= \frac{4}{1-2\delta}.
\end{aligned}
$$

The constant on the right-hand side of this last inequality is larger than 4 and can be made arbitrarily close to 4 by taking δ arbitrarily close to zero. $\triangle$

Finally, in this section we consider the nonhomogeneous periodic BVP

$$Lx = h(t), \tag{5.55}$$
$$x(a) = x(b), \quad x'(a) = x'(b). \tag{5.56}$$

The next theorem is the reason why the BVP (5.55), (5.56) is called periodic.

Theorem 5.118 *In this theorem, in addition to the standard assumptions on p, q, and h, assume these three functions are periodic on $\mathbb{R}$ with period*

$b - a$. *If the BVP* (5.55), (5.56) *has a solution* x, *then it is periodic with period* $b - a$.

Proof Assume x is a solution of the BVP (5.55), (5.56) and let

$$y(t) := x(t + b - a), \quad t \in \mathbb{R}.$$

Using the chain rule for differentiation and the periodicity of p, q, and h, it can be shown that y is a solution of (5.55). Also,

$$y(a) = x(b) = x(a),$$

and

$$y'(a) = x'(b) = x'(a).$$

Hence x and y are both solutions of the same IVP. It follows that

$$x(t) = y(t) = x(t + b - a),$$

for $t \in \mathbb{R}$. That is, x is periodic with period $b - a$. □

Theorem 5.119 *Assume that the homogeneous periodic BVP* $Lx = 0$, (5.56) *has only the trivial solution. Then the nonhomogeneous BVP* $Lx = h(t)$,

$$x(a) - x(b) = A, \quad x'(a) - x'(b) = B, \tag{5.57}$$

where A and B are given constants and h is a given continuous function on $[a, b]$, has a unique solution.

Proof Assume the homogeneous periodic BVP $Lx = 0$, (5.56), has only the trivial solution. Let x_1, x_2 be linearly independent solutions of $Lx = 0$; then

$$x(t) = c_1 x_1(t) + c_2 x_2(t)$$

defines a general solution of $Lx = 0$. Note that x satisfies the boundary conditions (5.56) iff c_1, c_2 are constants such that the two equations

$$c_1[x_1(a) - x_1(b)] + c_2[x_2(a) - x_2(b)] = 0 \tag{5.58}$$
$$c_1[x_1'(a) - x_1'(b)] + c_2[x_2'(a) - x_2'(b)] = 0 \tag{5.59}$$

hold. Since we are assuming that the BVP $Lx = 0$, (5.56) has only the trivial solution, it follows that the only solution of the system (5.58), (5.59) is

$$c_1 = c_2 = 0.$$

Therefore, the determinant of the coefficients in the system (5.58), (5.59) is different from zero; that is,

$$\begin{vmatrix} x_1(a) - x_1(b) & x_2(a) - x_2(b) \\ x_1'(a) - x_1'(b) & x_2'(a) - x_2'(b) \end{vmatrix} \neq 0. \tag{5.60}$$

Now we will show that the BVP $Lx = h(t)$, (5.57) has a unique solution. Let u_0 be a fixed solution of $Lu = h(t)$; then a general solution of $Lu = h(t)$ is defined by

$$u(t) = a_1 x_1(t) + a_2 x_2(t) + u_0(t).$$

It follows that u satisfies the BCs (5.57) iff a_1, a_2 are constants satisfying the system of equations

$$a_1[x_1(a) - x_1(b)] + a_2[x_2(a) - x_2(b)] = A - u_0(a) + u_0(b) \qquad (5.61)$$
$$a_1[x_1'(a) - x_1'(b)] + a_2[x_2'(a) - x_2'(b)] = B - u_0'(a) + u_0'(b). \qquad (5.62)$$

Since (5.60) holds, the system (5.61), (5.62) has a unique solution a_1, a_2. This implies that the BVP $Lx = h(t)$, (5.57) has a unique solution.

$\square$

Theorem 5.120 (Green's Function for Periodic BVP) *Assume the homogeneous BVP $Lx = 0$, (5.56) has only the trivial solution. For each fixed $s \in [a, b]$, let $u(\cdot, s)$ be the solution of the BVP*

$$Lu = 0, \qquad (5.63)$$
$$u(a, s) - u(b, s) = x(b, s), \qquad (5.64)$$
$$u'(a, s) - u'(b, s) = x'(b, s), \qquad (5.65)$$

where $x(\cdot, \cdot)$ is the Cauchy function for $Lx = 0$. Define the Green's function G for the BVP $Lx = 0$, (5.56) by

$$G(t, s) := \begin{cases} u(t, s), & \text{if } a \le t \le s \le b \\ v(t, s), & \text{if } a \le s \le t \le b, \end{cases} \qquad (5.66)$$

where $v(t, s) := u(t, s) + x(t, s)$. Assume h is continuous on $[a, b]$; then

$$x(t) := \int_a^b G(t, s)h(s) \, ds,$$

for $t \in [a, b]$, defines the unique solution x of the nonhomogeneous periodic BVP $Lx = h(t)$, (5.56). Furthermore, for each fixed $s \in [a, b]$, $v(\cdot, s)$ is a solution of $Lx = 0$ and $u(a, s) = v(b, s)$, $u'(a, s) = v'(b, s)$.

Proof The existence and uniqueness of $u(\cdot, \cdot)$ is guaranteed by Theorem 5.119. Since $v(t, s) := u(t, s) + x(t, s)$, we have for each fixed s that $v(\cdot, s)$ is a solution of $Lx = 0$. Using the boundary conditions (5.64), (5.65), it is easy to see that for each fixed s, $u(a, s) = v(b, s)$, $u'(a, s) = v'(b, s)$. Let

$G(t, s)$ be as in the statement of this theorem and consider

$$
\begin{aligned}
x(t) &= \int_a^b G(t, s)h(s)\, ds \\
&= \int_a^t G(t, s)h(s)\, ds + \int_t^b G(t, s)h(s)\, ds \\
&= \int_a^t v(t, s)h(s)\, ds + \int_t^b u(t, s)h(s)\, ds \\
&= \int_a^t [u(t, s) + x(t, s)]h(s)\, ds + \int_t^b u(t, s)h(s)\, ds \\
&= \int_a^b u(t, s)h(s)\, ds + \int_a^t x(t, s)h(s)\, ds \\
&= \int_a^b u(t, s)h(s)\, ds + z(t),
\end{aligned}
$$

where, by the variation of constants formula (Theorem 5.22), $z(t) :=$ $\int_a^t x(t, s)h(s)\, ds$ defines the solution of the IVP

$$
Lz = h(t), \quad z(a) = 0, \quad z'(a) = 0.
$$

Hence

$$
\begin{aligned}
Lx(t) &= \int_a^b Lu(t, s)h(s)\, ds + Lz(t) \\
&= \int_a^b Lu(t, s)h(s)\, ds + h(t) \\
&= h(t).
\end{aligned}
$$

Thus x is a solution of $Lx = h(t)$. Note that

$$
\begin{aligned}
x(a) &= \int_a^b G(a, s)h(s)\, ds \\
&= \int_a^b u(a, s)h(s)\, ds \\
&= \int_a^b v(b, s)h(s)\, ds \\
&= \int_a^b G(b, s)h(s)\, ds \\
&= x(b).
\end{aligned}
$$

Similarly,

$$
\begin{aligned}
x'(a) &= \int_a^b G'(a,s)h(s)\,ds \\
&= \int_a^b u'(a,s)h(s)\,ds \\
&= \int_a^b v'(b,s)h(s)\,ds \\
&= \int_a^b G'(b,s)h(s)\,ds \\
&= x'(b).
\end{aligned}
$$

Hence x satisfies the periodic boundary conditions (5.56). $\qquad\square$

Example 5.121 Using Theorem 5.120, we will solve the periodic BVP

$$x'' + x = \sin(2t), \tag{5.67}$$
$$x(0) = x(\pi), \quad x'(0) = x'(\pi). \tag{5.68}$$

It is easy to show that the homogeneous BVP

$$x'' + x = 0, \quad x(0) = x(\pi), \quad x'(0) = x'(\pi)$$

has only the trivial solution and hence we can use Theorem 5.120 to solve the BVP (5.67), (5.68). The Cauchy function for $x'' + x = 0$ is given [see Exercise 5.10, part (v)] by

$$x(t,s) = \sin(t-s).$$

By Theorem 5.120, the Green's function G is given by

$$
G(t,s) := \begin{cases} u(t,s), & \text{if } 0 \le t \le s \le \pi \\ v(t,s), & \text{if } 0 \le s \le t \le \pi, \end{cases} \tag{5.69}
$$

where for each fixed $s \in [0, \pi]$, $u(\cdot, s)$ is the solution of the BVP

$$u'' + u = 0, \tag{5.70}$$
$$u(0,s) - u(\pi,s) = x(\pi,s), \tag{5.71}$$
$$u'(0,s) - u'(\pi,s) = x'(\pi,s), \tag{5.72}$$

where $x(\cdot, \cdot)$ is the Cauchy function for $Lx = 0$ and $v(t,s) := u(t,s) + x(t,s)$. Using (5.70), we get

$$u(t,s) = A(s)\cos t + B(s)\sin t.$$

From the boundary conditions (5.71), (5.72), we get

$$2A(s) = \sin(\pi - s) = \sin s,$$
$$2B(s) = \cos(\pi - s) = -\cos s.$$

It follows that

$$u(t,s) = \frac{1}{2}\sin s \cos t - \frac{1}{2}\cos s \sin t = -\frac{1}{2}\sin(t-s).$$

Therefore,

$$v(t, s) = u(t, s) + x(t, s) = \frac{1}{2} \sin(t - s).$$

Hence by Theorem 5.120 the solution of the BVP (5.67), (5.68) is given by

$$
\begin{aligned}
x(t) &= \int_a^b G(t, s) h(s) \, ds \\
&= \int_0^\pi G(t, s) \sin(2s) \, ds \\
&= \int_0^t v(t, s) \sin(2s) \, ds + \int_t^\pi u(t, s) \sin(2s) \, ds \\
&= \frac{1}{2} \int_0^t \sin(t - s) \sin(2s) \, ds - \frac{1}{2} \int_t^\pi \sin(t - s) \sin(2s) \, ds \\
&= \frac{1}{2} \sin t \int_0^t \cos s \sin(2s) \, ds - \frac{1}{2} \cos t \int_0^t \sin(s) \sin(2s) \, ds \\
&\quad - \frac{1}{2} \sin t \int_t^\pi \cos s \sin(2s) \, ds + \frac{1}{2} \cos t \int_t^\pi \sin(s) \sin(2s) \, ds \\
&= \frac{1}{2} \sin t \left\{ -\frac{1}{3} \sin(2t) \sin t - \frac{2}{3} \cos(2t) \cos t + \frac{2}{3} \right\} \\
&\quad - \frac{1}{2} \cos t \left\{ \frac{1}{3} \cos t \sin(2t) - \frac{2}{3} \sin t \cos(2t) \right\} \\
&\quad - \frac{1}{2} \sin t \left\{ \frac{2}{3} + \frac{1}{3} \sin(2t) \sin t + \frac{2}{3} \cos(2t) \cos t \right\} \\
&\quad + \frac{1}{2} \cos t \left\{ -\frac{1}{3} \cos t \sin(2t) + \frac{2}{3} \sin t \cos(2t) \right\} \\
&= -\frac{1}{3} \sin^2 t \sin(2t) - \frac{2}{3} \sin t \cos t \cos(2t) \\
&\quad - \frac{1}{3} \cos^2 t \sin(2t) + \frac{2}{3} \sin t \cos t \cos(2t) \\
&= -\frac{1}{3} \sin(2t).
\end{aligned}
$$

The fact that this solution is periodic with period π as guaranteed by Theorem 5.118. (Note that this simple problem could be solved by the annihilator method). △

5.10 Exercises

5.1 Write each of the following differential equations in self-adjoint form:

    (i) (Legendre's Equation) $(1 - t^2)x'' - 2tx' + n(n + 1)x = 0$, $\quad t \in I := (-1, 1)$

    (ii) (Chebychev's Equation) $(1 - t^2)x'' - tx' + n^2 x = 0$, $\quad t \in I := (-1, 1)$

    (iii) (Laquerre's Equation) $tx'' + (1 - t)x' + ax = 0$, $\quad t \in I := (0, \infty)$

   (iv) (Hermite's Equation) $x'' - 2tx' + 2nx = 0$, $t \in I := (-\infty, \infty)$

   (v) $x'' + \frac{2}{3+t}x' + \frac{\lambda}{(3+t)^2}x = 0$, $t \in I := (-3, \infty)$

5.2 Find a homogeneous self-adjoint equation that has the function x defined by $x(t) = \sin(\frac{1}{t^n})$, $t \in I := (0, \infty)$, where n is a positive integer as a solution.

5.3 Prove Abel's formula directly by taking the derivative of

$$p(t)w[x(t), y(t)].$$

5.4 Let $a \in I$ and let $\tau(t) := \int_a^t \frac{1}{p(s)}ds$, for $t \in I$. Show that this defines $t = t(\tau)$. Then show that if x is a solution of $Lx = 0$, then $y(\tau) := x(t(\tau))$ defines a solution of $y'' + Q(\tau)y = 0$, where $Q(\tau) = p(t(\tau))q(t(\tau))$.

5.5 Complete the proof of Corollary 5.10.

5.6 Show that if $x_1, x_2, \cdots, x_k$ are functions that have $k-1$ derivatives on an interval I and if

$$w[x_1, x_2, \cdots, x_k](t_0) \neq 0,$$

where $t_0 \in I$, then $x_1, x_2, \cdots, x_k$ are linearly independent functions on I.

5.7 Let $x_1(t) = t^2, x_2(t) = t|t|$, for $t \in I := \mathbb{R}$. Show that x_1, x_2 are linearly independent on $\mathbb{R}$ but $w[x_1, x_2](0) = 0$. Are these two functions x_1, x_2 solutions of the same self-adjoint equation $Lx = 0$ on $\mathbb{R}$? Are x_1, x_2 linearly dependent or linearly independent on $\mathbb{R}^+ := [0, \infty)$?

5.8 Show that if x, $y \in \mathbb{D}$ and they satisfy the boundary conditions $\alpha z(a) - \beta z'(a) = 0$, $\gamma z(b) + \delta z'(b) = 0$, where $\alpha^2 + \beta^2 > 0$, $\gamma^2 + \delta^2 > 0$, then $< Lx, y >=< x, Ly >$.

5.9 Define the functions ch and sh to be the solutions of the differential equation $x'' - x = 0$ satisfying the initial conditions $ch(0) = 1$, $ch'(0) = 0$ and $sh(0) = 0$, $sh'(0) = 1$, respectively. State and prove a theorem like Theorem 5.13 giving properties of ch and sh.

5.10 Use the definition of the Cauchy function to find the Cauchy function for each of the following differential equations:

   (i) $(e^{-3t}x')' + 2e^{-3t}x = 0$, $I = (-\infty, \infty)$

   (ii) $(e^{-10t}x')' + 25e^{-10t}x = 0$, $I = (-\infty, \infty)$

   (iii) $(\frac{1}{t^4}x')' + \frac{6}{t^6}x = 0$, $I = (0, \infty)$

   (iv) $(e^{-2t}x')' = 0$, $I = (-\infty, \infty)$

   (v) $x'' + x = 0$, $I = (-\infty, \infty)$

In (v) write your answer as a single term.

5.11 Use your answers in Exercise 5.10 and the variation of constants formula to solve each of the following initial value problems:

   (i) $(e^{-3t}x')' + 2e^{-3t}x = e^{-t}$, $x(0) = 0$, $x'(0) = 0$

   (ii) $(e^{-10t}x')' + 25e^{-10t}x = e^{-5t}$, $x(0) = 0$, $x'(0) = 0$

   (iii) $(\frac{1}{t^4}x')' + \frac{6}{t^6}x = t$, $x(1) = 0$, $x'(1) = 0$

(iv) $(e^{-2t}x')' = t + 2,$ $x(0) = 0,$ $x'(0) = 0$

(v) $x'' + x = 4,$ $x(0) = 0,$ $x'(0) = 0$

5.12 Use Theorem 5.18 to find the Cauchy function for each of the differential equations in Exercise 5.10.

5.13 Use Theorem 5.18 to find the Cauchy function for the differential equation $\left(\frac{1}{t}x'\right)' + \frac{2}{t^3}x = 0, t > 0.$

5.14 Use Corollary 5.23 to solve the following IVPs:

(i) $\left(\frac{1}{t^6}x'\right)' + \frac{12}{t^8}x = \frac{2}{t^6},$ $x(1) = 1,$ $x'(1) = 2$

(ii) $x'' + 9x = 1,$ $x(0) = 2,$ $x'(0) = 0$

5.15 Find the Cauchy function for the differential equation

$$\left(\frac{x'}{1+t}\right)' = 0.$$

Use this Cauchy function and an appropriate variation of constants formula to solve the IVP

$$\left(\frac{x'}{1+t}\right)' = t^2,$$ $x(0) = x'(0) = 0.$

5.16 Find eigenpairs for the following Sturm-Liouville problems:

(i) $x'' = -\lambda x,$ $x(0) = 0,$ $x(\frac{\pi}{2}) = 0$

(ii) $x'' = -\lambda x,$ $x'(0) = 0,$ $x(\frac{\pi}{2}) = 0$

(iii) $(tx')' = -\frac{\lambda}{t}x,$ $x(1) = 0,$ $x(e) = 0$

(iv) $x'' = -\lambda x,$ $x'(-\pi) = 0,$ $x'(\pi) = 0$

(v) $(t^3x')' + \lambda tx = 0,$ $x(1) = 0,$ $x(e) = 0$

5.17 Find a constant α so that $x(t) = t,$ $y(t) = t + \alpha$ are orthogonal with respect to $r(t) = t + 1$ on $[0, 1].$

5.18 Prove that if λ_0, x_0 is an eigenpair for the SLP (5.5)–(5.7), then $\overline{\lambda_0}, \overline{x_0}$ is also an eigenpair.

5.19 Show that if $x_1,$ x_2 satisfy the boundary conditions (5.6), then

$$w[x_1(t), x_2(t)](a) = 0.$$

5.20 Find all eigenpairs for the periodic Sturm-Liouville problem:

$$x'' + \lambda x = 0,$$ $x\left(-\frac{\pi}{2}\right) = x\left(\frac{\pi}{2}\right),$ $x'\left(-\frac{\pi}{2}\right) = x'\left(\frac{\pi}{2}\right).$

5.21 Find all eigenpairs for the periodic Sturm-Liouville problem:

$$x'' + \lambda x = 0,$$ $x(-3) = x(3),$ $x'(-3) = x'(3).$

5.22 What do you get if you cross a hurricane with the Kentucky Derby?

5.23 Use separation of variables to find solutions of each of the following partial differential equations:

(i) $u_t = ku_{xx}$

(ii) $y_{tt} = ky_{xx}$

(iii) $u_{rr} + \frac{1}{r}u_r + \frac{1}{r^2}u_{\theta\theta} = 0$

5.24 (Periodic Boundary Conditions) Show that if, in addition to the standard assumptions on the coefficient functions p, q, we assume that these coefficient functions are periodic on $\mathbb{R}$ with period $b - a$, then any solution of $Lx = 0$ satisfying the boundary conditions $x(a) = x(b)$, $x'(a) = x'(b)$ is periodic with period $b - a$. Because of this fact we say that the boundary conditions $x(a) = x(b)$, $x'(a) = x'(b)$ are periodic boundary conditions.

5.25 (Sturm-Liouville Problem) Assume that $I = (a, b]$, and, in addition to the standard assumptions on the coefficient functions p, q, and r, that $\lim_{t \to a+} p(t) = 0$. Show that eigenfunctions corresponding to distinct eigenvalues for the singular Sturm-Liouville problem

$$Lx = -\lambda r(t)x,$$

$$x, \ x' \text{ are bounded on } I,$$

$$\gamma x(b) + \delta x'(b) = 0$$

are orthogonal with respect to the weight function r on I.

5.26 Show that if $q(t) \le 0$ on an interval I, then $x'' + q(t)x = 0$ is disconjugate on I.

5.27 Find a self-adjoint equation $Lx = 0$ and an interval I such that $Lx = 0$ is disconjugate on I but there is no positive solution on I (*Hint*: See Theorem 5.46.)

5.28 In the proof of Theorem 5.50, prove

$$\lim_{t \to b-} \left\{ \left(\frac{u(t)}{v(t)} \right) [p_1(t)u'(t)v(t) - p_2(t)u(t)v'(t)] \right\} = 0.$$

5.29 Show that the Euler–Cauchy differential equation

$$t^2 x'' + 2tx' + bx = 0$$

is oscillatory on $[1, \infty)$ if $b > \frac{1}{4}$ and disconjugate on $[1, \infty)$ if $b \le \frac{1}{4}$.

5.30 Complete the proof of Theorem 5.51.

5.31 Find a Polya factorization for each of the following differential equations on the given intervals J:

(i) $\left(e^{-5t}x' \right)' + 6e^{-5t}x = 0, \quad J = (-\infty, \infty)$

(ii) $x'' - x = 0, \quad J = (-\infty, \infty)$

(iii) $t^2 x'' - 5tx' + 8x = 0, \quad J = (0, \infty)$

(iv) $x'' - 6x' + 9x = 0, \quad J = (-\infty, \infty)$

5.32 Find a Trench factorization for each of the differential equations in Exercise 5.31 on the given intervals J.

5.33 For each of the following show that the given function u is a solution on the given interval J and solve the given differential equation on that interval:

(i) $\left(\frac{e^{2t}}{1+2t}x'\right)' - \frac{4e^{2t}}{(1+2t)^2}x = 0, \quad J = (-\frac{1}{2}, \infty), \quad u(t) = e^{-2t}$

(ii) $\left(\frac{e^t}{1+t}x'\right)' - \frac{e^t}{(1+t)^2}x = 0, \quad J = (0, \infty), \quad u(t) = t$

(iii) $\left(\frac{1}{(t^2+2t-1)}x'\right)' + \frac{2}{(t^2+2t-1)^2}x = 0, \quad J = [2, \infty), \quad u(t) = 1+t$

(iv) $tx'' - (1+t)x' + x = 0, \quad J = (0, \infty), \quad u(t) = e^t$

5.34 Show that $u(t) = \frac{\sin t}{\sqrt{t}}$ defines a nonzero solution of

$$(tx')' + \left(t - \frac{1}{4t}\right)x = 0,$$

on $J = (0, \pi)$, and then solve this differential equation.

5.35 Find a recessive and dominant solution on J for each of the following equations and for these two solutions show directly that the conclusions of Theorem 5.59 are true:

(i) $x'' - 5x' + 6x = 0, \quad J = [0, \infty)$

(ii) $t^2x'' - 5tx' + 9x = 0, \quad J = [1, \infty)$

(iii) $x'' - 4x' + 4x = 0, \quad J = [0, \infty)$

5.36 Prove the last statement in Theorem 5.59.

5.37 Find a Mammana factorization on J for each of the problems in Exercise 5.31.

5.38 Show that the Riccati operator $R : C^1(I) \to C(I)$, which is defined in Section 5.7, is not a linear operator (hence is a nonlinear operator).

5.39 Find a solution of some IVP for the Riccati equation $Rz = z' + z^2 = 0$ whose maximal interval of existence is not the whole interval $I = \mathbb{R}$. Give the maximal interval of existence for the solution you found.

5.40 Prove Corollary 5.75.

5.41 Solve the following Riccati equations:

(i) $z' - 3e^{2t} + e^{-2t}z^2 = 0$

(ii) $z' + 4e^{-4t} + e^{4t}z^2 = 0$

(iii) $z' - \frac{1}{t} + \frac{1}{t}z^2 = 0$

(iv) $z' + 4 + z^2 = 0;$

(v) $z' = -3e^{-4t} - e^{4t}z^2$

(vi) $z' + 16e^{-8t} + e^{8t}z^2 = 0$

5.42 For the Riccati equations in Exercises 5.41 (i), (ii), (iii), (v), and (vi), find the minimum solution that exists for all t sufficiently large. For the Riccati equation in (iv), find the minimum solution that exists for all $t < \frac{\pi}{2}$, sufficiently close.

5.43 Show if the following equations are oscillatory or not:

(i) $(t \log t x')' + \frac{1}{t \log t} x = 0$

(ii) $x'' + \frac{1}{t^{1.99}} x = 0$

5.44 Show that if $I = [2, \infty)$, and

$$\int_2^\infty \frac{t}{\log^{1+\beta} t} q(t) \, dt = \infty,$$

where $\beta > 0$, then $x'' + q(t)x = 0$ is oscillatory on $[2, \infty)$. [*Hint*: Use Theorem 5.81 with $u(t) = \frac{t^{\frac{1}{2}}}{\log^{\frac{1+\beta}{2}} t}$.]

5.45 Assume that for each of the following, I has a global minimum at x_0 and x_0 is twice continuously differentiable. Find $x_0(t)$.

(i) $I[x] = \int_0^{\frac{\pi}{2}} \{[x'(t)]^2 - x^2(t)\} \, dt$, $\quad x(0) = 1$, $\quad x(\frac{\pi}{2}) = 0$

(ii) $I[x] = \int_0^1 \{e^{-3t}[x'(t)]^2 - 2e^{-3t}x^2(t)\} \, dt$, $\quad x(0) = 1$, $\quad x(1) = 2$

(iii) $I[x] = \int_1^e \{\frac{1}{t}[x'(t)]^2 - \frac{1}{t^3}x^2(t)\} \, dt$, $\quad x(1) = 0$, $\quad x(e) = 1$

5.46 Do the proof of the case where $g(t_0) < 0$ in the proof of Lemma 5.87.

5.47 Assume that the calculus of variations problem in Example 5.83 has a global minimum at x_0 and $x_0 \in C^2[a, b]$. Find $x_0(t)$ by solving the appropriate Euler-Lagrange differential equation.

5.48 Assume that I has a global minimum at x_0 and x_0 is twice continuously differentiable. Find $x_0(t)$, given that

$$I[x] = \int_0^1 \left[e^{-5t}(x'(t))^2 - 6e^{-5t}x^2(t) \right] \, dt, \quad x(0) = 0, \quad x(1) = 1.$$

5.49 Show that the function x_1 defined by $x_1(t) := 1 - t$, $0 \le t \le 1$, is in the set D for the calculus of variations problem considered in Example 5.96 and show directly that $Q[x_1] > -1$.

5.50 Using Theorem 5.92 (Legendre's necessary condition), state what you can about the existence of local extrema for each of the simplest problems of the calculus of variations, where I is given by

(i) $I[x] = \int_0^1 \{e^{-t}x(t) + x^2(t) - t^2x(t)x'(t) - (1+t^2)(x'(t))^2\} \, dt$, $\quad x \in D$

(ii) $I[x] = \int_1^2 \{e^{-t} \sin(x(t)) - e^{-2t^2} x(t)x'(t) + (t-1)(x'(t))^2\} \, dt$, $\quad x \in D$

(iii) $I[x] = \int_0^\pi \{e^{-t}x'(t) - t^2x(t)x'(t) - \sin(2t)(x'(t))^2\} \, dt$, $\quad x \in D$

5.51 Show that if $f(t, u, v) = f(v)$, then The Euler–Lagrange equation reduces to

$$f_{vv}(x')x'' = 0.$$

Also show if $f(t, u, v) = f(t, v)$, then the Euler–Lagrange equation leads to the differential equation

$$f_v(t, x') = C$$

where C is a constant.

5.52 Solve the Euler–Lagrange equation given that

    (i) $f(t, u, v) = 2u^2 - v^2$

    (ii) $f(t, u, v) = \frac{\sqrt{1+v^2}}{u}$.

5.53 Find a possible extremum for the problem:

$$I[x] = \int_0^4 [tx'(t) - (x'(t))^2]\, dt$$

$$x(0) = 0, \quad x(4) = 3.$$

5.54 Prove Theorem 5.92 for the local maximum case.

5.55 Find the minimum value of

$$Q[x] = \int_1^2 \{t(x'(t))^2 + \frac{1}{t}x^2(t)\}\, dt,$$

subject to

$$x(1) = 3, \quad x(2) = 3.$$

Find $Q[3]$ and compare it to your minimum value.

5.56 Consider the simple pendulum problem considered in Example 3.10. Find the action integral. Assuming that Hamilton's principle holds, show that the angular displacement $\theta(t)$ of the pendulum at time t satisfies the pendulum equation

$$\theta'' + \frac{g}{L}\sin\theta = 0.$$

5.57 Prove Theorem 5.101.

5.58 (Boundary Conditions) Define $M : C^1[a, b] \to \mathbb{R}$ by

$$Mx = c_1 x(a) + c_2 x'(a) + c_3 x(b) + c_4 x'(b),$$

where c_i, $1 \le i \le 4$, are given constants. Show that M is a linear operator. Because of this any boundary condition of the form $Mx = A$, where A is a given constant, is called a *linear nonhomogeneous boundary condition*, whereas any boundary condition of the form $Mx = 0$ is called a *linear homogeneous boundary condition*. If a boundary condition is not linear it is called a *nonlinear boundary condition*. Classify each of the following boundary conditions:

    (i) $2x(a) - 3x'(a) = 0$

    (ii) $x(a) = x(b)$

    (iii) $x(a) = 0$

    (iv) $x(a) + 6 = x(b)$

    (v) $x(a) + 2x^2(b) = 0$

5.59 Use Theorem 5.102 to find the Green's function for the left focal BVP

$$(p(t)x')' = 0,$$

$$x'(a) = 0, \quad x(b) = 0.$$

5.60 Show that if the boundary value problem (5.32)–(5.34) has only the trivial solution, then the Green's function for the BVP (5.32)–(5.34) satisfies $G(s+, s) = G(s-, s)$ and satisfies the *jump condition*

$$G'(s+, s) - G'(s-, s) = \frac{1}{p(s)},$$

for $s \in (a, b)$. Here $G(s+, s) := \lim_{t \to s+} G(t, s)$.

5.61 Use Theorem 5.102 to find the Green's function for the BVP

$$x'' = 0, \quad x(0) - x'(0) = 0, \quad x(1) + x'(1) = 0.$$

5.62 For each of the following, find an appropriate Green's function and solve the given BVP:

    (i) $x'' = t^2, \quad x(0) = 0, \quad x(1) = 0$

    (ii) $\left(e^{2t}x'\right)' = e^{3t}, \quad x(0) = 0, \quad x(\log(2)) = 0$

    (iii) $\left(e^{-5t}x'\right)' + 6e^{-5t}x = e^{3t}, \quad x(0) = 0, \quad x(\log(2)) = 0$

5.63 Prove Corollary 5.103.

5.64 Use Corollary 5.103 to solve the BVPs

    (i) $x'' - 3x' + 2x = e^{3t}, \quad x(0) = 1, \quad x'(\log(2)) = 2$

    (ii) $x'' + 4x = 5e^t, \quad x(0) = 1, \quad x'(\frac{\pi}{8}) = 0$

    (iii) $t^2 x'' - 6tx' + 12x = 2t^5, \quad x(1) = 0, \quad x'(2) = 88$

5.65 Using an appropriate Green's function solve the BVP

$$x'' = t, \quad x(0) = 0, \quad x(1) = 1.$$

5.66 Let $G(\cdot, \cdot)$ be the Green's function for the right focal BVP

$$x'' = 0, \quad x(a) = x'(b) = 0.$$

Show that

$$-(b - a) \le G(t, s) \le 0, \quad \text{for} \quad a \le t, s \le b,$$

$$\int_a^b |G(t, s)| \, ds \le \frac{(b - a)^2}{2}, \quad \text{for} \quad a \le t \le b,$$

and

$$\int_a^b |G'(t, s)| \, ds \le b - a, \quad \text{for} \quad a \le t \le b.$$

5.67 Find the Green's function for the BVP

$$\left(\frac{x'}{1 + t}\right)' = 0, \quad x(0) = 0, \quad x(1) = 0.$$

Use this Green's function and an appropriate variation of constants formula to solve the BVP

$$\left(\frac{x'}{1 + t}\right)' = t, \quad x(0) = 0, \quad x'(1) = 1.$$

5.68 In each of the following use Theorem 5.114 to find T as large as possible so that the given differential equation is disconjugate on $[0, T]$.

 (i) $x'' + t^2 x = 0$

 (ii) $x'' + \left(t^2 + \frac{1}{3}\right) x = 0$

 (iii) $x'' + \left(\frac{5}{3} + 3t^2\right) x = 0$

 (iv) $x'' + \frac{3}{2}t^2 x = 0$

 (v) $x'' + 4(t - 1)x = 0$

In part (iii) use your calculator to solve the inequality in T that you got.

5.69 Use Theorem 5.114 to prove that if q is a continuous function on $[a, \infty)$, then there is a λ_0, $0 \le \lambda_0 \le \infty$ such that $x'' + \lambda q(t)x = 0$ is oscillatory on $[a, \infty)$ if $\lambda > \lambda_0$ and nonoscillatory on $[a, \infty)$ if $0 \le \lambda < \lambda_0$. Also show that if $\int_a^\infty q(t)\, dt = \infty$, then $\lambda_0 = 0$.

5.70 For each of the following, find an appropriate Green's function and solve the given periodic BVP (note that by Theorem 5.118 the solution you will find is periodic with period $b - a$):

 (i) $x'' + x = 4$, $x(0) = x(\pi)$, $x'(0) = x'(\pi)$

 (ii) $x'' + x = 2$, $x(0) = x(\frac{\pi}{2})$, $x'(0) = x'(\frac{\pi}{2})$

 (iii) $x'' + x = \cos(4t)$, $x(0) = x(\frac{\pi}{2})$, $x'(0) = x'(\frac{\pi}{2})$

 (iv) $x'' - x = \sin t$, $x(0) = x(\pi)$, $x'(0) = x'(\pi)$

Chapter 6

Linear Differential Equations of Order n

6.1 Basic Results

In this chapter we are concerned with the nth-order linear differential equation

$$y^{(n)} + p_{n-1}(t)y^{(n-1)} + \cdots + p_0(t)y = h(t), \tag{6.1}$$

where we assume $p_i : I \to \mathbb{R}$ is continuous for $0 \le i \le n-1$, and $h : I \to \mathbb{R}$ is continuous, where I is a subinterval of $\mathbb{R}$.

Definition 6.1 Let

$$C(I) := \{x : I \to \mathbb{R} : x \text{ is continuous on } I\}$$

and

$$C^n(I) := \{x : I \to \mathbb{R} : x \text{ has } n \text{ continuous derivatives on } I\},$$

and define $L_n : C^n(I) \to C(I)$ by

$$L_n x(t) = x^{(n)}(t) + p_{n-1}(t)x^{(n-1)}(t) + \cdots + p_0(t)x(t),$$

for $t \in I$.

Then it can be shown (Exercise 6.1) that L_n is a linear operator. Note that the differential equation (6.1) can be written in the form $L_n y = h$. If h is not the trivial function on I, then $L_n y = h$ is said to be nonhomogeneous, whereas the differential equation $L_n x = 0$ is said to be homogeneous.

Theorem 6.2 (Existence-Uniqueness Theorem) *Assume $p_i : I \to \mathbb{R}$ are continuous for $0 \le i \le n-1$, and $h : I \to \mathbb{R}$ is continuous. Assume that $t_0 \in I$ and $A_i \in \mathbb{R}$ for $0 \le i \le n-1$. Then the IVP*

$$L_n y = h(t), \quad y^{(i)}(t_0) = A_i, \quad 0 \le i \le n-1, \tag{6.2}$$

has a unique solution and this solution exists on the whole interval I.

Proof The existence and uniqueness of solutions of the IVP (6.2) follows from Corollary 8.19. The fact that solutions exist on the whole interval I follows from Theorem 8.65. $\square$

W.G. Kelley and A.C. Peterson, *The Theory of Differential Equations: Classical and Qualitative*, Universitext 278, DOI 10.1007/978-1-4419-5783-2_6, © Springer Science+Business Media, LLC 2010

Definition 6.3 Assume that $x_1, x_2, \cdots, x_k$ are k times differentiable functions on I, then we define the *Wronskian determinant* of these k functions by

$$W(t) = W[x_1(t), x_2(t), \cdots, x_k(t)]$$

$$= \begin{vmatrix} x_1(t) & x_2(t) & \cdots & x_k(t) \\ x_1'(t) & x_2'(t) & \cdots & x_k'(t) \\ \cdots & \cdots & \cdots & \cdots \\ x_1^{(k-1)}(t) & x_2^{(k-1)}(t) & \cdots & x_k^{(k-1)}(t) \end{vmatrix},$$

for $t \in I$.

Theorem 6.4 (Liouville's Theorem) *Suppose $x_1, x_2, \ldots, x_n$ are solutions of the differential equation $L_n x = 0$, and let $W(t)$ be the Wronskian determinant at t of $x_1, x_2, \ldots, x_n$. Then*

$$W(t) = e^{-\int_{t_0}^{t} p_{n-1}(\tau) d\tau} W(t_0), \tag{6.3}$$

for all $t, t_0 \in I$.

Proof Note that

$$W'(t) = \begin{vmatrix} x_1(t) & x_2(t) & \cdots & x_n(t) \\ x_1'(t) & x_2'(t) & \cdots & x_n'(t) \\ \cdots & \cdots & \cdots & \cdots \\ x_1^{(n-2)}(t) & x_2^{(n-2)}(t) & \cdots & x_n^{(n-2)}(t) \\ x_1^{(n)}(t) & x_2^{(n)}(t) & \cdots & x_n^{(n)}(t) \end{vmatrix}.$$

Using the fact that $x_1, x_2, \ldots, x_n$ are solutions of equation $L_n x = 0$ and properties of determinants, we get that $W'(t)$ is the determinant

$$\begin{vmatrix} x_1(t) & x_2(t) & \cdots & x_n(t) \\ x_1'(t) & x_2'(t) & \cdots & x_n'(t) \\ \cdots & \cdots & \cdots & \cdots \\ x_1^{(n-2)}(t) & x_2^{(n-2)}(t) & \cdots & x_n^{(n-2)}(t) \\ -p_{n-1}(t)x_1^{(n-1)}(t) & -p_{n-1}(t)x_2^{(n-1)}(t) & \cdots & -p_{n-1}(t)x_n^{(n-1)}(t) \end{vmatrix}.$$

Hence $W(t)$ solves the differential equation

$$W' = -p_{n-1}(t)W.$$

Solving this differential equation, we get Liouville's formula (6.3). $\square$

The proof of the following corollary is Exercise 6.4.

Corollary 6.5 *Suppose $x_1, x_2, \ldots, x_n$ are solutions of the differential equation $L_n x = 0$ and let $W(t)$ be the Wronskian determinant of $x_1, x_2, \ldots, x_n$ at t Then either $W(t) = 0$ for all $t \in I$ or $W(t) \neq 0$ for all $t \in I$. In the first case $x_1, x_2, \ldots, x_n$ are linearly dependent on I and in the second case $x_1, x_2, \ldots, x_n$ are linearly independent on I.*

Note that by Exercise 6.6 we know that $L_n x = 0$ has n linearly independent solutions on I. With this in mind we state and prove the following theorem.

Theorem 6.6 *If $x_1, x_2, \cdots, x_n$ are n linearly independent solutions of $L_n x = 0$ on I, then*

$$x = c_1 x_1 + c_2 x_2 + \cdots + c_n x_n$$

is a general solution of $L_n x = 0$. If, in addition, y_0 is a solution of the nonhomogeneous differential equation $L_n y = h(t)$, then

$$y = c_1 x_1 + c_2 x_2 + \cdots + c_n x_n + y_0$$

is a general solution of $L_n y = h(t)$.

Proof Assume $x_1, x_2, \cdots, x_n$ are solutions of $L_n x = 0$, then if $x = c_1 x_1 + c_2 x_2 + \cdots + c_n x_n$ it follows that

$$L_n x = c_1 L_n x_1 + c_2 L_n x_2 + \cdots + c_n L_n x_n = 0.$$

Hence any linear combination of solutions is a solution. Now assume that $x_1, x_2, \cdots, x_n$ are n linearly independent solutions of $L_n x = 0$, and $t_0 \in I$. Then by Corollary 6.5, $W(t_0) \neq 0$, where $W(t_0)$ is the Wronskian of $x_1, x_2, \cdots, x_n$ at t_0. Now let u be an arbitrary but fixed solution of $L_n x = 0$ and let

$$u_i := u^{(i)}(t_0), \quad 0 \leq i \leq n - 1.$$

Since $W(t_0) \neq 0$, there are constants $a_1, a_2, \cdots, a_n$ such that

$$
\begin{aligned}
a_1 x_1(t_0) + a_2 x_2(t_0) + \cdots + a_n x_n(t_0) &= u_0 \\
a_1 x_1'(t_0) + a_2 x_2'(t_0) + \cdots + a_n x_n'(t_0) &= u_1 \\
&\cdots \\
a_1 x_1^{(n-1)}(t_0) + a_2 x_2^{(n-1)}(t_0) + \cdots + a_n x_n^{(n-1)}(t_0) &= u_{n-1}.
\end{aligned}
$$

Set

$$v = a_1 x_1 + a_2 x_2 + \cdots + a_n x_n,$$

then v is a solution of $L_n x = 0$ and v satisfies the same initial conditions as u at t_0. So by the uniqueness theorem (Theorem 6.2)

$$u = v = a_1 x_1 + a_2 x_2 + \cdots + a_n x_n.$$

The proof of the last statement in this theorem is Exercise 6.7. $\square$

6.2 Variation of Constants Formula

In this section we state and prove the variation of constants formula (Theorem 6.11) for $L_n x = h(t)$. Before we do that we define the Cauchy function for $L_n x = 0$ and we show how to find this Cauchy function.

Definition 6.7 We define the Cauchy function $x : I \times I \to \mathbb{R}$ for the linear differential equation $L_n x = 0$ to be for each fixed $s \in I$ the solution of the IVP $L_n x = 0$,

$$x^{(i)}(s, s) = 0, \quad 0 \le i \le n - 2, \quad x^{(n-1)}(s, s) = 1$$

[here we let $x^{(i)}(t, s)$ denote the ith partial derivative of $x(t, s)$ with respect to t].

Example 6.8 Note that

$$x(t, s) := \frac{(t - s)^{n-1}}{(n - 1)!}$$

gives the Cauchy function for $x^{(n)} = 0$. $\triangle$

Theorem 6.9 *Assume $x_1, x_2, \ldots, x_n$ are linearly independent solutions of the differential equation $L_n x = 0$ on I, then the Cauchy function is given by*

$$x(t, s) = \frac{W(t, s)}{W(s)},$$

where

$$W(t, s) = \begin{vmatrix} x_1(s) & \cdots & x_n(s) \\ x_1'(s) & \cdots & x_n'(s) \\ \vdots & & \vdots \\ x_1^{(n-2)}(s) & \cdots & x_n^{(n-2)}(s) \\ x_1(t) & \cdots & x_n(t) \end{vmatrix};$$

that is, $W(t, s)$ is $W(s)$ with the last row replaced by $(x_1(t), \ldots, x_n(t))$.

Proof Since $x_1, x_2, \ldots, x_n$ are linearly independent solutions of the differential equation $L_n x = 0$ on I, $W(t) \ne 0$ on I, and so we can let

$$y(t, s) := \frac{W(t, s)}{W(s)},$$

for $t, s \in I$. Expanding the numerator of $y(t, s)$ along the last row of $W(t, s)$ we see that for each fixed s, $y(t, s)$ is a linear combination of $x_1(t), x_2(t), \ldots, x_n(t)$ and hence $y(\cdot, s)$ is a solution of $L_n x = 0$. Also, it is easy to see that for each fixed s, $y(t, s)$ satisfies the initial conditions

$$y^{(i)}(s, s) = 0, \quad 0 \le i \le n - 2, \quad y^{(n-1)}(s, s) = 1.$$

Hence we see that for each fixed s, $y(\cdot, s)$ and the Cauchy function $x(\cdot, s)$ satisfy the same initial conditions. By the existence-uniqueness theorem (Theorem 6.2) we get for each fixed $s \in I$,

$$x(t, s) = y(t, s) = \frac{W(t, s)}{W(s)},$$

and since $s \in I$ is arbitrary we are done. $\square$

Example 6.10 Use Theorem 6.9 to find the Cauchy function for

$$x''' - 2x'' = 0. \tag{6.4}$$

Three linearly independent solutions of the differential equation (6.4) are given by $x_1(t) = 1$, $x_2(t) = t$, $x_3(t) = e^{2t}$. The Wronskian determinant of these three solutions is

$$W(t) = \begin{vmatrix} 1 & t & e^{2t} \\ 0 & 1 & 2e^{2t} \\ 0 & 0 & 4e^{2t} \end{vmatrix} = 4e^{2t}.$$

Also,

$$W(t, s) = \begin{vmatrix} 1 & s & e^{2s} \\ 0 & 1 & 2e^{2s} \\ 1 & t & e^{2t} \end{vmatrix} = e^{2t} - e^{2s}(2t - 2s + 1).$$

It follows from Theorem 6.9 that

$$x(t, s) = \frac{1}{4}(e^{2(t-s)} - 2t + 2s - 1).$$

$\triangle$

In the following result we will see why the Cauchy function is important.

Theorem 6.11 (Variation of Constants) *The solution of the IVP*

$$L_n y = h(t), \quad y^{(i)}(t_0) = 0, \quad 0 \le i \le n - 1,$$

is given by

$$y(t) = \int_{t_0}^{t} x(t, s) h(s) \, ds,$$

where $x(\cdot, \cdot)$ is the Cauchy function for $L_n x = 0$.

Proof Let

$$y(t) := \int_{t_0}^{t} x(t, s) h(s) \, ds, \quad t \in I,$$

where $x(\cdot, \cdot)$ is the Cauchy function for $L_n x = 0$. Then

$$y^{(i)}(t) = \int_{t_0}^{t} x^{(i)}(t, s) h(s) \, ds + x^{(i-1)}(t, t) h(t) = \int_{t_0}^{t} x^{(i)}(t, s) h(s) \, ds,$$

for $0 \le i \le n - 1$, and

$$y^{(n)}(t) = \int_{t_0}^{t} x^{(n)}(t, s) h(s) \, ds + x^{(n-1)}(t, t) h(t)$$

$$= \int_{t_0}^{t} x^{(n)}(t, s) h(s) \, ds + h(t).$$

It follows from these equations that

$$y^{(i)}(t_0) = 0, \quad 0 \le i \le n - 1,$$

and

$$Ly(t) = \int_{t_0}^{t} Lx(t, s)h(s)\, ds + h(t) = h(t),$$

and the proof is complete.   □

Example 6.12 Use the variation of constants formula in Theorem 6.11 to solve the IVP

$$y''' - 2y'' = 4t, \quad y(0) = y'(0) = y''(0) = 0. \tag{6.5}$$

By Example 6.10, the Cauchy function for $x''' - 2x'' = 0$ is given by

$$x(t, s) = \frac{1}{4}(e^{2(t-s)} - 2t + 2s - 1).$$

Hence by the variation of constants formula in Theorem 6.11, the solution x of the IVP (6.5) is given by

$$
\begin{aligned}
y(t) &= \int_0^t x(t, s)4s\, ds \\
&= \int_0^t (se^{2(t-s)} - 2ts + 2s^2 - s)\, ds \\
&= e^{2t}\int_0^t se^{-2s}\, ds - 2t\int_0^t s\, ds + \int_0^t (2s^2 - s)\, ds \\
&= e^{2t}(-\frac{1}{2}te^{-2t} - \frac{1}{4}e^{-2t} + \frac{1}{4}) - t^3 + \frac{2}{3}t^3 - \frac{1}{2}t^2 \\
&= \frac{1}{4}e^{2t} - \frac{1}{3}t^3 - \frac{1}{2}t^2 - \frac{1}{2}t - \frac{1}{4}.
\end{aligned}
$$

△

The proof of the following corollary is Exercise 6.12

Corollary 6.13 *The solution of the IVP*

$$L_n y = h(t), \quad y^{(i)}(t_0) = A_i, \quad 0 \le i \le n - 1,$$

where A_i, $0 \le i \le n - 1$, are given constants, is given by

$$y(t) = u(t) + \int_{t_0}^{t} x(t, s)h(s)\, ds,$$

where $x(\cdot, \cdot)$ is the Cauchy function for $L_n x = 0$ and u solves the IVP

$$L_n u = 0, \quad u^{(i)}(t_0) = A_i, \quad 0 \le i \le n - 1.$$

Example 6.14 Use Corollary 6.13 to solve the nonhomogeneous IVP

$$y''' - 2y'' = 4t, \quad y(0) = 5, \quad y'(0) = 5, \quad y''(0) = 12. \tag{6.6}$$

By Corollary 6.13 the solution of this IVP is given by

$$y(t) = u(t) + \int_0^t x(t, s)h(s)\, ds.$$

Since u solves the IVP

$$u''' - 2u'' = 0, \quad u(0) = 5, \quad u'(0) = 5, \quad u''(0) = 12,$$

we get that

$$u(t) = 2 - t + 3e^{2t}.$$

In Example 6.12 we found that

$$\int_0^t x(t,s)h(s)\, ds = \int_0^t x(t,s)4s\, ds = \frac{1}{4}e^{2t} - \frac{1}{3}t^3 - \frac{1}{2}t^2 - \frac{1}{2}t - \frac{1}{4}.$$

Hence

$$
\begin{aligned}
y(t) &= u(t) + \int_0^t x(t,s)h(s)\, ds \\
&= 2 - t + 3e^{2t} + \frac{1}{4}e^{2t} - \frac{1}{3}t^3 - \frac{1}{2}t^2 - \frac{1}{2}t - \frac{1}{4}. \\
&= \frac{13}{4}e^{2t} - \frac{1}{3}t^3 - \frac{1}{2}t^2 - \frac{3}{2}t + \frac{7}{4}.
\end{aligned}
$$

$\triangle$

6.3 Green's Functions

In this section we study Green's functions for various BVPs for the nth-order linear equation $L_n x = 0$. The comparison theorem (Theorem 6.26) shows why it is important to know sign conditions on Green's functions.

Theorem 6.15 (Existence-Uniqueness Theorem) *Assume that $I = [a, b]$, $1 \le k \le n-1$ and that the homogeneous $(k, n-k)$ conjugate BVP $L_n x = 0$,*

$$x^{(i)}(a) = 0, \quad 0 \le i \le k - 1, \quad x^{(j)}(b) = 0, \quad 0 \le j \le n - k - 1 \quad (6.7)$$

has only the trivial solution. Then the nonhomogeneous $(k, n-k)$ conjugate BVP $L_n y = h(t)$,

$$y^{(i)}(a) = A_i, \quad 0 \le i \le k - 1, \quad y^{(j)}(b) = B_j, \quad 0 \le j \le n - k - 1, \quad (6.8)$$

where $A_i, B_j \in \mathbb{R}$ for $0 \le i \le k - 1, 0 \le j \le n - k - 1$, has a unique solution.

Proof Let $x_1, x_2, \ldots, x_n$ be n linearly independent solutions on $[a, b]$ of the homogeneous differential equation $L_n x = 0$. Then a general solution of $L_n x = 0$ is given by

$$x = c_1 x_1 + c_2 x_2 + \ldots + c_n x_n.$$

Then x satisfies the boundary conditions in (6.7) iff

$$
\begin{aligned}
c_1 x_1(a) + c_2 x_2(a) + \cdots + c_n x_n(a) &= 0 \\
c_1 x_1'(a) + c_2 x_2'(a) + \cdots + c_n x_n'(a) &= 0 \\
&\cdots \quad \cdots \\
c_1 x_1^{(k-1)}(a) + c_2 x_2^{(k-1)}(a) + \cdots + c_n x_n^{(k-1)}(a) &= 0 \\
c_1 x_1(b) + c_2 x_2(b) + \cdots + c_n x_n(b) &= 0 \\
c_1 x_1'(b) + c_2 x_2'(b) + \cdots + c_n x_n'(b) &= 0 \\
&\cdots \quad \cdots \\
c_1 x_1^{(n-k-1)}(b) + c_2 x_2^{(n-k-1)}(b) + \cdots + c_n x_n^{(n-k-1)}(b) &= 0.
\end{aligned}
$$

We can write this system as the vector equation

$$Mc = 0,$$

where

$$
M = \begin{pmatrix}
x_1(a) & x_2(a) & \cdots & x_n(a) \\
x_1'(a) & x_2'(a) & \cdots & x_n'(a) \\
\cdots & \cdots & \cdots & \cdots \\
x_1^{(k-1)}(a) & x_2^{(k-1)}(a) & \cdots & x_n^{(k-1)}(a) \\
x_1(b) & x_2(b) & \cdots & x_n(b) \\
x_1'(b) & x_2'(b) & \cdots & x_n'(b) \\
\cdots & \cdots & \cdots & \cdots \\
x_1^{(n-k-1)}(b) & x_2^{(n-k-1)}(b) & \cdots & x_n^{(n-k-1)}(b)
\end{pmatrix}, \quad
c = \begin{pmatrix} c_1 \\ c_2 \\ c_3 \\ \vdots \\ c_n \end{pmatrix}.
$$

Since the homogeneous BVP (6.7) has only the trivial solution, we get that

$$c_1 = c_2 = \cdots = c_n = 0$$

is the unique solution of the above linear system. Hence

$$\det M \neq 0. \tag{6.9}$$

Now let w be a solution of the nonhomogeneous differential equation $L_n y = h(t)$. Then a general solution of $L_n y = h(t)$ is

$$y = d_1 x_1 + d_2 x_2 + \cdots + d_n x_n + w.$$

Then the boundary value problem $L_n y = h(t)$, (6.8) has a solution iff the vector equation

$$
Md = \begin{pmatrix}
A_0 - w(a) \\
\vdots \\
A_{k-1} - w^{(k-1)}(a) \\
B_0 - w(b) \\
\vdots \\
B_{n-k-1} - w^{(n-k-1)}(b)
\end{pmatrix},
$$

where $d = \begin{pmatrix} d_1 \\ d_2 \\ \vdots \\ d_n \end{pmatrix}$, has a solution. Using (6.9), we get that this system has a unique solution $d_1, d_2, \cdots, d_n$ and this completes the proof. $\square$

The function $G(\cdot, \cdot)$ in Theorem 6.16 is called the *Green's function* for the $(k, n - k)$ conjugate BVP $L_n x = 0$, (6.7).

Theorem 6.16 *Assume* $I = [a, b]$, $1 \le k \le n - 1$, *and that the* $(k, n - k)$ *conjugate BVP* $L_n x = 0$, (6.7) *has only the trivial solution. For each fixed* $s \in I$, *let* $u(\cdot, s)$ *be the solution of the BVP* $L_n u = 0$,

$$u^{(i)}(a, s) = 0, \quad u^{(j)}(b, s) = -x^{(j)}(b, s), \qquad (6.10)$$

for $0 \le i \le k-1, 0 \le j \le n-k-1$, *respectively, where* $x(\cdot, \cdot)$ *is the Cauchy function for* $L_n x = 0$. *Define*

$$G(t, s) := \begin{cases} u(t, s), & if \ a \le t \le s \le b \\ v(t, s), & if \ a \le s \le t \le b, \end{cases} \qquad (6.11)$$

where $v(t, s) := u(t, s) + x(t, s)$. *Assume* h *is continuous on* $[a, b]$, *then*

$$y(t) := \int_a^b G(t, s)h(s) \, ds$$

is the unique solution of the nonhomogeneous $(k, n - k)$ *conjugate BVP* $L_n y = h(t)$, (6.7). *Furthermore, for each fixed* $s \in [a, b]$, $v(\cdot, s)$ *is a solution of* $Lx = 0$ *satisfying the second set of boundary conditions in* (6.7).

Proof By Theorem 6.15 the BVP $L_n u = 0$, (6.10) has, for each fixed s, a unique solution and hence the $u(t, s)$ in the statement of this theorem is well defined. Let $G(t, s)$ be defined by (6.11), where $v(t, s) := u(t, s) + x(t, s)$. Let

$$y(t) := \int_a^b G(t, s)h(s) \, ds,$$

then

$$
\begin{aligned}
y(t) &= \int_a^t G(t,s)h(s)\,ds + \int_t^b G(t,s)h(s)\,ds \\
&= \int_a^t v(t,s)h(s)\,ds + \int_t^b u(t,s)h(s)\,ds \\
&= \int_a^t [u(t,s) + x(t,s)]h(s)\,ds + \int_t^b u(t,s)h(s)\,ds \\
&= \int_a^b u(t,s)h(s)\,ds + \int_a^t x(t,s)h(s)\,ds. \\
&= \int_a^b u(t,s)h(s)\,ds + z(t),
\end{aligned}
$$

where, by the variation of constants formula (Theorem 6.11),

$$
z(t) := \int_a^t x(t,s)h(s)\,ds
$$

is the solution of the nonhomogeneous differential equation $L_n y = h(t)$ satisfying the initial conditions $z^{(i)}(a) = 0$, $0 \le i \le n-1$. It then follows that

$$
\begin{aligned}
L_n y(t) &= \int_a^b L_n u(t,s)h(s)\,ds + L_n z(t) \\
&= h(t).
\end{aligned}
$$

Also,

$$
y^{(i)}(a) = \int_a^b u^{(i)}(a,s)h(s)\,ds + z^{(i)}(a) = 0,
$$

for $0 \le i \le k-1$, since for each fixed s, $u(t,s)$ satisfies the first set of boundary conditions in (6.10). Since $v(t,s) = u(t,s) + x(t,s)$, we have for each fixed s, $v(\cdot,s)$ is a solution of $L_n x = 0$. Using the fact that $u(t,s)$ satisfies the second set of boundary conditions in (6.10), we get that $v(t,s)$ satisfies the second set of boundary conditions in (6.7). To see that y

satisfies the second set of boundary conditions in (6.7), note that

$$
\begin{aligned}
y(t) &= \int_a^t G(t,s)h(s)\,ds + \int_t^b G(t,s)h(s)\,ds \\
&= \int_a^t v(t,s)h(s)\,ds + \int_t^b u(t,s)h(s)\,ds \\
&= \int_a^t v(t,s)h(s)\,ds + \int_t^b [v(t,s) - x(t,s)]h(s)\,ds \\
&= \int_a^b v(t,s)h(s)\,ds - \int_t^b x(t,s)h(s)\,ds \\
&= \int_a^b v(t,s)h(s)\,ds + \int_b^t x(t,s)h(s)\,ds \\
&= \int_a^b v(t,s)h(s)\,ds + w(t),
\end{aligned}
$$

where, by the variation of constants formula (Theorem 6.11),

$$
w(t) := \int_b^t x(t,s)h(s)\,ds
$$

solves the initial conditions $w^{(j)}(b) = 0$, $0 \le j \le n-1$. Hence

$$
y^{(j)}(b) = \int_a^b v^{(j)}(b,s)h(s)\,ds + w^{(j)}(b) = 0,
$$

for $0 \le j \le n-k-1$.   □

In the standard way (see Exercise 6.14) you can prove the following corollary.

Corollary 6.17 *Assume that $I = [a,b]$ and that the $(k, n-k)$ conjugate BVP $L_n x = 0$, (6.7) has only the trivial solution. For each fixed $s \in I$, let $u(\cdot, s)$ be the solution of the BVP $L_n u = 0$,*

$$
u^{(i)}(a,s) = 0, \quad u^{(j)}(b,s) = -x^{(j)}(b,s), \tag{6.12}
$$

for $0 \le i \le k-1$, $0 \le j \le n-k-1$, respectively, where $x(\cdot, \cdot)$ is the Cauchy function for $L_n x = 0$. Define $G(t,s)$ by (6.11), where $v(t,s) := u(t,s) + x(t,s)$. Assume h is continuous on $[a,b]$, then

$$
y(t) := w(t) + \int_a^b G(t,s)h(s)\,ds,
$$

where w is the unique solution of the BVP $L_n w = 0$,

$$
w^{(i)}(a) = A_i, \quad 0 \le i \le k-1, \quad w^{(j)}(b) = B_j, \quad 0 \le j \le n-k-1,
$$

is the unique solution of the nonhomogeneous BVP $L_n y = h(t)$, (6.8). Furthermore, for each fixed $s \in [a,b]$, $v(\cdot, s)$ is a solution of $Lx = 0$ satisfying the second set of boundary conditions in (6.7).

Definition 6.18 The solutions $x_0(\cdot, t_0), x_1(\cdot, t_0), \cdots, x_{n-1}(\cdot, t_0)$ of $L_n x = 0$ satisfying the initial conditions

$$x_i^{(j)}(t_0, t_0) = \delta_{ij}, \quad 0 \le i, j \le n - 1,$$

where δ_{ij} is the Kroneker delta function (i.e., $\delta_{ij} = 0$ if $i \ne j$ and $\delta_{ij} = 1$ if $i = j$), are called the *normalized solutions* of $L_n x = 0$ at $t = t_0$.

Example 6.19 The normalized solutions of $x^{(n)} = 0$ at t_0 are given by

$$x_i(t, t_0) := \frac{(t - t_0)^i}{i!}, \quad t \in \mathbb{R},$$

$0 \le i \le n - 1$.

$\triangle$

Theorem 6.20 *Assume that $I = [a, b]$ and that the homogeneous $(k, n - k)$ conjugate BVP $L_n x = 0$, (6.7) has only the trivial solution. Then the Green's function for the $(k, n - k)$ conjugate BVP $L_n x = 0$, (6.7) is given by (6.11), where*

$$u(t, s) = \frac{1}{D} \begin{vmatrix} 0 & x_k(t, a) & \cdots & x_{n-1}(t, a) \\ x(b, s) & x_k(b, a) & \cdots & x_{n-1}(b, a) \\ x'(b, s) & x_k'(b, a) & \cdots & x_{n-1}'(b, a) \\ \cdots & \cdots & \cdots & \cdots \\ x^{(n-k-1)}(b, s) & x_k^{(n-k-1)}(b, a) & \cdots & x_{n-1}^{(n-k-1)}(b, a) \end{vmatrix},$$

and $v(t, s)$ is the preceding expression for $u(t, s)$ with the 0 in the upper left hand corner of the determinant replaced by $x(t, s)$ and

$$D := W[x_k(t, a), x_{k+1}(t, a), \cdots, x_{n-1}(t, a)](b).$$

Proof Since the homogeneous $(k, n - k)$ conjugate BVP $L_n x = 0$, (6.7) has only the trivial solution we have by Exercise 6.18 that $D \ne 0$. Let

$$w(t, s) := \frac{1}{D} \begin{vmatrix} 0 & x_k(t, a) & \cdots & x_{n-1}(t, a) \\ x(b, s) & x_k(b, a) & \cdots & x_{n-1}(b, a) \\ x'(b, s) & x_k'(b, a) & \cdots & x_{n-1}'(b, a) \\ \cdots & \cdots & \cdots & \cdots \\ x^{(n-k-1)}(b, s) & x_k^{(n-k-1)}(b, a) & \cdots & x_{n-1}^{(n-k-1)}(b, a) \end{vmatrix}.$$

Expanding $w(t, s)$ along the first row, we see that for each fixed s, $w(t, s)$ is a linear combination of $x_k(t, a), x_{k+1}(t, a), \cdots, x_{n-1}(t, a)$ and so we conclude that for each fixed s, $w(\cdot, s)$ is a solution of $L_n x = 0$ and $w^{(i)}(a, s) = 0$ for $0 \le i \le k - 1$. Now let $z(t, s)$ to be the expression used to define $w(t, s)$ with the 0 in the upper left-hand corner of the determinant replaced by

$x(t, s)$. Then

$$
z(t, s)
$$

$$
= \frac{1}{D} \begin{vmatrix} x(t, s) & x_k(t, a) & \cdots & x_{n-1}(t, a) \\ x(b, s) & x_k(b, a) & \cdots & x_{n-1}(b, a) \\ x'(b, s) & x_k'(b, a) & \cdots & x_{n-1}'(b, a) \\ \cdots & \cdots & \cdots & \cdots \\ x^{(n-k-1)}(b, s) & x_k^{(n-k-1)}(b, a) & \cdots & x_{n-1}^{(n-k-1)}(b, a) \end{vmatrix}
$$

$$
= x(t, s) + \frac{1}{D} \begin{vmatrix} 0 & x_k(t, a) & \cdots & x_{n-1}(t, a) \\ x(b, s) & x_k(b, a) & \cdots & x_{n-1}(b, a) \\ x'(b, s) & x_k'(b, a) & \cdots & x_{n-1}'(b, a) \\ \cdots & \cdots & \cdots & \cdots \\ x^{(n-k-1)}(b, s) & x_k^{(n-k-1)}(b, a) & \cdots & x_{n-1}^{(n-k-1)}(b, a) \end{vmatrix}
$$

$$
= x(t, s) + w(t, s).
$$

It follows that

$$
z^{(j)}(b, s) = x^{(j)}(b, s) + w^{(j)}(b, s) = 0, \quad 0 \le j \le n - k - 1.
$$

This implies that

$$
w^{(j)}(b, s) = -x^{(j)}(b, s).
$$

Since for each fixed s, $u(t, s)$ and $w(t, s)$ satisfy the same $(k, n-k)$ conjugate BVP, we have (using Theorem 6.15) the desired result

$$
u(t, s) = w(t, s)
$$

$$
= \frac{1}{D} \begin{vmatrix} 0 & x_k(t, a) & \cdots & x_{n-1}(t, a) \\ x(b, s) & x_k(b, a) & \cdots & x_{n-1}(b, a) \\ x'(b, s) & x_k'(b, a) & \cdots & x_{n-1}'(b, a) \\ \cdots & \cdots & \cdots & \cdots \\ x^{(n-k-1)}(b, s) & x_k^{(n-k-1)}(b, a) & \cdots & x_{n-1}^{(n-k-1)}(b, a) \end{vmatrix}
$$

Also,

$$
v(t, s) = x(t, s) + u(t, s) = x(t, s) + w(t, s) = z(t, s),
$$

and so $v(t, s)$ has the desired form. $\qquad \square$

Example 6.21 Use Theorem 6.20 to find the Green's function for the $(n - 1, 1)$ conjugate BVP

$$
x^{(n)} = 0, \quad x^{(i)}(a) = 0, \quad 0 \le i \le n - 2, \quad x(b) = 0.
$$

First,

$$
u(t, s) = \frac{1}{x_{n-1}(b, a)} \begin{vmatrix} 0 & x_{n-1}(t, a) \\ x(b, s) & x_{n-1}(b, a) \end{vmatrix} = -\frac{(t - a)^{n-1}(b - s)^{n-1}}{(n - 1)!(b - a)^{n-1}}.
$$

Also,

$$v(t,s) = \frac{1}{x_{n-1}(b,a)} \begin{vmatrix} x(t,s) & x_{n-1}(t,a) \\ x(b,s) & x_{n-1}(b,a) \end{vmatrix} = \frac{(t-s)^{n-1}}{(n-1)!}$$
$$- \frac{(t-a)^{n-1}(b-s)^{n-1}}{(n-1)!(b-a)^{n-1}}.$$

Hence

$$G(t,s) = \begin{cases} -\frac{(t-a)^{n-1}(b-s)^{n-1}}{(n-1)!(b-a)^{n-1}}, & \text{if } a \le t \le s \le b \\ \frac{(t-s)^{n-1}}{(n-1)!} - \frac{(t-a)^{n-1}}{(n-1)!}\frac{(b-s)^{n-1}}{(n-1)!}, & \text{if } a \le s \le t \le b. \end{cases}$$

$\triangle$

Example 6.22 Using an appropriate Green's function, solve the BVP

$$y''' = t, \quad y(0) = y'(0) = 0, \quad y(1) = 0.$$

From Example 6.21 we get that the Green's function for

$$x''' = 0, \quad x(0) = x'(0) = 0, \quad x(1) = 0$$

is given by

$$G(t,s) = \begin{cases} -\frac{t^2(1-s)^2}{2}, & \text{if } 0 \le t \le s \le 1 \\ \frac{(t-s)^2}{2} - \frac{t^2(1-s)^2}{2}, & \text{if } 0 \le s \le t \le 1. \end{cases}$$

Hence

$$\begin{aligned} y(t) &= \int_0^1 G(t,s)h(s)\,ds = \int_0^1 G(t,s)s\,ds \\ &= \int_0^t \left(\frac{(t-s)^2}{2} - \frac{t^2(1-s)^2}{2} \right) s\,ds + \int_t^1 \left(-\frac{t^2(1-s)^2}{2} \right) s\,ds \\ &= -\frac{t^2}{2} \int_0^1 (1-s)^2 s\,ds + \frac{1}{2} \int_0^t (t-s)^2 s\,ds \\ &= -\frac{1}{24} t^2 + \frac{1}{24} t^4. \end{aligned}$$

$\triangle$

Definition 6.23 We say $L_n x = 0$ is *disconjugate* on I provided no non-trivial solution of $L_n x = 0$ has n (or more) zeros, counting multiplicities, in I.

Example 6.24 It is easy to see that $x^{(n)} = 0$ is disconjugate on $\mathbb{R}$. $\triangle$

Theorem 6.25 *If $L_n x = 0$ is disconjugate on $I = [a,b]$, then the Green's function $G(t,s)$ for the BVP $L_n x = 0$, (6.7) satisfies*

$$(-1)^{n-k} G(t,s) \ge 0,$$

for $t,s \in [a,b]$.

Proof For a proof of this result see Hartman [18]. $\square$

Consistency of the sign of a Green's function is very important. The following theorem is an application of this fact.

Theorem 6.26 (Comparison Theorem) *Assume $L_n x = 0$ is disconjugate on $[a, b]$ and $u, v \in C^n[a, b]$ satisfy*

$$L_n u(t) \geq L_n v(t), \quad t \in [a, b], \tag{6.13}$$

and

$$u^{(i)}(a) = v^{(i)}(a), \quad 0 \leq i \leq k - 1, \quad u^{(j)}(b) = v^{(j)}(b), \quad 0 \leq j \leq n - k - 1.$$

Then

$$(-1)^{n-k} u(t) \geq (-1)^{n-k} v(t), \quad t \in [a, b].$$

Proof Let

$$h(t) := L_n u(t) - L_n v(t) \geq 0, \quad t \in [a, b],$$

where we have used (6.13). Then let

$$w(t) := u(t) - v(t), \quad t \in [a, b].$$

It follows that w is the solution of the $(k, n - k)$ conjugate BVP

$$L_n w = h(t), \quad w^{(i)}(a) = 0, \quad 0 \leq i \leq k-1, \quad w^{(j)}(b) = 0, \quad 0 \leq i \leq n-k-1.$$

This implies that

$$(-1)^{n-k} w(t) = \int_a^b (-1)^{n-k} G(t, s) h(s) \, ds \geq 0,$$

for $t \in [a, b]$. This implies the desired result

$$(-1)^{n-k} u(t) \geq (-1)^{n-k} v(t), \quad t \in [a, b].$$

$\square$

We end this section by stating the analogues of Theorems 6.15, 6.16, and 6.20 (see Exercises 6.15, 6.16 and 6.17) for the $(k, n - k)$ right-focal BVP $L_n x = 0$,

$$x^{(i)}(a) = 0, \quad 0 \leq i \leq k - 1, \quad x^{(j)}(b) = 0, \quad k \leq j \leq n - 1, \tag{6.14}$$

where $1 \leq k \leq n - 1$. In Example 7.16 we are interested in how tall a vertical thin rod can be before it starts to bend. This problem leads to a right-focal BVP.

Theorem 6.27 (Existence-Uniqueness Theorem) *Assume that $I = [a, b]$, $1 \leq k \leq n-1$ and that the homogeneous $(k, n-k)$ right-focal BVP $L_n x = 0$,*

$$x^{(i)}(a) = 0, \quad 0 \leq i \leq k - 1, \quad x^{(j)}(b) = 0, \quad k \leq j \leq n - 1, \tag{6.15}$$

has only the trivial solution. Then the nonhomogeneous $(k, n-k)$ right-focal BVP $L_n y = h(t)$,

$$y^{(i)}(a) = A_i, \quad 0 \leq i \leq k - 1, \quad y^{(j)}(b) = B_j, \quad k \leq j \leq n - 1, \tag{6.16}$$

where $A_i, B_j \in \mathbb{R}$ for $0 \leq i \leq k - 1$, $k \leq j \leq n - 1$, has a unique solution.

Theorem 6.28 *Assume that $I = [a, b]$, $1 \leq k \leq n-1$ and that the $(k, n-k)$ right-focal BVP $L_n x = 0$, (6.15) has only the trivial solution. For each fixed $s \in I$, let $u(\cdot, s)$ be the solution of the BVP $L_n u = 0$,*

$$u^{(i)}(a, s) = 0, \quad u^{(j)}(b, s) = -x^{(j)}(b, s), \tag{6.17}$$

for $0 \leq i \leq k - 1$, $k \leq j \leq n - 1$, respectively, where $x(\cdot, \cdot)$ is the Cauchy function for $L_n x = 0$. Define $G(t,s)$ by (6.11), where $v(t, s) := u(t, s) + x(t, s)$. Assume h is continuous on $[a, b]$; then

$$y(t) := \int_a^b G(t, s) h(s) \, ds$$

is the unique solution of the nonhomogeneous $(k, n-k)$ right-focal BVP $L_n y = h(t)$, (6.15). Furthermore, for each fixed $s \in [a, b]$, $v(\cdot, s)$ is a solution of $Lx = 0$ satisfying the second set of boundary conditions in (6.15).

Theorem 6.29 *Assume $I = [a, b]$ and that the homogeneous $(k, n - k)$ right-focal BVP $L_n x = 0$, (6.15) has only the trivial solution. Then the Green's function for the $(k, n-k)$ right-focal BVP $L_n x = 0$, (6.15) is given by (6.11), where*

$$u(t, s) = \frac{1}{D} \begin{vmatrix} 0 & x_k(t, a) & \cdots & x_{n-1}(t, a) \\ x^{(k)}(b, s) & x_k^{(k)}(b, a) & \cdots & x_{n-1}^{(k)}(b, a) \\ x^{(k+1)}(b, s) & x_k^{(k+1)}(b, a) & \cdots & x_{n-1}^{(k+1)}(b, a) \\ \cdots & \cdots & \cdots & \cdots \\ x^{(n-1)}(b, s) & x_k^{(n-1)}(b, a) & \cdots & x_{n-1}^{(n-1)}(b, a) \end{vmatrix},$$

and $v(t, s)$ is the preceding expression for $u(t, s)$ with the 0 in the upper left-hand corner of the determinant replaced by $x(t, s)$ and

$$D := W[x_k^{(k)}(t, a), x_{k+1}^{(k)}(t, a), \cdots, x_{n-1}^{(k)}(t, a)](b).$$

Example 6.30 Use Theorem 6.29 to find the Green's function for the $(n - 1, 1)$ right-focal BVP

$$x^{(n)} = 0, \quad x^{(i)}(a) = 0, \quad 0 \leq i \leq n - 2, \quad x^{(n-1)}(b) = 0.$$

First,

$$u(t, s) = \frac{1}{x_{n-1}^{(n-1)}(b, a)} \begin{vmatrix} 0 & x_{n-1}(t, a) \\ x^{(n-1)}(b, s) & x_{n-1}^{(n-1)}(b, a) \end{vmatrix} = -\frac{(t - a)^{n-1}}{(n - 1)!}.$$

Also,

$$v(t, s) =$$

$$\frac{1}{x_{n-1}^{(n-1)}(b, a)} \begin{vmatrix} x(t, s) & x_{n-1}(t, a) \\ x^{(n-1)}(b, s) & x_{n-1}^{(n-1)}(b, a) \end{vmatrix} = \frac{(t - s)^{n-1}}{(n - 1)!} - \frac{(t - a)^{n-1}}{(n - 1)!}.$$

Hence

$$G(t, s) = \begin{cases} -\frac{(t-a)^{n-1}}{(n-1)!}, & \text{if } a \leq t \leq s \leq b \\ \frac{(t-s)^{n-1}}{(n-1)!} - \frac{(t-a)^{n-1}}{(n-1)!}, & \text{if } a \leq s \leq t \leq b. \end{cases}$$

$\triangle$

Example 6.31 Using an appropriate Green's function, solve the (2,1) right-focal BVP

$$y''' = t, \quad y(0) = y'(0) = 0, \quad y''(1) = 0.$$

From Example 6.30 we get that the Green's function for

$$x''' = 0, \quad x(0) = x'(0) = 0, \quad x''(1) = 0$$

is given by

$$G(t, s) = \begin{cases} -\frac{t^2}{2}, & \text{if } 0 \le t \le s \le 1 \\ \frac{(t-s)^2}{2} - \frac{t^2}{2}, & \text{if } 0 \le s \le t \le 1. \end{cases}$$

Hence

$$\begin{aligned} y(t) &= \int_0^1 G(t, s) h(s) \, ds = \int_0^1 G(t, s) s \, ds \\ &= \int_0^t \left(\frac{(t-s)^2}{2} - \frac{t^2}{2} \right) s \, ds + \int_t^1 \left(-\frac{t^2}{2} \right) s \, ds \\ &= -\frac{t^2}{2} \int_0^1 s \, ds + \frac{1}{2} \int_0^t (t-s)^2 s \, ds \\ &= -\frac{1}{4} t^2 + \frac{1}{24} t^4. \end{aligned}$$

$\triangle$

6.4 Factorizations and Principal Solutions

In this section we will derive the Polya and Trench factorizations and use the Trench factorization to get the existence of "principal solutions" of $L_n x = 0$.

Theorem 6.32 *Assume that $x_1, x_2, \cdots, x_n$ are n linearly independent solutions of $L_n x = 0$ on I. If $x \in C^n[I]$, then*

$$L_n[x](t) = \frac{w[x_1(t), x_2(t), \cdots, x_n(t), x(t)]}{w[x_1(t), x_2(t), \cdots, x_n(t)]},$$

for $t \in I$, where $w[x_1(t), x_2(t), \cdots, x_n(t)]$ is the Wronskian determinant of

$$x_1(t), x_2(t) \cdots, x_n(t).$$

Proof Consider

$$w[x_1(t), x_2(t), \cdots, x_n(t), x(t)]$$

$$= \begin{vmatrix} x_1(t) & x_2(t) & \cdots & x_n(t) & x(t) \\ x_1'(t) & x_2'(t) & \cdots & x_n'(t) & x'(t) \\ \cdots & \cdots & \cdots & \cdots & \cdots \\ x_1^{(n)}(t) & x_2^{(n)}(t) & \cdots & x_n^{(n)}(t) & x^{(n)}(t) \end{vmatrix}.$$

Taking $p_0(t)$ times the first row, $p_1(t)$ times the second row, and so forth until we take $p_{n-1}(t)$ times the second to last row and adding to the last row we get

$$w[x_1(t), x_2(t), \cdots, x_n(t), x(t)]$$

$$= \begin{vmatrix} x_1(t) & x_2(t) & \cdots & x_n(t) & x(t) \\ x_1'(t) & x_2'(t) & \cdots & x_n'(t) & x'(t) \\ \cdots & \cdots & \cdots & \cdots & \cdots \\ L_n x_1(t) & L_n x_2(t) & \cdots & L_n x_n(t) & L_n x(t) \end{vmatrix}$$

$$= \begin{vmatrix} x_1(t) & x_2(t) & \cdots & x_n(t) & x(t) \\ x_1'(t) & x_2'(t) & \cdots & x_n'(t) & x'(t) \\ \cdots & \cdots & \cdots & \cdots & \cdots \\ 0 & 0 & \cdots & 0 & L_n x(t) \end{vmatrix}$$

$$= w[x_1(t), x_2(t), \cdots, x_n(t)] L_n x(t). \quad \square$$

Corollary 6.33 *Assume* $x, x_1, x_2, \cdots, x_k \in C^k[I]$, $k \geq 2$ *and*

$$w_i(t) := w[x_1(t), x_2(t), \cdots, x_i(t)] \neq 0,$$

for $i = k - 1, k$ *and for* $t \in I$; *then*

$$\left\{ \frac{w[x_1(t), x_2(t), \cdots, x_{k-1}(t), x(t)]}{w_k(t)} \right\}' = \frac{w_{k-1}(t) w[x_1(t), \cdots, x_k(t), x(t)]}{w_k^2(t)},$$

for $t \in I$.

Proof Define an operator M_k on $C^k[I]$ by

$$M_k[x](t) := \frac{w_k(t)}{w_{k-1}(t)} \left\{ \frac{w[x_1(t), x_2(t), \cdots, x_{k-1}(t), x(t)]}{w_k(t)} \right\}',$$

for $t \in I$. Expanding out the right-hand side, we get

$$M_k[x](t) = x^{(k)}(t) + q_{k-1}(t) x^{(k-1)}(t) + \cdots + q_0(t) x(t),$$

where $q_0(t), \cdots, q_{k-1}(t)$ are defined appropriately. Note that $x_1, x_2, \cdots, x_k$ are k linearly independent solutions of $M_k x = 0$. It follows from Theorem 6.32 with $n = k$ that

$$M_k[x](t) = \frac{w_k(t)}{w_{k-1}(t)} \left\{ \frac{w[x_1(t), x_2(t), \cdots, x_{k-1}(t), x(t)]}{w_k(t)} \right\}'$$

$$= \frac{w[x_1(t), \cdots, x_k(t), x(t)]}{w_k(t)},$$

for $t \in I$, which implies the desired result. $\quad \square$

We now use Corollary 6.33 to prove the Polya factorization theorem.

Theorem 6.34 (Polya Factorization) *Assume* $x_1, x_2, \cdots, x_n$ *are solutions of* $L_n x = 0$ *such that* $w_k(t) := w[x_1(t), x_2(t), \cdots, x_k(t)] \neq 0$ *for* $t \in I$, $1 \leq k \leq n$. *Then for* $x \in C^n[I]$,

$$L_n x(t) = \rho_n(t) \left\{ \rho_{n-1}(t) \left\{ \cdots \left\{ \rho_0(t) x(t) \right\}' \cdots \right\}' \right\}',$$

for $t \in I$, *where*

$$\rho_n(t) = \frac{w_n(t)}{w_{n-1}(t)}, \quad \rho_k(t) = \frac{w_k^2(t)}{w_{k-1}(t)w_{k+1}(t)}, \quad \rho_0(t) = \frac{1}{x_1(t)},$$

for $t \in I$, $1 \leq k \leq n-1$, *where* $w_0(t) := 1$.

Proof Since $w_n(t) \neq 0$, we have by Theorem 6.32 that

$$L_n[x](t) = \frac{w[x_1(t), x_2(t), \cdots, x_n(t), x(t)]}{w_n(t)},$$

for $t \in I$. Since $w_{n-1}(t) \neq 0, w_n(t) \neq 0$ for $t \in I$, we get, using Corollary 6.33, that

$$
\begin{aligned}
L_n[x](t) &= \frac{w_n(t)}{w_{n-1}(t)} \left\{ \frac{w[x_1(t), \cdots, x_{n-1}(t), x(t)]}{w_n(t)} \right\}' \\
&= \rho_n(t) \left\{ \frac{w[x_1(t), \cdots, x_{n-1}(t), x(t)]}{w_n(t)} \right\}',
\end{aligned}
$$

for $t \in I$. Since $w_{n-2}(t) \neq 0, w_{n-1}(t) \neq 0$ for $t \in I$, we get, using Corollary 6.33, that

$$
\begin{aligned}
L_n[x](t) &= \rho_n(t) \left\{ \frac{w_{n-1}^2(t)}{w_{n-2}(t)w_n(t)} \left\{ \frac{w[x_1(t), \cdots, x_{n-2}(t), x(t)]}{w_{n-1}(t)} \right\}' \right\}' \\
&= \rho_n(t) \left\{ \rho_{n-1}(t) \left\{ \frac{w[x_1(t), \cdots, x_{n-2}(t), x(t)]}{w_{n-1}(t)} \right\}' \right\}',
\end{aligned}
$$

for $t \in I$. Continuing in this fashion (using mathematical induction), we obtain

$$L_n[x](t) = \rho_n(t) \left\{ \rho_{n-1}(t) \left\{ \cdots \left\{ \rho_2(t) \left\{ \frac{w[x_1(t), x(t)]}{w_2(t)} \right\}' \right\}' \cdots \right\}' \right\}',$$

for $t \in I$. Finally, using

$$\{\rho_0(t)x(t)\}' = \left\{ \frac{x(t)}{x_1(t)} \right\}' = \frac{w[x_1(t), x(t)]}{x_1^2(t)},$$

for $t \in I$ we get the desired result. $\square$

Example 6.35 Find a Polya factorization for

$$x''' - 2x'' - x' + 2x = 0.$$

The corresponding characteristic equation is

$$\lambda^3 - 2\lambda^2 - \lambda + 2 = (\lambda - 1)(\lambda - 2)(\lambda + 1) = 0,$$

and hence the characteristic roots are

$$\lambda_1 = 1, \quad \lambda_2 = 2, \quad \lambda_3 = -1,$$

and therefore

$$x_1 = e^t, \quad x_2 = e^{2t}, \quad x_3 = e^{-t},$$

are solutions. Note that

$$w_1(t) = x_1(t) = e^t, \quad w_2(t) = w[e^t, e^{2t}] = e^{3t},$$

and

$$w_3(t) = w[e^t, e^{2t}, e^{-t}] = 6e^{2t}.$$

By Theorem 6.34, we get a Polya factorization where

$$\rho_0(t) = e^{-t}, \quad \rho_1(t) = e^{-t}, \quad \rho_2(t) = \frac{1}{6}e^{3t}, \quad \rho_3(t) = 6e^{-t}.$$

Hence

$$
\begin{aligned}
x''' - 2x'' - x' + 2x &= \rho_3(t)\{\rho_2(t)[\rho_1(t)(\rho_0(t)x)']']'\}' \\
&= 6e^{-t}\{\tfrac{1}{6}e^{3t}[e^{-t}(e^{-t}x(t))')']'\}' \\
&= e^{-t}\{e^{3t}[e^{-t}(e^{-t}x(t))')']'\}' = 0
\end{aligned}
$$

is a Polya factorization.   △

Corollary 6.36 *If $L_n x = 0$ is disconjugate on I, then $L_n x = 0$ has a Poyla factorization on I.*

Proof We will only give the proof for the case where $I = [a, b]$. Let $x_i(t, a)$, $0 \le i \le n - 1$ be the normalized solutions of $L_n x = 0$ at a. Then, using the disconjugacy of $L_n x = 0$ on $[a, b]$, we have, by Exercise 6.18, that

$$w[x_k(t, a), x_{k+1}(t, a), \cdots, x_{n-1}(t, a)] \ne 0,$$

for $t \in (a, b]$, $1 \le k \le n - 1$ (why is this also true for $k = 0$?). Next let

$$u_i(t) := (-1)^{i+1} x_{n-i}(t, a),$$

for $t \in [a, b]$, $1 \le k \le n$; then it follows that

$$w[u_1(t), u_2(t), \cdots, u_k(t)] \ne 0,$$

for $t \in (a, b]$, $1 \le k \le n$. Now let $x_{i,\epsilon}$ be the solution of the IVP $L_n x = 0$,

$$x_{i,\epsilon}^{(j)}(a) = (-1)^{i+1} \frac{d^j}{d\epsilon^j} \frac{\epsilon^{n-i}}{(n-i)!};$$

then for $\epsilon > 0$, sufficiently small, we get that

$$w[x_{1,\epsilon}(t), x_{2,\epsilon}(t), \cdots, x_{k,\epsilon}(t)] > 0,$$

for $t \in [a, b]$, $1 \le k \le n$. The final result then follows from Theorem 6.34.   □

Theorem 6.37 (Trench Factorization) *Assume $L_n x = 0$ is disconjugate on $I := [a, \infty)$; then there exist positive functions γ_i, $0 \leq i \leq n$, such that for $x \in C^n[I]$*

$$L_n x(t) = \gamma_n(t) \left\{ \gamma_{n-1}(t) \left\{ \cdots \left\{ \gamma_0(t) x(t) \right\}' \cdots \right\}' \right\}',$$

for $t \in I$, where

$$\int_a^\infty \frac{1}{\gamma_i(t)} dt = \infty,$$

for $1 \leq k \leq n - 1$.

Proof For a proof of this result see Trench [49].   □

Theorem 6.38 (Principal Solutions) *Assume $L_n x = 0$ is disconjugate on $I := [a, \infty)$; then there exist solutions u_i, $1 \leq i \leq n$ of $L_n x = 0$ such that*

$$\lim_{t \to \infty} \frac{u_i(t)}{u_{i+1}(t)} = 0,$$

for $1 \leq i \leq n - 1$. We call u_i an ith principal solution of $L_n x = 0$ at ∞.

Proof Since $L_n x = 0$ is disconjugate on $[a, \infty)$, we have from Theorem 6.38 that there exist positive functions γ_i such that the differential equation $L_n x = 0$ is equivalent to the differential equation

$$L_n x = \gamma_n(t) \left\{ \gamma_{n-1}(t) \left\{ \cdots \left\{ \gamma_0(t) x \right\}' \cdots \right\}' \right\}' = 0,$$

for $t \in I$, where

$$\int_a^\infty \frac{1}{\gamma_i(t)} dt = \infty,$$

for $1 \leq k \leq n - 1$. Let $u_1(t) := \frac{1}{\gamma_0(t)}$ and

$$u_{i+1}(t) := \frac{1}{\gamma_0(t)} \int_a^t \frac{1}{\gamma_1(s_1)} \int_a^{s_2} \frac{1}{\gamma_2(s_2)} \cdots \int_a^{s_{i-1}} \frac{1}{\gamma_i(s_i)} ds_i \cdots ds_2 \, ds_1,$$

$1 \leq i \leq n - 1$. Then each u_i, $1 \leq i \leq n$ is a solution of $L_n x = 0$ and, applying l'Hôpital's rule $i - 1$ times, we get

$$\lim_{t \to \infty} \frac{u_i(t)}{u_{i+1}(t)} = \lim_{t \to \infty} \frac{1}{\int_a^t \frac{1}{\gamma_i(s)} ds} = 0.$$

□

Example 6.39 Find a Trench factorization for

$$x''' - 2x'' - x' + 2x = 0,$$

and a set of principal solutions.

Note that the Poyla factorization in Example 6.35 is not a Trench factorization since

$$\int_0^\infty \frac{1}{\rho_2(t)} \, dt = 6 \int_0^\infty e^{-3t} \, dt < \infty.$$

This time let
$$x_1 = e^{-t}, \quad x_2 = e^t, \quad x_3 = e^{2t}$$
denote three solutions. In this case
$$w_1(t) = x_1(t) = e^{-t}, \quad w_2(t) = w[e^{-t}, e^t] = 2,$$
and
$$w_3(t) = w[e^{-t}, e^t, e^{2t}] = 6e^{2t}.$$
By Theorem 6.34, we get a Polya factorization, where
$$\gamma_0(t) = e^t, \quad \gamma_1(t) = \frac{1}{2}e^{-2t}, \quad \gamma_2(t) = \frac{2}{3}e^{-t}, \quad \gamma_3(t) = 3e^{2t}.$$
Hence
$$
\begin{aligned}
x''' - 2x'' - x' + 2x &= \gamma_3(t)\{\gamma_2(t)[\gamma_1(t)(\gamma_0(t)x(t))']'\}' \\
&= 3e^{2t}\{\tfrac{2}{3}e^{-t}[\tfrac{1}{2}e^{-2t}(e^t x(t))']'\}' \\
&= e^{2t}\{e^{-t}[e^{-2t}(e^t x(t))']'\}'
\end{aligned}
$$
is a Polya factorization. Since
$$\int_0^\infty \frac{1}{\gamma_1(t)}\,dt = 2\int_0^\infty e^{2t}\,dt = \infty,$$
and
$$\int_0^\infty \frac{1}{\gamma_2(t)}\,dt = \frac{3}{2}\int_0^\infty e^t\,dt = \infty,$$
this Polya factorization is actually a Trench factorization. A set of principal solutions is given by
$$\{u_1(t) = e^{-t}, u_2(t) = e^t, u_3(t) = e^{2t}\}.$$

$$\triangle$$

6.5 Adjoint Equation

In this section we discuss the (formal) adjoint differential equation of the vector differential equation $y' = A(t)y$ and we use this to motivate the definition of the (formal) adjoint differential equation of the nth-order scalar differential equation $L_n x = 0$.

The (formal) adjoint equation of the vector equation $y' = A(t)y$, where A (with complex-valued entries) is assumed to be a continuous $n \times n$ matrix function on I, is defined to be the vector differential equation

$$z' = -A^*(t)z,$$

where $A^*(t)$ is the conjugate transpose of $A(t)$. Likewise, we say that the matrix differential equation

$$Z' = -A^*(t)Z$$

is the (formal) adjoint of the matrix differential equation $Y' = A(t)Y$. We say that the vector differential equation $y' = A(t)y$ is (formally) self-adjoint provided $A(t)$ is skew Hermitian on I [i.e., $A(t) = -A^*(t)$, for

$t \in I$]. Before we prove an important relation between the vector equation $y' = A(t)y$ and its adjoint $z' = -A^*(t)z$, we prove the following important lemma.

Lemma 6.40 *Assume Y is a nonsingular and differentiable matrix function on the interval I. Then the inverse matrix function Y^{-1} is differentiable on I and*

$$\left(Y^{-1}\right)'(t) = -Y^{-1}(t)Y'(t)Y^{-1}(t),$$

for $t \in I$.

Proof Since Y is differentiable and nonsingular, we get from a formula for calculating the inverse that Y^{-1} is differentiable. Differentiating both sides of

$$Y(t)Y^{-1}(t) = I,$$

for $t \in I$, we get that

$$Y'(t)Y^{-1}(t) + Y(t)\left(Y^{-1}\right)'(t) = 0,$$

for $t \in I$. Solving for $\left(Y^{-1}\right)'(t)$, we get the desired result. $\quad\square$

Theorem 6.41 *The matrix function Y is a fundamental matrix for the vector differential equation $y' = A(t)y$ on I iff $Z := (Y^*)^{-1} = \left(Y^{-1}\right)^*$ is a fundamental matrix for the adjoint equation $z' = -A^*(t)z$ on I.*

Proof Assume Y is a fundamental matrix for $y' = A(t)y$. Since $Y(t)$ is nonsingular on I, $Z(t) := (Y^*)^{-1}(t)$ is well defined and nonsingular on I. Also,

$$Y'(t) = A(t)Y(t), \quad t \in I$$

implies that (using Lemma 6.40)

$$\left(Y^{-1}(t)\right)' = -Y^{-1}(t)Y'(t)Y^{-1}(t) = -Y^{-1}(t)A(t).$$

Taking the conjugate transpose of both sides, we get the desired result

$$Z'(t) = -A^*(t)Z(t), \quad t \in I.$$

The converse follows easily from the fact that the adjoint of $z' = -A^*(t)z$ is $y' = A(t)y$. $\quad\square$

Theorem 6.42 *If $Y(t)$ is a fundamental matrix of the vector equation $y' = A(t)y$ and $Z(t)$ is a fundamental matrix of the adjoint equation $z' = -A^*(t)z$, then*

$$Y^*(t)Z(t) = C,$$

for $t \in I$, where C is a nonsingular $n \times n$ constant matrix.

Proof By the product rule,

$$
\begin{aligned}
\{Y^*(t)Z(t)\}' &= (Y'(t))^* Z(t) + Y^*(t)Z'(t) \\
&= (A(t)Y(t))^* Z(t) + Y^*(t)(-A^*(t)Z(t)) \\
&= Y^*(t)A^*(t)Z(t) - Y^*(t)A^*(t)Z(t) \\
&= 0,
\end{aligned}
$$

for $t \in I$. This implies that

$$
Y^*(t)Z(t) = C, \quad t \in I,
$$

where C is an $n \times n$ constant matrix. It follows from this last equation that C is nonsingular. □

For the remainder of this section we will assume that the coefficient functions in the equation $L_n u = 0$ are continuous complex-valued functions on a real interval I. We now motivate what we call the (formal) adjoint of $L_n u = 0$. The equation $L_n u = 0$ is equivalent to the vector equation $y' = A(t)y$, where

$$
A(t) = \begin{pmatrix}
0 & 1 & 0 & \cdots & 0 \\
0 & 0 & 1 & \cdots & 0 \\
\cdots & \cdots & \cdots & \cdots & \cdots \\
-p_0(t) & -p_1(t) & -p_2(t) & \cdots & -p_{n-1}(t)
\end{pmatrix}.
$$

Hence the adjoint of $y' = A(t)y$ is $z' = -A^*(t)z$, where

$$
-A^*(t) = \begin{pmatrix}
0 & 0 & 0 & \cdots & \overline{p_0(t)} \\
-1 & 0 & 0 & \cdots & \overline{p_1(t)} \\
0 & -1 & 0 & \cdots & \overline{p_2(t)} \\
\cdots & \cdots & \cdots & \cdots & \cdots \\
0 & 0 & 0 & \cdots & \overline{p_{n-1}(t)}
\end{pmatrix}.
$$

Letting $z_1, z_2, \cdots, z_n$ be the components of z, we get that

$$
\begin{aligned}
z_1' &= \overline{p_0(t)}z_n \\
z_2' &= -z_1 + \overline{p_1(t)}z_n \\
z_3' &= -z_2 + \overline{p_2(t)}z_n \\
&\cdots \\
z_{n-1}' &= -z_{n-2} + \overline{p_{n-2}(t)}z_n \\
z_n' &= -z_{n-1} + \overline{p_{n-1}(t)}z_n.
\end{aligned}
$$

It then follows that

$$
\begin{aligned}
z_n'' &= -z_{n-1}' + (\overline{p_{n-1}(t)}z_n)' \\
&= z_{n-2} - \overline{p_{n-2}(t)}z_n + (\overline{p_{n-1}(t)}z_n)' \\
z_n''' &= z_{n-2}' - (\overline{p_{n-2}(t)}z_n)' + (\overline{p_{n-1}(t)}z_n)'' \\
&= -z_{n-3} + \overline{p_{n-3}(t)}z_n - (\overline{p_{n-2}(t)}z_n)' + (\overline{p_{n-1}(t)}z_n)''
\end{aligned}
$$

$$\cdots$$

$$
z_n^{(n-1)} = (-1)^{n+1}\left\{ z_1 - \overline{p_1(t)}z_n + (\overline{p_2(t)}z_n)' - \cdots + (\overline{p_{n-1}(t)}z_n)^{(n-2)} \right\}
$$

hold. Finally, we get that

$$
\begin{aligned}
& z_n^{(n)} \\
&= (-1)^{n+1}\left\{ z_1' - (\overline{p_1(t)}z_n)' + (\overline{p_2(t)}z_n)'' - \cdots + (\overline{p_{n-1}(t)}z_n)^{(n-1)} \right\} \\
&= (-1)^{n+1}\left\{ \overline{p_0(t)}z_n - (\overline{p_1(t)}z_n)' + \cdots + (\overline{p_{n-1}(t)}z_n)^{(n-1)} \right\}.
\end{aligned}
$$

Hence if z is a solution of the adjoint vector equation $z' = -A^*(t)z$, then its n-th component $v = z_n$ is a solution of the nth-order linear differential equation

$$
L_n^* v = (-1)^n v^{(n)} + \sum_{k=1}^{n} (-1)^{n-k} (\overline{p_{n-k}(t)}v)^{(n-k)} = 0.
$$

The differential equation $L_n^* v = 0$ is called the *(formal) adjoint* of the differential equation $L_n u = 0$. If $L_n u = 0$ and $L_n^* u = 0$ are the same equation, then we say that $L_n u = 0$ is (formally) self-adjoint. An important relation between the nth-order linear differential equation and its (formal) adjoint is given by the next theorem.

Theorem 6.43 (Lagrange Identity) *Assume u, v have n continuous derivatives on I. Then*

$$
\overline{v(t)}L_n u(t) - u(t)\overline{L_n^* v(t)} = \{v(t); u(t)\}',
$$

for $t \in I$, where the Lagrange bracket $\{v(t); u(t)\}$ is defined by

$$
\{v(t); u(t)\} = \sum_{k=1}^{n}\sum_{j=0}^{k-1} (-1)^j u^{(k-j-1)}(t)(p_k(t)\overline{v(t)})^{(j)},
$$

for $t \in I$, where $p_n(t) \equiv 1$.

Proof We leave the proof of this theorem to the reader. We now show the proof for the special case when $n = 2$. Assume u, v are twice differentiable on I and consider (note that when $n = 2$, $p_2(t) \equiv 1$)

$$
\begin{aligned}
\{v(t); u(t)\}' &= \left\{\overline{v(t)}u'(t) - \overline{v'(t)}u(t) + p_1(t)\overline{v(t)}u(t)\right\}' \\
&= \overline{v(t)}u''(t) - \overline{v''(t)}u(t) + (p_1(t)\overline{v(t)})'u(t) + p_1(t)\overline{v(t)}u'(t) \\
&= \overline{v(t)}\left\{u''(t) + p_1(t)u'(t)\right\} - u(t)\left\{\overline{v''(t)} - (p_1(t)\overline{v(t)})'\right\} \\
&= \overline{v(t)}\left\{u''(t) + p_1(t)u'(t) + p_0(t)u(t)\right\} \\
&\quad - u(t)\left\{\overline{v''(t)} - (p_1(t)\overline{v(t)})' + p_0(t)\overline{v(t)}\right\} \\
&= \overline{v(t)}L_2 u(t) - u(t)\overline{L_2^* v(t)},
\end{aligned}
$$

for $t \in I$.   □

The following result follows immediately from the Lagrange identity.

Corollary 6.44 (Abel's Formula) *If u, v are complex-valued solutions of $L_n u = 0$ and $L_n^* v = 0$, respectively, on I, then*

$$\{v(t); u(t)\} = C, \quad t \in I,$$

where C is a constant.

Definition 6.45 For complex-valued, continuous functions u, v defined on $[a, b]$, define the *inner product* of u and v by

$$< u, v > = \int_a^b u(t)\overline{v(t)} \, dt.$$

Theorem 6.46 (Green's Theorem) *If u, v are n times continuously differentiable complex-valued functions on $I = [a, b]$, then*

$$< L_n u, v > - < u, L_n^* v > = \{v(t); u(t)\}_a^b. \tag{6.18}$$

Proof By the Lagrange identity,

$$\overline{v(t)}L_n u(t) - u(t)\overline{L_n^* v(t)} = \{v(t); u(t)\}',$$

for $t \in I = [a, b]$, where $\{v(t); u(t)\}$ is the Lagrange bracket. Integrating both sides from a to b then gives Green's formula (6.18).   □

Note that if u, v are n times continuously differentiable functions on $[a, b]$ satisfying

$$\{v(t); u(t)\}_a^b = 0,$$

then

$$< L_n u, v > = < u, L_n^* v > .$$

6.6 Exercises

6.1 Show that the operator L_n defined in Definition 6.1 is a linear operator.

6.2 Show that if in the definition of the linear operator L_n, p_{n-1} is the zero function, then the Wronskian determinant of any n solutions of $L_n x = 0$ is constant on I.

6.3 Use Theorem 2.23 to prove Theorem 6.4.

6.4 Prove Corollary 6.5.

6.5 Prove Theorem 6.4 directly for the special case $n = 3$.

6.6 Prove that the nth-order linear homogeneous differential equation $L_n x = 0$ has n linearly independent solutions on I.

6.7 Prove the last statement in Theorem 6.6.

6.8 Find the Cauchy function for each of the following differential equations:

    (i) $x''' - x'' = 0$
    (ii) $x''' - x' = 0$
    (iii) $x'' - \frac{4}{t}x' + \frac{6}{t^2}x = 0$
    (iv) $x''' - \frac{2}{t}x'' + \frac{2}{t^2}x' = 0$

6.9 Use Theorem 6.9 and the solutions $x_i(t) = t^{i-1}$, $1 \le i \le n$, to find the Cauchy function in Example 6.8.

6.10 Use the variation of constants formula in Theorem 6.11 to solve the following IVPs:

    (i) $y''' - y'' = 2t$, $y(0) = y'(0) = y''(0) = 0$
    (ii) $y''' - y' = e^{3t}$, $y(0) = y'(0) = y''(0) = 0$
    (iii) $y'' - \frac{4}{t}y' + \frac{6}{t^2}y = t$; $y(1) = y'(1) = 0$
    (iv) $y''' - \frac{2}{t}y'' + \frac{2}{t^2}y' = 5$, $y(1) = y'(1) = y''(1) = 0$

6.11 Use Theorem 6.11 to solve the IVP
$$x''' - 2x'' = 2, \quad x(0) = x'(0) = x''(0) = 0.$$

6.12 Prove Corollary 6.13.

6.13 (Taylor's Theorem) Use Corollary 6.13 to prove the following version of Taylor's theorem: If $f \in C^{n+1}(I)$ and $t_0 \in I$, then
$$f(t) = \sum_{k=0}^{n} f^{(k)}(t_0)\frac{(t - t_0)^k}{k!} + \int_{t_0}^{t} \frac{(t - s)^n}{n!} f^{(n+1)}(s)\, ds,$$
for $t \in I$.

6.14 Prove Corollary 6.17.

6.15 Prove Theorem 6.27.

6.16 Prove Theorem 6.28.

6.17 Prove Theorem 6.29.

6.18 Show that if the $(k, n - k)$ conjugate BVP $L_n x = 0$, (6.7), has only the trivial solution, then $D \neq 0$, where D is defined in Theorem 6.20.

6.19 Find the normalized solutions $\{x_k(t, 0)\}_{k=0}^{1}$ at zero for the differential equation $x'' + x = 0$.

6.20 Find the normalized solutions $\{x_k(t, 0)\}_{k=0}^{2}$ at zero for the differential equation $x''' - x'' = 0$.

6.21 Find the Green's functions for each of the following boundary value problems:

    (i) $x^{(4)} = 0, \quad x(0) = x'(0) = x''(0) = 0 = x(1)$
    (ii) $x'' + x = 0, \quad x(0) = 0 = x(\frac{\pi}{2})$
    (iii) $x^{(4)} = 0, \quad x(0) = x'(0) = x''(0) = 0 = x'''(1)$
    (iv) $x'' + x = 0, \quad x(0) = 0 = x'(\frac{\pi}{4})$

6.22 Use the Green's functions you found in Exercise 6.21 to solve each of the following boundary value problems:

    (i) $y^{(4)} = 24, \quad y(0) = y'(0) = y''(0) = 0 = y(1)$
    (ii) $y'' + y = 2e^t, \quad y(0) = 0 = y(\frac{\pi}{2})$
    (iii) $y^{(4)} = 24, \quad y(0) = y'(0) = y''(0) = 0 = y'''(1)$
    (iv) $y'' + y = t, \quad y(0) = 0 = y'(\frac{\pi}{4})$

6.23 Find a Polya factorization of each of the following:

    (i) $x''' - 6x'' + 11x' - 6x = 0$
    (ii) $x''' - x'' - 4x' + 4x = 0$
    (iii) $x''' + \frac{1}{t}x'' - \frac{2}{t^2}x' + \frac{2}{t^3}x = 0$
    (iv) $x^{(4)} - 4x''' + 6x'' - 4x' + x = 0$

6.24 Find a Trench factorization for each of the differential equations in Exercise 6.23 and find a set of principal solutions.

6.25 Show that $y' = A(t)y$ has a unitary fundamental matrix X on I, i.e., $X(t) = (X^*)^{-1}(t)$ for $t \in I$, iff $y' = A(t)y$ is (formally) self-adjoint. In this case show that if y is a solution of $y' = A(t)y$, then $\|y(t)\| \equiv c, t \in I$, where c is a constant and $\| \cdot \|$ is the Euclidean norm.

6.26 Find conditions on p_0 and p_1 so that $L_2 x = 0$ is (formally) self-adjoint.

6.27 Prove Theorem 6.43 for $n = 3$.

6.28 Assume that $u, v \in C^2[a, b]$. Show that if either both u, v satisfy the boundary conditions $x(a) = 0 = x(b)$, or if both u, v satisfy the boundary conditions

$$\alpha x(a) - \beta x'(a) = 0, \quad \gamma x(b) + \delta x'(b) = 0,$$

where $\alpha^2 + \beta^2 > 0$, $\gamma^2 + \delta^2 > 0$, and $p_1(a) = 0 = p_1(b)$, then

$$< L_2 u, v >=< u, L_2^* v > .$$

Chapter 7

BVPs for Nonlinear Second-Order DEs

7.1 Contraction Mapping Theorem (CMT)

Our discussion of differential equations up to this point has focused on solutions of initial value problems. We have made good use of the fact that these problems have unique solutions under mild conditions. We have also encountered boundary value problems in a couple of chapters. These problems require the solution or its derivative to have prescribed values at two or more points. Such problems arise in a natural way when differential equations are used to model physical problems. An interesting complication arises in that BVPs may have many solutions or no solution at all. Consequently, we will present some results on the existence and uniqueness of solutions of BVPs. A useful device will be the contraction mapping theorem (Theorem 7.5), which is introduced by the definitions below. One of the goals of this chapter is to show how the contraction mapping theorem is commonly used in differential equations.

Definition 7.1 A linear (vector) space $\mathbb{X}$ is called a normed linear space (NLS) provided there is a function $\|\cdot\| : \mathbb{X} \to \mathbb{R}$, called a norm, satisfying

(i) $\|x\| \geq 0$ for all $x \in \mathbb{X}$ and $\|x\| = 0$ iff $x = 0$,
(ii) $\|\lambda x\| = |\lambda|\|x\|$, for all $\lambda \in \mathbb{R}$ and $x \in \mathbb{X}$,
(iii) $\|x + y\| \leq \|x\| + \|y\|$, for all $x, y \in \mathbb{X}$.

Definition 7.2 We say that $T : \mathbb{X} \to \mathbb{X}$ is a *contraction mapping* on a NLS $\mathbb{X}$ provided there is a constant $\alpha \in (0,1)$ such that

$$\|Tx - Ty\| \leq \alpha\|x - y\|,$$

for all $x, y \in \mathbb{X}$. We say $\bar{x} \in \mathbb{X}$ is a *fixed point* of T provided $T\bar{x} = \bar{x}$.

Definition 7.3 We say that $\{x_n\} \subset \mathbb{X}$ is a *Cauchy sequence* provided given any $\varepsilon > 0$ there is a positive integer N such that $\|x_n - x_m\| < \varepsilon$ for all $n, m \geq N$.

Definition 7.4 We say $\mathbb{X}$ is a *Banach space* provided it is a NLS and every Cauchy sequence in $\mathbb{X}$ converges to an element in $\mathbb{X}$.

Theorem 7.5 (Contraction Mapping Theorem) *If T is a contraction mapping on a Banach space $\mathbb{X}$ with contraction constant α, with $0 < \alpha < 1$, then*

W.G. Kelley and A.C. Peterson, *The Theory of Differential Equations:*
Classical and Qualitative, Universitext 278, DOI 10.1007/978-1-4419-5783-2_7,
© Springer Science+Business Media, LLC 2010

T has a unique fixed point $\overline{x}$ in $\mathbb{X}$. *Also if* $x_0 \in \mathbb{X}$ *and we set* $x_{n+1} = Tx_n$, *for* $n \geq 0$, *then*

$$\lim_{n \to \infty} x_n = \overline{x}.$$

Furthermore,

$$\|x_n - \overline{x}\| \leq \frac{\alpha^n}{1 - \alpha}\|x_1 - x_0\|, \tag{7.1}$$

for $n \geq 1$.

Proof Let $x_0 \in \mathbb{X}$ and define the sequence $\{x_n\}$ of "Picard iterates" by $x_{n+1} = Tx_n$, for $n \geq 0$. Consider

$$
\begin{aligned}
\|x_{m+1} - x_m\| &= \|Tx_m - Tx_{m-1}\| \\
&\leq \alpha\|x_m - x_{m-1}\| \\
&= \alpha\|Tx_{m-1} - Tx_{m-2}\| \\
&\leq \alpha^2\|x_{m-1} - x_{m-2}\| \\
&\cdots \\
&\leq \alpha^m\|x_1 - x_0\|.
\end{aligned}
$$

Next consider

$$
\begin{aligned}
&\|x_{n+k} - x_n\| \\
&\leq \|x_{n+k} - x_{n+k-1}\| + \|x_{n+k-1} - x_{n+k-2}\| + \cdots + \|x_{n+1} - x_n\| \\
&\leq (\alpha^{n+k-1} + \cdots + \alpha^n)\|x_1 - x_0\| \\
&= \alpha^n(1 + \alpha + \cdots + \alpha^{k-1})\|x_1 - x_0\| \\
&\leq \alpha^n(1 + \alpha + \alpha^2 + \cdots)\|x_1 - x_0\| \\
&= \frac{\alpha^n}{1 - \alpha}\|x_1 - x_0\|.
\end{aligned}
$$

This last inequality implies that $\{x_n\}$ is a Cauchy sequence. Hence there is an $\overline{x} \in \mathbb{X}$ such that

$$\lim_{n \to \infty} x_n = \overline{x}.$$

Letting $k \to \infty$ in the last inequality above, we get that (7.1) holds. Since T is a contraction mapping on $\mathbb{X}$, T is uniformly continuous on $\mathbb{X}$ and hence is continuous on $\mathbb{X}$. Therefore,

$$\overline{x} = \lim_{n \to \infty} x_{n+1} = \lim_{n \to \infty} Tx_n = T\overline{x},$$

and so $\overline{x}$ is a fixed point of T. Assume $\overline{\overline{x}} \in \mathbb{X}$ is a fixed point of T. Then

$$\|\overline{x} - \overline{\overline{x}}\| = \|T\overline{x} - T\overline{\overline{x}}\| \leq \alpha\|\overline{x} - \overline{\overline{x}}\|,$$

which implies that

$$(1 - \alpha)\|\overline{x} - \overline{\overline{x}}\| \leq 0,$$

which implies that $\overline{x} = \overline{\overline{x}}$, so $\overline{x}$ is the only fixed point of T. $\square$

7.2 Application of the CMT to a Forced Equation

First we use the contraction mapping theorem to study the forced second-order self-adjoint equation

$$[p(t)x']' + q(t)x = f(t), \tag{7.2}$$

where p, q, f are continuous on $[a, \infty)$ and $p(t) > 0$ on $[a, \infty)$. In the next theorem we will give conditions on p, q, f that ensure that (7.2) has a zero tending solution.

Theorem 7.6 *Assume*

    (i) $p(t) > 0$, $q(t) \geq 0$, *for all* $t \in [a, \infty)$
    (ii) $\int_a^\infty \frac{1}{p(t)} dt < \infty$
    (iii) $\int_a^\infty q(t) P(t) \, dt < \infty$, *where* $P(t) := \int_t^\infty \frac{1}{p(s)} ds$
    (iv) $\int_a^\infty f(t) \, dt < \infty$

hold; then equation (7.2) has a solution that converges to zero as $t \to \infty$.

Proof We will use the contraction mapping theorem to prove this theorem. Since $\int_a^\infty q(t) P(t) \, dt < \infty$, we can choose $b \in [a, \infty)$ sufficiently large so that

$$\alpha := \int_b^\infty q(\tau) P(\tau) \, d\tau < 1. \tag{7.3}$$

Let $\mathbb{X}$ be the Banach space (see Exercise 7.2) of all continuous functions $x : [b, \infty) \to \mathbb{R}$ that converge to zero as $t \to \infty$ with the norm $\| \cdot \|$ defined by

$$\|x\| = \max_{t \in [b, \infty)} |x(t)|.$$

Define an operator T on $\mathbb{X}$ by

$$Tx(t) = K(t) + P(t) \int_b^t q(\tau) x(\tau) \, d\tau + \int_t^\infty q(\tau) x(\tau) P(\tau) \, d\tau,$$

for $t \in [b, \infty)$, where

$$K(t) := \int_t^\infty \frac{F(s)}{p(s)} ds, \quad F(t) := \int_t^\infty f(s) \, ds, \quad \text{and} \quad P(t) := \int_t^\infty \frac{1}{p(s)} ds,$$

for $t \in [b, \infty)$. Clearly, Tx is continuous on $[b, \infty)$. Hence to show $T : \mathbb{X} \to \mathbb{X}$ it remains to show that $\lim_{t \to \infty} Tx(t) = 0$. To show this it suffices to show that if

$$y(t) := P(t) \int_b^t q(\tau) x(\tau) \, d\tau,$$

then $\lim_{t \to \infty} y(t) = 0$. To see this let $\epsilon > 0$ be given. Since $\lim_{t \to \infty} x(t) = 0$, there is a $c \geq b$ such that

$$|x(t)| < \epsilon \quad \text{for} \quad t \in [c, \infty).$$

Since $\lim_{t\to\infty} P(t) = 0$, there is a $d \geq c$ such that

$$P(t) \int_b^c q(\tau)|x(\tau)|\, d\tau < \epsilon,$$

for $t \geq d$. Consider for $t \geq d$

$$
\begin{aligned}
|y(t)| \ &\leq\ P(t) \int_b^t q(\tau)|x(\tau)|\, d\tau \\
&=\ P(t) \int_b^c q(\tau)|x(\tau)|\, d\tau + P(t) \int_c^t q(\tau)|x(\tau)|\, d\tau \\
&\leq\ \epsilon + \epsilon P(t) \int_c^t q(\tau)\, d\tau \\
&\leq\ \epsilon + \epsilon \int_c^t q(\tau) P(\tau)\, d\tau \\
&\leq\ \epsilon + \epsilon\alpha \leq 2\epsilon.
\end{aligned}
$$

Hence $\lim_{t\to\infty} y(t) = 0$ and so $T : \mathbb{X} \to \mathbb{X}$.

Next we show that T is a contraction mapping on X. Let $x, y \in X$, $t \geq b$, and consider

$$
\begin{aligned}
&|Tx(t) - Ty(t)| \\
\leq\ & P(t) \int_b^t q(\tau)|x(\tau) - y(\tau)|\, d\tau + \int_t^\infty q(\tau)P(\tau)|x(\tau) - y(\tau)|\, d\tau \\
\leq\ & \left\{ P(t) \int_b^t q(\tau)\, d\tau + \int_t^\infty q(\tau)P(\tau)\, d\tau \right\} \|x - y\| \\
\leq\ & \left\{ \int_b^t q(\tau)P(\tau)\, d\tau + \int_t^\infty q(\tau)P(\tau)\, d\tau \right\} \|x - y\| \\
=\ & \int_b^\infty q(\tau)P(\tau)\, d\tau \|x - y\| \\
=\ & \alpha \|x - y\|,
\end{aligned}
$$

where we have used the fact that P is decreasing and we have also used (7.3). Therefore, T is a contraction mapping on X. Hence by the contraction mapping theorem (Theorem 7.5), T has a unique fixed point $x \in X$. This implies that

$$x(t) = K(t) + P(t) \int_b^t q(\tau)x(\tau)\, d\tau + \int_t^\infty q(\tau)P(\tau)x(\tau)\, d\tau, \qquad (7.4)$$

for all $t \geq b$. Since $x \in \mathbb{X}$,

$$\lim_{t\to\infty} x(t) = 0.$$

It remains to show that x given by (7.4) is a solution of (7.2). Differentiating both sides of (7.4), we get

$$x'(t) = K'(t) + P(t)q(t)x(t) + P'(t)\int_b^t q(\tau)x(\tau)\,d\tau - q(t)P(t)x(t)$$

$$= -\frac{F(t)}{p(t)} - \frac{1}{p(t)}\int_b^t q(\tau)x(\tau)\,d\tau,$$

and so

$$p(t)x'(t) = -F(t) - \int_b^t q(\tau)x(\tau)\,d\tau,$$

and therefore

$$(p(t)x'(t))' = f(t) - q(t)x(t).$$

So we have x is a solution of the forced equation (7.2).   □

7.3 Applications of the CMT to BVPs

We begin our study of the nonlinear BVP

$$x'' = f(t,x), \quad x(a) = A, \quad x(b) = B, \tag{7.5}$$

where we assume $a < b$ and $A, B \in \mathbb{R}$, by using the contraction mapping theorem to establish the existence of a unique solution for a class of functions f.

Theorem 7.7 *Assume* $f : [a,b] \times \mathbb{R} \to \mathbb{R}$ *is continuous and satisfies a uniform Lipschitz condition with respect to* x *on* $[a,b] \times \mathbb{R}$ *with Lipschitz constant* K; *that is,*

$$|f(t,x) - f(t,y)| \le K|x - y|,$$

for all $(t,x), (t,y) \in [a,b] \times \mathbb{R}$. *If*

$$b - a < \frac{2\sqrt{2}}{\sqrt{K}}, \tag{7.6}$$

then the BVP (7.5) has a unique solution.

Proof Let $\mathbb{X}$ be the Banach space of continuous functions on $[a,b]$ with norm (called the max norm) defined by

$$\|x\| = \max\{|x(t)| : a \le t \le b\}.$$

Note that x is a solution of the BVP (7.5) iff x is a solution of the linear, nonhomogeneous BVP

$$x'' = h(t) := f(t,x(t)), \quad x(a) = A, \quad x(b) = B. \tag{7.7}$$

But the BVP (7.7) has a solution iff the integral equation

$$x(t) = z(t) + \int_a^b G(t,s)f(s,x(s))\,ds$$

has a solution, where z is the solution of the BVP

$$z'' = 0, \quad z(a) = A, \quad z(b) = B$$

and G is the Green's function for the BVP

$$x'' = 0, \quad x(a) = 0, \quad x(b) = 0.$$

Define $T : \mathbb{X} \to \mathbb{X}$ by

$$Tx(t) = z(t) + \int_a^b G(t, s) f(s, x(s)) \, ds,$$

for $t \in [a, b]$. Hence the BVP (7.5) has a unique solution iff the operator T has a unique fixed point.

We will use the contraction mapping theorem (Theorem 7.5) to show that T has a unique fixed point in $\mathbb{X}$. Let $x, y \in \mathbb{X}$ and consider

$$
\begin{aligned}
|Tx(t) - Ty(t)| &\leq \int_a^b |G(t, s)| |f(s, x(s)) - f(s, y(s))| \, ds \\
&\leq \int_a^b |G(t, s)| K |x(s) - y(s)| \, ds \\
&\leq K \int_a^b |G(t, s)| \, ds \, \|x - y\| \\
&\leq K \frac{(b - a)^2}{8} \|x - y\|,
\end{aligned}
$$

for $t \in [a, b]$, where for the last inequality we used Theorem 5.109. It follows that

$$\|Tx - Ty\| \leq \alpha \|x - y\|,$$

where from (7.6)

$$\alpha := K \frac{(b - a)^2}{8} < 1.$$

Hence T is a contraction mapping on $\mathbb{X}$, and by the contraction mapping theorem (Theorem 7.5) we get the desired conclusion. $\square$

In the proof of Theorem 7.7 we used the contraction mapping theorem to prove the existence of the solution of the BVP (7.7). In this proof the operator T was defined by

$$Tx(t) = z(t) + \int_a^b G(t, s) f(s, x(s)) \, ds,$$

for $t \in [a, b]$. Hence, in this case, if x_0 is any continuous function on $[a, b]$, then recall that the Picard iterates are defined by

$$x_{n+1}(t) = Tx_n(t),$$

for $t \in [a, b]$, $n = 0, 1, 2, \cdots$. Hence

$$x_{n+1}(t) = z(t) + \int_a^b G(t, s) f(s, x_n(s)) \, ds,$$

for $t \in [a, b]$, $n = 0, 1, 2, \cdots$. It follows from this that the Picard iterate x_{n+1} solves the BVP

$$x_{n+1}'' = f(t, x_n(t)),$$
$$x(a) = A, \quad x(b) = B,$$

which is usually easy to solve.

Example 7.8 Consider the BVP

$$x'' = -1 - \sin x, \quad x(0) = 0, \quad x(1) = 0. \tag{7.8}$$

Here $f(t, x) = -1 - \sin x$ and so

$$|f_x(t, x)| = |\cos x| \leq K := 1.$$

Since

$$b - a = 1 < \frac{2\sqrt{2}}{\sqrt{K}} = 2\sqrt{2},$$

we have by Theorem 7.7 that the BVP (7.8) has a unique solution. If we take as our initial approximation $x_0(t) \equiv 0$, then the first Picard iterate x_1 solves the BVP

$$x_1'' = f(t, x_0(t)) = f(t, 0) = -1,$$
$$x_1(0) = 0, \quad x_1(1) = 0.$$

Solving this BVP, we get that the first Picard iterate is given by

$$x_1(t) = -\frac{1}{2}t(t - 1), \quad t \in [0, 1].$$

Next we want to use (7.1) to see how good an approximation x_1 is to the actual solution x of the BVP (7.8). Recall in the proof of Theorem 7.7 the norm $\| \cdot \|$ was the max norm and the contraction constant α is given by

$$\alpha = K \frac{(b - a)^2}{8} = \frac{1}{8}.$$

Hence by (7.1)

$$\|x - x_1\| \leq \frac{\alpha}{1 - \alpha} \|x_1 - x_0\|$$
$$= \frac{1}{7} \|x_1\| = \frac{1}{56}.$$

It follows that

$$|x(t) - x_1(t)| \leq \frac{1}{56}, \quad t \in [0, 1],$$

and we see that $x_1(t) = \frac{1}{2}t(1 - t)$ is a very good approximation of the actual solution x of the BVP (7.8). $\triangle$

Next we see if we use a slightly more complicated norm than in the proof of Theorem 7.7 we can prove the following theorem, which is a better theorem in the sense that it allows the length of the interval $[a, b]$ to be larger. However, since the norm is more complicated it makes the application of the inequality (7.1) more complicated.

Theorem 7.9 *Assume $f : [a, b] \times \mathbb{R} \to \mathbb{R}$ is continuous and satisfies a uniform Lipschitz condition with respect to x on $[a, b] \times \mathbb{R}$ with Lipschitz constant K; that is,*

$$|f(t, x) - f(t, y)| \leq K|x - y|,$$

for all $(t, x), (t, y) \in [a, b] \times \mathbb{R}$. If

$$b - a < \frac{\pi}{\sqrt{K}}, \tag{7.9}$$

then the BVP (7.5) has a unique solution.

Proof Let $\mathbb{X}$ be the Banach space (see Exercise 7.3) of continuous functions on $[a, b]$ with norm defined by

$$\|x\| = \max\left\{ \frac{|x(t)|}{\rho(t)} : a \leq t \leq b \right\},$$

where ρ is a positive continuous function on $[a, b]$ that we will choose later. Note that x is a solution of the BVP (7.5) iff x is a solution of the linear, nonhomogeneous BVP

$$x'' = h(t) := f(t, x(t)), \quad x(a) = A, \quad x(b) = B. \tag{7.10}$$

But the BVP (7.10) has a solution iff the integral equation

$$x(t) = z(t) + \int_a^b G(t, s) f(s, x(s)) \, ds$$

has a solution, where z is the solution of the BVP

$$z'' = 0, \quad z(a) = A, \quad z(b) = B,$$

and G is the Green's function for the BVP

$$x'' = 0, \quad x(a) = 0, \quad x(b) = 0.$$

Define $T : \mathbb{X} \to \mathbb{X}$ by

$$Tx(t) = z(t) + \int_a^b G(t, s) f(s, x(s)) \, ds,$$

for $t \in [a, b]$. Hence the BVP (7.5) has a unique solution iff the operator T has a unique fixed point.

We will use the contraction mapping theorem (Theorem 7.5) to show that T has a unique fixed point in $\mathbb{X}$. Let $x, y \in \mathbb{X}$ and consider

$$
\begin{aligned}
\frac{|Tx(t) - Ty(t)|}{\rho(t)} &\leq \frac{1}{\rho(t)} \int_a^b |G(t,s)||f(s,x(s)) - f(s,y(s))| \, ds \\
&\leq \frac{1}{\rho(t)} \int_a^b |G(t,s)|K|x(s) - y(s)| \, ds \\
&= \frac{1}{\rho(t)} \int_a^b |G(t,s)|K\rho(s)\frac{|x(s) - y(s)|}{\rho(s)} ds \\
&\leq \frac{K}{\rho(t)} \int_a^b |G(t,s)|\rho(s) \, ds \, \|x - y\|,
\end{aligned}
$$

for $t \in [a, b]$. If we could choose a positive continuous function ρ such that there is a constant α for which

$$
K\frac{1}{\rho(t)} \int_a^b |G(t,s)|\rho(s) \, ds \leq \alpha < 1, \tag{7.11}
$$

for $a \leq t \leq b$, then it would follow that T is a contraction mapping on $\mathbb{X}$, and by the contraction mapping theorem (Theorem 7.5) we would get the desired conclusion. Hence it remains to show that we can pick a positive continuous function ρ such that (7.11) holds.

Since $b - a < \pi/\sqrt{K}$, the function defined by

$$
\rho(t) = \sin(\sqrt{K/\alpha}(t - c))
$$

is positive on the interval $[a, b]$, for $\alpha < 1$ near 1 and $c < a$ near a. Now ρ satisfies

$$
\rho'' + \frac{K}{\alpha}\rho = 0,
$$

so

$$
\begin{aligned}
\rho(t) &= \frac{\rho(b)(t - a) + \rho(a)(b - t)}{b - a} + \int_a^b G(t,s)\left(-\frac{K}{\alpha}\rho(s)\right) \, ds \\
&> \int_a^b G(t,s)\left(-\frac{K}{\alpha}\rho(s)\right) \, ds.
\end{aligned}
$$

Since the Green's function $G(t, s) \leq 0$, we have

$$
\rho(t) > \int_a^b |G(t,s)|[\frac{K}{\alpha}\rho(s)] \, ds.
$$

Then (7.11) holds, and the proof is complete.   □

The next example shows that Theorem 7.9 is sharp.

Example 7.10 Consider the BVP

$$
x'' + Kx = 0, \quad x(a) = 0, \quad x(b) = B,
$$

where K is a positive constant.

By Theorem 7.9, if $b - a < \frac{\pi}{\sqrt{K}}$ the preceding BVP has a unique solution. Note, however, that if $b - a = \frac{\pi}{\sqrt{K}}$, then the given BVP does not have a solution if $B \neq 0$ and has infinitely many solutions if $B = 0$. $\triangle$

Theorem 7.11 *Assume $f : [a, b] \times \mathbb{R}^2 \to \mathbb{R}$ is continuous and satisfies the uniform Lipschitz condition with respect to x and x',*

$$|f(t, x, x') - f(t, y, y')| \leq K|x - y| + L|x' - y'|,$$

for $(t, x, x'), (t, y, y') \in [a, b] \times \mathbb{R}^2$, where $L \geq 0, K > 0$ are constants. If

$$K\frac{(b - a)^2}{8} + L\frac{b - a}{2} < 1, \tag{7.12}$$

then the BVP

$$x'' = f(t, x, x'), \quad x(a) = A, \quad x(b) = B \tag{7.13}$$

has a unique solution.

Proof Let $\mathbb{X}$ be the Banach space of continuously differentiable functions on $[a, b]$ with norm $\| \cdot \|$ defined by

$$\|x\| = \max\{K|x(t)| + L|x'(t)| : a \leq t \leq b\}.$$

Note that x is a solution of the BVP (7.13) iff x is a solution of the BVP

$$x'' = h(t) := f(t, x(t), x'(t)), \quad x(a) = A, \quad x(b) = B. \tag{7.14}$$

But the BVP (7.14) has a solution iff the integral equation

$$x(t) = z(t) + \int_a^b G(t, s)f(s, x(s), x'(s)) \, ds$$

has a solution, where z is the solution of the BVP

$$z'' = 0, \quad z(a) = A, \quad z(b) = B,$$

and G is the Green's function for the BVP

$$x'' = 0, \quad x(a) = 0, \quad x(b) = 0.$$

Define $T : \mathbb{X} \to \mathbb{X}$ by

$$Tx(t) = z(t) + \int_a^b G(t, s)f(s, x(s), x'(s)) \, ds,$$

for $t \in [a, b]$. Then the BVP (7.13) has a unique solution iff the operator T has a unique fixed point. We will use the contraction mapping theorem (Theorem 7.5) to show that T has a unique fixed point. Let $x, y \in \mathbb{X}$ and consider

$$K|Tx(t) - Ty(t)|$$

$$\leq K \int_a^b |G(t,s)||f(s,x(s),x'(s)) - f(s,y(s),y'(s))| \, ds$$

$$\leq K \int_a^b |G(t,s)| \, [K|x(s) - y(s)| + L|x'(s) - y'(s)|] \, ds$$

$$\leq K \int_a^b |G(t,s)| \, ds \, \|x - y\|$$

$$\leq K \frac{(b-a)^2}{8} \, \|x - y\|,$$

for $t \in [a,b]$, where we have used Theorem 5.109. Similarly,

$$L|(Tx)'(t) - (Ty)'(t)|$$

$$\leq L \int_a^b |G_t(t,s)||f(s,x(s),x'(s)) - f(s,y(s),y'(s))| \, ds$$

$$\leq L \int_a^b |G_t(t,s)| \, [K|x(s) - y(s)| + L|x'(s) - y'(s)|] \, ds$$

$$\leq L \int_a^b |G_t(t,s)| \, ds \, \|x - y\|$$

$$\leq L \frac{(b-a)}{2} \, \|x - y\|,$$

for $t \in [a,b]$, where we have used Theorem 5.109. It follows that

$$\|Tx - Ty\| \leq \alpha \|x - y\|,$$

where

$$\alpha := K \frac{(b-a)^2}{8} + L \frac{b-a}{2} < 1$$

by (7.12). From the contraction mapping theorem (Theorem 7.5) we get that T has a unique fixed point in $\mathbb{X}$. This implies that the BVP (7.13) has a unique solution. $\square$

Theorem 7.12 *Assume* $f : [a,b] \times \mathbb{R}^2 \to \mathbb{R}$ *is continuous and there are nonnegative continuous functions* p, q *on* $[a,b]$ *such that*

$$|f(t,x,x') - f(t,y,y')| \leq p(t)|x - y| + q(t)|x' - y'|,$$

for $(t,x,x'), (t,y,y') \in [a,b] \times \mathbb{R}^2$. *Let* u *be the solution of the IVP*

$$u'' + q(t)u' + p(t)u = 0, \quad u(a) = 0, \quad u'(a) = 1; \qquad (7.15)$$

then if $u'(t) > 0$ *on* $[a,b]$ *it follows that the right-focal BVP*

$$x'' = f(t,x,x'), \quad x(a) = A, \quad x'(b) = m, \qquad (7.16)$$

where A *and* m *are constants, has a unique solution.*

Proof Let $\mathbb{X}$ be the Banach space of continuously differentiable functions on $[a, b]$ with norm $\| \cdot \|$ defined by

$$\|x\| = \max \left\{ \max_{t \in [a,b]} \frac{|x(t)|}{v(t)}, \max_{t \in [a,b]} \frac{|x'(t)|}{w(t)} \right\},$$

where v, w are positive continuous functions that we will choose later in this proof. Then $\mathbb{X}$ is a Banach space. Define $T : \mathbb{X} \to \mathbb{X}$ by

$$Tx(t) = z(t) + \int_a^b G(t, s) f(s, x(s), x'(s)) \, ds,$$

for $t \in [a, b]$, where z is the solution of the BVP

$$z'' = 0, \quad z(a) = A, \quad z'(b) = m,$$

and G is the Green's function for the BVP

$$x'' = 0, \quad x(a) = 0, \quad x'(b) = 0.$$

Then the BVP (7.16) has a unique solution iff the operator T has a unique fixed point. We will use the contraction mapping theorem (Theorem 7.5) to show that T has a unique fixed point. Let $x, y \in \mathbb{X}$ and consider

$$\frac{|Tx(t) - Ty(t)|}{v(t)}$$

$$\leq \frac{1}{v(t)} \int_a^b |G(t, s)| |f(s, x(s), x'(s)) - f(s, y(s), y'(s))| \, ds$$

$$\leq \frac{1}{v(t)} \int_a^b |G(t, s)| \left[p(s)|x(s) - y(s)| + q(s)|x'(s) - y'(s)| \right] ds$$

$$\leq \frac{1}{v(t)} \int_a^b |G(t, s)| \left[p(s)v(s)\frac{|x(s) - y(s)|}{v(s)} + q(s)w(s)\frac{|x'(s) - y'(s)|}{w(s)} \right] ds$$

$$\leq \frac{1}{v(t)} \int_a^b |G(t, s)| \left[p(s)v(s) + q(s)w(s) \right] ds \, \|x - y\|,$$

for $t \in [a, b]$. Similarly,

$$\frac{|(Tx)'(t) - (Ty)'(t)|}{w(t)}$$

$$\leq \frac{1}{w(t)} \int_a^b |G_t(t, s)||f(s, x(s), x'(s)) - f(s, y(s), y'(s))| \, ds$$

$$\leq \frac{1}{w(t)} \int_a^b |G_t(t, s)| \left[p(s)|x(s) - y(s)| + q(s)|x'(s) - y'(s)| \right] \, ds$$

$$\leq \frac{1}{w(t)} \int_a^b |G_t(t, s)| \left[p(s)v(s) \frac{|x(s) - y(s)|}{v(s)} \right.$$

$$\left. + q(s)w(s) \frac{|x'(s) - y'(s)|}{w(s)} \right] ds$$

$$\leq \frac{1}{w(t)} \int_a^b |G_t(t, s)| \left[p(s)v(s) + q(s)w(s) \right] ds \, \|x - y\|,$$

for $t \in [a, b]$. Thus we see that we want to find positive continuous functions v, w and a constant α such that

$$\frac{1}{v(t)} \int_a^b |G(t, s)| \left[p(s)v(s) + q(s)w(s) \right] ds \leq \alpha < 1, \tag{7.17}$$

and

$$\frac{1}{w(t)} \int_a^b |G_t(t, s)| \left[p(s)v(s) + q(s)w(s) \right] ds \leq \alpha < 1, \tag{7.18}$$

for $t \in [a, b]$. If we can do this, then T is a contraction mapping on $\mathbb{X}$ and the conclusion of this theorem follows from the contraction mapping theorem (Theorem 7.5). By hypothesis the solution u of the IVP (7.15) satisfies $u'(t) > 0$ on $[a, b]$. Let v be the solution of the IVP

$$v'' = -\frac{1}{\alpha}[p(t)v + q(t)v'], \quad v(a) = \epsilon, \quad v'(a) = 1,$$

where we pick $\epsilon > 0$ sufficient close to zero and $\alpha < 1$ sufficiently close to one so that

$$v(t) > 0, \quad v'(t) > 0, \quad \text{on} \quad [a, b].$$

Let $w(t) := v'(t)$. If $B := v'(b) > 0$, it follows that v solves the BVP

$$v'' = -\frac{1}{\alpha}[p(t)v + q(t)v'], \quad v(a) = \epsilon, \quad v'(b) = B. \tag{7.19}$$

Let z_1 be the solution of the BVP

$$z_1'' = 0, \quad z_1(a) = \epsilon > 0, \quad z_1'(b) = B > 0.$$

It follows that $z_1(t) > 0$ and $z_1'(t) > 0$ on $[a, b]$. Since v solves the BVP (7.19),

$$
\begin{aligned}
v(t) &= z_1(t) + \frac{-1}{\alpha} \int_a^b G(t, s) \left[p(s)v(s) + q(s)v'(s) \right] ds \quad (7.20) \\
&> \frac{-1}{\alpha} \int_a^b G(t, s) \left[p(s)v(s) + q(s)v'(s) \right] ds \\
&= \frac{1}{\alpha} \int_a^b |G(t, s)| \left[p(s)v(s) + q(s)w(s) \right] ds,
\end{aligned}
$$

for $t \in [a, b]$. From this we get the desired inequality (7.17), for $t \in [a, b]$. Differentiating both sides of (7.20), we get

$$
\begin{aligned}
w(t) &= v'(t) = z_1'(t) + \frac{-1}{\alpha} \int_a^b G_t(t, s) \left[p(s)v(s) + q(s)v'(s) \right] ds \\
&> \frac{-1}{\alpha} \int_a^b G_t(t, s) \left[p(s)v(s) + q(s)v'(s) \right] ds \\
&= \frac{1}{\alpha} \int_a^b |G_t(t, s)| \left[p(s)v(s) + q(s)w(s) \right] ds,
\end{aligned}
$$

for $t \in [a, b]$, but this gives us the other desired inequality (7.18), for $t \in [a, b]$. $\qquad \square$

Now we deduce some useful information about the linear differential equation with constant coefficients

$$u'' + Lu' + Ku = 0. \quad (7.21)$$

Definition 7.13 Assume that u is the solution of the IVP (7.21),

$$u(0) = 0, \quad u'(0) = 1.$$

We define $\rho(K, L)$ as follows: If $u'(t) > 0$ on $[0, \infty)$, then $\rho(K, L) := \infty$. Otherwise $\rho(K, L)$ is the first point to the right of 0 where $u'(t)$ is zero.

In Exercise 7.14 you are asked to show that if u is a nontrivial solution of (7.21) with $u(0) = 0$, then if $u'(t) \neq 0$ on $[0, \infty)$, then $\rho(K, L) := \infty$. Otherwise $\rho(K, L)$ is the first point to the right of 0 where $u'(t)$ is zero. We will use this fact in the proof of the next theorem.

Theorem 7.14 *Assume $K, L \geq 0$; then*

$$
\rho(K, L) = \begin{cases}
\frac{2}{L}, & \text{if } L^2 - 4K = 0, \ L > 0, \\
\frac{2}{\sqrt{L^2 - 4K}} \operatorname{arccosh}\left(\frac{L}{2\sqrt{K}} \right), & \text{if } L^2 - 4K > 0, \ K > 0, \\
\frac{2}{\sqrt{4K - L^2}} \arccos\left(\frac{L}{2\sqrt{K}} \right), & \text{if } L^2 - 4K < 0, \\
\infty, & \text{if } L \geq 0, \ K = 0.
\end{cases}
$$

Proof We will only consider the case $L^2 - 4K > 0$, $K > 0$. The other cases are contained in Exercise 7.15. In the case $L^2 - 4K > 0$, a general solution of (7.21) is

$$u(t) = Ae^{-\frac{L}{2}t} \cosh\left(\frac{\sqrt{L^2 - 4K}}{2}t\right) + Be^{-\frac{L}{2}t} \sinh\left(\frac{\sqrt{L^2 - 4K}}{2}t\right).$$

The boundary condition $u(0) = 0$ gives us that $A = 0$ and hence

$$u(t) = Be^{-\frac{L}{2}t} \sinh\left(\frac{\sqrt{L^2 - 4K}}{2}t\right).$$

Differentiating, we get

$$u'(t) = -\frac{L}{2}Be^{-\frac{L}{2}t} \sinh\left(\frac{\sqrt{L^2 - 4K}}{2}t\right)$$

$$+ \; B\frac{\sqrt{L^2 - 4K}}{2}e^{-\frac{L}{2}t} \cosh\left(\frac{\sqrt{L^2 - 4K}}{2}t\right).$$

Setting $u'(t) = 0$, dividing by $Be^{-\frac{L}{2}t}$, and simplifying, we get

$$\frac{L}{2} \sinh\left(\frac{\sqrt{L^2 - 4K}}{2}t\right) = \frac{\sqrt{L^2 - 4K}}{2} \cosh\left(\frac{\sqrt{L^2 - 4K}}{2}t\right).$$

Squaring both sides, we get

$$\frac{L^2}{4} \sinh^2\left(\frac{\sqrt{L^2 - 4K}}{2}t\right) = \frac{L^2 - 4K}{4} \cosh^2\left(\frac{\sqrt{L^2 - 4K}}{2}t\right).$$

Using the identity $\cosh^2\theta - \sinh^2\theta = 1$ and simplifying, we get

$$\cosh^2\left(\frac{\sqrt{L^2 - 4K}}{2}t\right) = \frac{L^2}{4K}.$$

Solving for t, we get the desired result

$$t = \frac{2}{\sqrt{L^2 - 4K}} \operatorname{arccosh}\left(\frac{L}{2\sqrt{K}}\right).$$

$\square$

Theorem 7.15 *Assume $f : [a,b] \times \mathbb{R}^2 \to \mathbb{R}$ is continuous and there are constants K, L such that*

$$|f(t, x, x') - f(t, y, y')| \leq K|x - y| + L|x' - y'|,$$

for $(t, x, x'), (t, y, y') \in [a,b] \times \mathbb{R}^2$. If

$$b - a < \rho(K, L),$$

then it follows that the right-focal BVP

$$x'' = f(t, x, x'), \quad x(a) = A, \quad x'(b) = m, \tag{7.22}$$

where A and m are constants, has a unique solution.

Proof By Theorem 7.14, the solution of the IVP

$$v'' + Lv' + Kv = 0, \quad v(0) = 0, \quad v'(0) = 1$$

satisfies $v'(t) > 0$ on $[0, d]$, if $d < \rho(K, L)$. But this implies that $u(t) := v(t - a)$ is the solution of the IVP

$$u'' + Lu' + Ku = 0, \quad u(a) = 0, \quad u'(a) = 1$$

and satisfies $u'(t) > 0$ on $[a, b]$ if $b - a < \rho(K, L)$. It then follows from Theorem 7.12 that the BVP (7.22) has a unique solution. $\square$

In the next example we give an interesting application of Theorem 7.15 (see [**3**]).

Example 7.16 We are interested in how tall a vertical thin rod can be before it starts to bend (see Figure 1). Assume that a rod has length T, weight W, and constant of flexual rigidity $B > 0$. Assume that the rod is constrained to remain vertical at the bottom end of the rod, which is located at the origin in the st-plane shown in Figure 1. Further, assume that the rod has uniform material and constant cross section. If we let $x(t) = \frac{ds}{dt}(t)$, then it turns out that x satisfies the differential equation

$$x'' + \frac{W}{B}\frac{T - t}{T}x = 0. \tag{7.23}$$

If the rod does not bend, then the BVP (7.23),

$$x(0) = 0, \quad x'(T) = 0$$

has only the trivial solution. If the rod bends, then this BVP has a non-trivial solution (so solutions of this BVP are not unique). We now apply Theorem 7.15 to this BVP. Here $f(t, x, x') = -\frac{W}{B}\frac{T-t}{T}x$ and so

$$|f_x(t, x, x')| = \frac{W}{B}\frac{T - t}{T} \le K := \frac{W}{B},$$

and

$$|f_{x'}(t, x, x')| = 0 =: L.$$

Since

$$L^2 - 4K = -4\frac{W}{B} < 0,$$

we have by Theorem 7.14,

$$\rho(K, L) = \frac{2}{\sqrt{4K - L^2}} \arccos\left(\frac{L}{2\sqrt{K}}\right) = \frac{\pi}{2}\sqrt{\frac{B}{W}}.$$

Thus if $T < \frac{\pi}{2}\sqrt{\frac{B}{W}}$, then the rod is stable (does not bend). It was pointed out in [**3**] that it has been shown that if $T < 2.80\sqrt{\frac{B}{W}}$, then the rod is stable. Note that the upper bound on T is larger if B is larger and smaller if W is larger, which is what we expect intuitively. $\triangle$

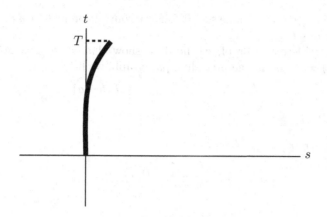

FIGURE 1. Vertical rod.

7.4 Lower and Upper Solutions

In this section, we will take a different approach to the study of BVPs, namely we will define functions called upper and lower solutions that, not only imply the existence of a solution of a BVP but also provide bounds on the location of the solution. The following result is fundamental:

Theorem 7.17 *Assume that* $f : [a, b] \times \mathbb{R}^2 \to \mathbb{R}$ *is continuous and bounded. Then the BVP*

$$x'' = f(t, x, x'),$$
$$x(a) = A, \quad x(b) = B,$$

where A and B are constants, has a solution.

Proof We can give a straightforward and more intuitive proof if we assume (as in Chapter 3) that f has continuous partials with respect to x and x'. Then IVP's have unique solutions that are continuous functions of their initial values. For a proof of the more general result, see Hartman [**18**].

Let $|f(t, x, x')| \le P$, for $(t, x, x') \in [a, b] \times \mathbb{R}^2$. Define $\phi(t, A, v)$ to be the solution of $x'' = f(t, x, x')$ satisfying the initial conditions $x(a) = A$, $x'(a) = v$. Then ϕ is continuous.

We first show that for each v, $\phi(t, A, v)$ exists on $[a, b]$. Since $|\phi''| \le P$, integration yields

$$|\phi'(t, A, v) - \phi'(a, A, v)| \le P|t - a|,$$

so that

$$|\phi'(t, A, v)| \le P(b - a) + |v|,$$

and ϕ' is bounded on the intersection of its maximal interval of existence with $[a, b]$. By integrating again, we get that

$$|\phi(t, A, v)| \le |A| + P(b - a)^2 + |v|(b - a),$$

so ϕ is bounded on the same set. It follows from Theorem 3.1 that $\phi(t, A, v)$ exists for $a \leq t \leq b$.

To complete the proof, we need to show that there is a w so that $\phi(b, A, w) = B$. As in the preceding paragraph, we have

$$\phi'(a, A, v) - \phi'(t, A, v) \leq P(b - a),$$

so

$$\phi'(t, A, v) \geq v - P(b - a),$$

for $a \leq t \leq b$. Choose

$$v > \frac{B - A}{b - a} + P(b - a).$$

Then

$$\phi'(t, A, v) > \frac{B - A}{b - a},$$

and by integrating both sides of the last inequality from a to b, we have

$$\phi(b, A, v) > B.$$

Similarly, we can find a value of v for which $\phi(b, A, v) < B$. The intermediate value theorem implies the existence of an intermediate value w so that $\phi(b, A, w) = B$. □

The method outlined in the previous proof of obtaining a solution of the BVP by varying the derivative of solutions at the initial point is called the *shooting method* and is one of the best ways to find numerical solutions of BVPs. Otherwise, Theorem 7.17 appears to be a result of little utility since one rarely deals with differential equations of the form $x'' = f(t, x, x')$ in which the function f is bounded. However, we shall see that this theorem is very useful indeed. A given equation with an unbounded, continuous right-hand side can be modified to obtain an equation in which the right-hand side is bounded and continuous and then Theorem 7.17 is applied to get a solution of the modified equation. It often happens that the solution of the modified equation is also a solution of the original equation.

First we define lower and upper solutions of $x'' = f(t, x, x')$.

Definition 7.18 We say that α is a *lower solution* of $x'' = f(t, x, x')$ on an interval I provided

$$\alpha''(t) \geq f(t, \alpha(t), \alpha'(t)),$$

for $t \in I$. Similarly, we say β is an *upper solution* of $x'' = f(t, x, x')$ on an interval I provided

$$\beta''(t) \leq f(t, \beta(t), \beta'(t)),$$

for $t \in I$.

In two of the main results (Theorem 7.20 and Theorem 7.34) in this chapter, we assume that

$$\alpha(t) \leq \beta(t), \quad t \in [a, b],$$

which is the motivation for calling α a lower solution and β an upper solution.

A very simple motivating example for what follows is given in the next example.

Example 7.19 Consider the differential equation

$$x'' = 0. \tag{7.24}$$

Note that a lower solution α of (7.24) satisfies $\alpha''(t) \geq 0$ and hence is concave upward, while an upper solution β of (7.24) satisfies $\beta''(t) \leq 0$ and hence is concave downward. To be specific, note that $\alpha(t) := t^2$ and $\beta(t) := -t^2 + 3$ are lower and upper solutions, respectively, of (7.24) satisfying $\alpha(t) \leq \beta(t)$ on $[-1, 1]$. Assume that A and B are constants satisfying,

$$\alpha(-1) = 1 \leq A \leq 2 = \beta(-1), \quad \text{and} \quad \alpha(1) = 1 \leq B \leq 2 = \beta(1).$$

Note that solutions of (7.24) are straight lines so the solution of the BVP

$$x'' = 0, \quad x(-1) = A, \quad x(1) = B$$

satisfies

$$\alpha(t) \leq x(t) \leq \beta(t),$$

for $t \in [-1, 1]$. (Compare this example to Theorems 7.20 and 7.34.) $\triangle$

We begin with the case that the differential equation does not contain the first derivative.

Theorem 7.20 *Assume f is continuous on $[a, b] \times \mathbb{R}$ and α, β are lower and upper solutions of $x'' = f(t, x)$, respectively, with $\alpha(t) \leq \beta(t)$ on $[a, b]$. If A and B are constants such that $\alpha(a) \leq A \leq \beta(a)$ and $\alpha(b) \leq B \leq \beta(b)$, then the BVP*

$$x'' = f(t, x),$$

$$x(a) = A, \quad x(b) = B$$

has a solution x satisfying

$$\alpha(t) \leq x(t) \leq \beta(t)$$

on $[a, b]$.

Proof First we define on $[a, b] \times \mathbb{R}$ the modification F of f as follows:

$$F(t, x) := \begin{cases} f(t, \beta(t)) + \frac{x - \beta(t)}{1 + |x|}, & \text{if } x \geq \beta(t) \\ f(t, x), & \text{if } \alpha(t) \leq x \leq \beta(t) \\ f(t, \alpha(t)) + \frac{x - \alpha(t)}{1 + |x|}, & \text{if } x \leq \alpha(t). \end{cases}$$

Note that F is continuous and bounded on $[a, b] \times \mathbb{R}$, and $F(t, x) = f(t, x)$, if $t \in [a, b]$, $\alpha(t) \leq x \leq \beta(t)$. Then by Theorem 7.17 the BVP

$$x'' = F(t, x),$$

$$x(a) = A, \quad x(b) = B,$$

has a solution x on $[a, b]$. We claim that $x(t) \leq \beta(t)$ on $[a, b]$. Assume not; then $w(t) := x(t) - \beta(t) > 0$ at some points in $[a, b]$. Since $w(a) \leq 0$ and $w(b) \leq 0$, we get that $w(t)$ has a positive maximum at some point $t_0 \in (a, b)$. Hence

$$w(t_0) > 0, \quad w'(t_0) = 0, \quad w''(t_0) \leq 0.$$

This implies that

$$x(t_0) > \beta(t_0), \quad x'(t_0) = \beta'(t_0), \quad x''(t_0) \leq \beta''(t_0).$$

But

$$
\begin{aligned}
x''(t_0) - \beta''(t_0) &\geq F(t_0, x(t_0)) - f(t_0, \beta(t_0)) \\
&= f(t_0, \beta(t_0)) + \frac{x(t_0) - \beta(t_0)}{1 + |x(t_0)|} - f(t_0, \beta(t_0)) \\
&= \frac{x(t_0) - \beta(t_0)}{1 + |x(t_0)|} \\
&> 0,
\end{aligned}
$$

which is a contradiction. Hence $x(t) \leq \beta(t)$ on $[a, b]$. It is left as an exercise (see Exercise 7.18) to show that $\alpha(t) \leq x(t)$ on $[a, b]$. Since

$$\alpha(t) \leq x(t) \leq \beta(t),$$

for $t \in [a, b]$, it follows that x is a solution of $x'' = f(t, x)$ on $[a, b]$.   □

Example 7.21 By Theorem 7.17 the BVP

$$x'' = -\cos x,$$

$$x(0) = 0 = x(1)$$

has a solution.

We now use Theorem 7.20 to find bounds on a solution of this BVP. It is easy to check that $\alpha(t) := 0$ and $\beta(t) := \frac{t(1-t)}{2}$, for $t \in [0, 1]$ are lower and upper solutions, respectively, of $x'' = f(t, x, x')$ on $[0, 1]$ satisfying the conditions of Theorem 7.20. Hence there is a solution x of this BVP satisfying

$$0 \leq x(t) \leq \frac{t(1-t)}{2},$$

for $t \in [0, 1]$.   △

Example 7.22 A BVP that comes up in the study of the viability of patches of plankton (microscopic plant and animal organisms) is the BVP

$$x'' = -rx\left(1 - \frac{1}{K}x\right),$$

$$x(0) = 0, \quad x(p) = 0,$$

where p is the width of the patch of plankton, $x(t)$ is the density of the plankton t units from one end of the patch of plankton, and $r > 0$, $K > 0$ are constants. It is easy to see that

$$\beta(t) = K$$

is an upper solution on $[0, p]$. We look for a lower solution of this equation of the form

$$\alpha(t) = a \sin\left(\frac{\pi t}{p}\right),$$

for some constant a. Consider

$$\alpha''(t) + r\alpha(t)\left[1 - \frac{1}{K}\alpha(t)\right]$$

$$= -a\frac{\pi^2}{p^2}\sin\left(\frac{\pi t}{p}\right) + ra\sin\left(\frac{\pi t}{p}\right)\left[1 - \frac{a}{K}\sin\left(\frac{\pi t}{p}\right)\right]$$

$$= a\sin\left(\frac{\pi t}{p}\right)\left\{-\frac{\pi^2}{p^2} + r\left[1 - \frac{a}{K}\sin\left(\frac{\pi t}{p}\right)\right]\right\}.$$

Hence if

$$r > \frac{\pi^2}{p^2},$$

and $0 < a < K$ is sufficiently small, then $\alpha(t) = a\sin\left(\frac{\pi t}{p}\right)$ is a lower solution of our given differential equation satisfying

$$\alpha(t) \le \beta(t).$$

The hypotheses of Theorem 7.20 are satisfied; hence we conclude that if the width p of the patch of plankton satisfies

$$p > \frac{\pi}{\sqrt{r}},$$

then our given BVP has a solution x satisfying

$$a\sin\left(\frac{\pi t}{p}\right) \le x(t) \le K,$$

for $t \in [0, p]$. This implies that if the width of the patch of plankton satisfies

$$p > \frac{\pi}{\sqrt{r}},$$

then the patch is viable. $\triangle$

Theorem 7.23 (Uniqueness Theorem) *Assume that $f : [a, b] \times \mathbb{R}^2 \to \mathbb{R}$ is continuous and for each fixed $(t, x') \in [a, b] \times \mathbb{R}$, $f(t, x, x')$ is strictly increasing with respect to x. Then the BVP*

$$x'' = f(t, x, x'),$$

$$x(a) = A, \quad x(b) = B$$

has at most one solution.

Proof Assume x and y are distinct solutions of the given BVP. Without loss of generality, we can assume that there are points $t \in (a, b)$ such that $x(t) > y(t)$. Let

$$w(t) := x(t) - y(t),$$

for $t \in [a, b]$. Since $w(a) = w(b) = 0$ and $w(t) > 0$ at some points in (a, b), w has a positive maximum at some point $d \in (a, b)$. Hence

$$w(d) > 0, \quad w'(d) = 0, \quad w''(d) \le 0.$$

But

$$
\begin{aligned}
w''(d) &= x''(d) - y''(d) \\
&= f(d, x(d), x'(d)) - f(d, y(d), y'(d)) \\
&= f(d, x(d), x'(d)) - f(d, y(d), x'(d)) \\
&> 0,
\end{aligned}
$$

where we have used that for each fixed $(t, x') \in [a, b] \times \mathbb{R}$, $f(t, x, x')$ is strictly increasing with respect to x. This is a contradiction and the proof is complete. $\qquad \square$

In the next example we show that in Theorem 7.23 you cannot replace *strictly increasing* in the statement of Theorem 7.23 by *nondecreasing*.

Example 7.24 Consider the differential equation

$$x'' = f(t, x, x') := |x'|^p,$$

where $0 < p < 1$.

Note that $f : [-a, a] \times \mathbb{R}^2 \to \mathbb{R}$ is continuous and for each fixed $(t, x') \in [-a, a] \times \mathbb{R}$, $f(t, x, x')$ is nondecreasing with respect to x. From Exercise 7.22 there is a solution of this differential equation of the form $x(t) = \kappa |t|^\alpha$, for some $\kappa > 0$ and some $\alpha > 2$. Since all constant functions are also solutions, we get that there are BVPs that do not have at most one solution. $\qquad \triangle$

The function $f(t, x, x') := |x'|^p$ in Example 7.24 does not satisfy a Lipschitz condition with respect to x' on $[-a, a] \times \mathbb{R}^2$. The next theorem, which we state without proof, shows that we do get uniqueness if we assume f satisfies a Lipschitz condition with respect to x'.

Theorem 7.25 (Uniqueness Theorem) *Assume that $f : [a, b] \times \mathbb{R}^2 \to \mathbb{R}$ is continuous and for each fixed $(t, x') \in [a, b] \times \mathbb{R}$, $f(t, x, x')$ is nondecreasing with respect to x. Further assume that f satisfies a Lipschitz condition with respect to x' on each compact subset of $[a, b] \times \mathbb{R}^2$. Then the BVP*

$$x'' = f(t, x, x'),$$

$$x(a) = A, \quad x(b) = B$$

has at most one solution.

Theorem 7.26 *Assume that $f(t,x)$ is continuous on $[a,b] \times \mathbb{R}$ and for each fixed $t \in [a,b]$, $f(t,x)$ is nondecreasing in x. Then the BVP*

$$x'' = f(t,x),$$

$$x(a) = A, \quad x(b) = B$$

has a unique solution.

Proof Let L be the line segment joining (a,A) and (b,B), and let M and m be, respectively, the maximum and minimum values of f on L. Since f is nondecreasing in x, if (t,x) is above L, then $f(t,x) \geq m$, and if (t,x) is below L, then $f(t,x) \leq M$. Define

$$\beta(t) = \frac{m}{2}t^2 + p, \quad \alpha(t) = \frac{M}{2}t^2 + q,$$

where p is chosen to be large enough that β is above L on $[a,b]$ and q is small enough that α is below L on $[a,b]$. Then $\beta(t) > \alpha(t)$, for $a \leq t \leq b$, $\beta(a) > A > \alpha(a)$, $\beta(b) > B > \alpha(b)$, and

$$\beta''(t) = m \leq f(t,\beta(t)), \quad \alpha''(t) = M \geq f(t,\alpha(t)),$$

for $a \leq t \leq b$. We can apply Theorem 7.20 to obtain a solution x of the BVP so that $\alpha(t) \leq x(t) \leq \beta(t)$, for $a \leq t \leq b$. Uniqueness follows from Theorem 7.25. □

Example 7.27 Consider the BVP

$$x'' = c(t)x + d(t)x^3 + e(t),$$

$$x(a) = A, \quad x(b) = B,$$

where $c(t)$, $d(t)$, and $e(t)$ are continuous on $[a,b]$. If further $c(t) \geq 0$ and $d(t) \geq 0$, then by Theorem 7.26 this BVP has a unique solution. △

Example 7.28 The following BVP serves as a simple model in the theory of combustion (see Williams [54]):

$$\epsilon x'' = x^2 - t^2,$$

$$x(-1) = x(1) = 1.$$

In the differential equation, ϵ is a positive parameter related to the speed of the reaction, x is a positive variable associated with mass, and t measures directed distance from the flame. Note that the right-hand side of the differential equation is increasing in x only for $x > 0$, so that Theorem 7.26 cannot be immediately applied. However, if we observe that $\alpha \equiv 0$ is a lower solution, then the proof of Theorem 7.26 yields a positive upper solution, and we can conclude that the BVP has a unique positive solution. The uniqueness of this positive solution follows by the proof of Theorem 7.23. △

Definition 7.29 Assume α and β are continuous functions on $[a, b]$ with $\alpha(t) \leq \beta(t)$ on $[a, b]$, and assume $c > 0$ is a given constant; then we say that $F(t, x, x')$ is the modification of $f(t, x, x')$ associated with the triple $\alpha(t)$, $\beta(t)$, c provided

$$F(t, x, x') := \begin{cases} g(t, \beta(t), x') + \frac{x - \beta(t)}{1 + |x|}, & \text{if } x \geq \beta(t), \\ g(t, x, x'), & \text{if } \alpha(t) \leq x \leq \beta(t), \\ g(t, \alpha(t), x') + \frac{x - \alpha(t)}{1 + |x|}, & \text{if } x \leq \alpha(t), \end{cases}$$

where

$$g(t, x, x') := \begin{cases} f(t, x, c), & \text{if } x' \geq c, \\ f(t, x, x'), & \text{if } |x'| \leq c, \\ f(t, x, -c), & \text{if } x' \leq -c. \end{cases}$$

We leave it to the reader to show that g, F are continuous on $[a, b] \times \mathbb{R}^2$, F is bounded on $[a, b] \times \mathbb{R}^2$, and

$$F(t, x, x') = f(t, x, x'),$$

if $t \in [a, b]$, $\alpha(t) \leq x \leq \beta(t)$, and $|x'| \leq c$.

Theorem 7.30 *Assume f is continuous on $[a, b] \times \mathbb{R}^2$ and α, β are lower and upper solutions of $x'' = f(t, x, x')$, respectively, with $\alpha(t) \leq \beta(t)$ on $[a, b]$. Further assume that solutions of IVPs for $x'' = f(t, x, x')$ are unique. If there is a $t_0 \in [a, b]$ such that*

$$\alpha(t_0) = \beta(t_0), \quad \alpha'(t_0) = \beta'(t_0),$$

then $\alpha(t) \equiv \beta(t)$ on $[a, b]$.

Proof Assume α, β, and t_0 are as in the statement of this theorem and it is not true that $\alpha(t) \equiv \beta(t)$ on $[a, b]$. We will only consider the case where there are points t_1, t_2 such that $a \leq t_0 \leq t_1 < t_2 \leq b$,

$$\alpha(t_1) = \beta(t_1), \quad \alpha'(t_1) = \beta'(t_1),$$

and

$$\alpha(t) < \beta(t), \quad \text{on } (t_1, t_2].$$

Pick $c > 0$ so that

$$|\alpha'(t)| < c, \quad |\beta'(t)| < c,$$

for $t \in [t_1, t_2]$. Let F be the modification of f with respect to the triple α, β, c for the interval $[t_1, t_2]$. By Theorem 7.17, the BVP

$$x'' = F(t, x, x'), \quad x(t_1) = \alpha(t_1), \quad x(t_2) = x_2,$$

where $\alpha(t_2) < x_2 < \beta(t_2)$ has a solution x. We claim that $x(t) \leq \beta(t)$ on $[t_1, t_2]$. Assume not; then there is a $d \in (t_1, t_2)$ such that $w(t) := x(t) - \beta(t)$ has a positive maximum on $[t_1, t_2]$ at d. It follows that

$$w(d) > 0, \quad w'(d) = 0, \quad w''(d) \leq 0,$$

and so
$$x(d) > \beta(d), \quad x'(d) = \beta'(d), \quad x''(d) \le \beta''(d).$$
But
$$\begin{aligned}
w''(d) &= x''(d) - \beta''(d) \\
&\ge F(d, x(d), x'(d)) - f(d, \beta(d), \beta'(d)) \\
&= f(d, \beta(d), \beta'(d)) + \frac{x(d) - \beta(d)}{1 + |\beta(d)|} - f(d, \beta(d), \beta'(d)) \\
&= \frac{x(d) - \beta(d)}{1 + |\beta(d)|} \\
&> 0,
\end{aligned}$$

which is a contradiction. Hence $x(t) \le \beta(t)$ on $[t_1, t_2]$. Similarly, $\alpha(t) \le x(t)$ on $[t_1, t_2]$. Thus
$$\alpha(t) \le x(t) \le \beta(t),$$
on $[t_1, t_2]$. Pick $t_3 \in [t_1, t_2)$ so that
$$x(t_3) = \alpha(t_3), \quad x'(t_3) = \alpha'(t_3),$$
and
$$\alpha(t) < x(t),$$
on $(t_3, t_2]$. Since
$$|x'(t_3)| = |\alpha'(t_3)| < c,$$
we can pick $t_4 \in (t_3, t_2]$ so that
$$|x'(t)| < c,$$
on $[t_3, t_4]$. Then x is a solution of $x'' = f(t, x, x')$ on $[t_3, t_4]$ and hence x is an upper solution of $x'' = f(t, x, x')$ on $[t_3, t_4]$. Let F_1 be the modification of f with respect to the triple α, x, c for the interval $[t_3, t_4]$. Let $x_4 \in (\alpha(t_4), x(t_4))$, then by Theorem 7.17 the BVP
$$x'' = F_1(t, x, x'), \quad x(t_3) = \alpha(t_3), \quad x(t_4) = x_4,$$
has a solution y on $[t_3, t_4]$. By a similar argument we can show that
$$\alpha(t) \le y(t) \le x(t),$$
on $[t_3, t_4]$. Now we can pick $t_5 \in (t_3, t_4]$ so that x and y differ at some points in $(t_3, t_5]$ and
$$|y'(t)| < c,$$
for $t \in [t_3, t_5]$. Then y is a solution of $x'' = f(t, x, x')$ on $[t_3, t_5]$. But now we have that x, y are distinct solutions of the same IVP (same initial conditions at t_3), which contradicts the uniqueness of solutions of IVPs. $\square$

Corollary 7.31 *Assume that the linear differential equation*
$$x'' + p(t)x' + q(t)x = 0 \tag{7.25}$$
has a positive upper solution β on $[a, b]$. Then (7.25) is disconjugate on $[a, b]$.

Proof Assume that (7.25) has a positive upper solution β on $[a, b]$ but that (7.25) is not disconjugate on $[a, b]$. It follows that there is a solution x of (7.25) and points $a \leq t_1 < t_2 \leq b$ such that $x(t_1) = x(t_2) = 0$ and $x(t) > 0$ on (t_1, t_2). But then there is a constant $\lambda > 0$ so that if $\alpha(t) := \lambda x(t)$, then

$$\alpha(t) \leq \beta(t), \quad t \in [t_1, t_2],$$

and $\alpha(c) = \beta(c)$ for some $c \in (a, b)$. It follows that $\alpha'(c) = \beta'(c)$. Since α is a solution, it is also a lower solution. Applying Theorem 7.30, we get that $\alpha(t) \equiv \beta(t)$ on $[t_1, t_2]$, which is a contradiction. □

7.5 Nagumo Condition

In this section we would like to extend Theorem 7.20 to the case where the nonlinear term is of the form $f(t, x, x')$; that is, also depends on x'. In Exercise 7.24 you are asked to show that $x(t) := 4 - \sqrt{4 - t}$ is a solution of the differential equation $x'' = 2(x')^3$ on $[0, 4)$. Note that this solution is bounded on $[0, 4)$, but its derivative is unbounded on $[0, 4)$. This is partly due to the fact that the nonlinear term $f(t, x, x') = 2(x')^3$ grows too fast with respect to the x' variable. We now define a Nagumo condition, which is a growth condition on $f(t, x, x')$ with respect to x'. In Theorem 7.33 we will see that this Nagumo condition will give us an a priori bound on the derivative of solutions of $x'' = f(t, x, x')$ that satisfy

$$\alpha(t) \leq x(t) \leq \beta(t),$$

for $t \in [a, b]$. We then will use this result to prove Theorem 7.34, which extends Theorem 7.20 to the case where f also depends on x' but satisfies a Nagumo condition. We end this section by giving several applications of Theorem 7.34.

Definition 7.32 We say that $f : [a, b] \times \mathbb{R}^2 \to \mathbb{R}$ satisfies a *Nagumo condition* with respect to the pair $\alpha(t), \beta(t)$ on $[a, b]$ provided $\alpha, \beta : [a, b] \to \mathbb{R}$ are continuous, $\alpha(t) \leq \beta(t)$ on $[a, b]$, and there is a function $h : [0, \infty) \to (0, \infty)$ such that

$$|f(t, x, x')| \leq h(|x'|),$$

for all $t \in [a, b]$, $\alpha(t) \leq x \leq \beta(t)$, $x' \in \mathbb{R}$ with

$$\int_\lambda^\infty \frac{s \, ds}{h(s)} > \max_{a \leq t \leq b} \beta(t) - \min_{a \leq t \leq b} \alpha(t),$$

where

$$\lambda := \max \left\{ \frac{|\beta(b) - \alpha(a)|}{b - a}, \frac{|\alpha(b) - \beta(a)|}{b - a} \right\}.$$

Theorem 7.33 *Assume that $f : [a, b] \times \mathbb{R}^2 \to \mathbb{R}$ is continuous and satisfies a Nagumo condition with respect to $\alpha(t), \beta(t)$ on $[a, b]$, where $\alpha(t) \leq \beta(t)$ on $[a, b]$. Assume x is a solution of $x'' = f(t, x, x')$ satisfying*

$$\alpha(t) \leq x(t) \leq \beta(t), \quad on \quad [a, b];$$

then there exists a constant $N > 0$, independent of x such that $|x'(t)| \leq N$ on $[a, b]$.

Proof Let λ be as in Definition 7.32. Choose $N > \lambda$ so that

$$\int_\lambda^N \frac{s \, ds}{h(s)} > \max_{t \in [a,b]} \beta(t) - \min_{t \in [a,b]} \alpha(t).$$

Let x be a solution of $x'' = f(t, x, x')$ satisfying

$$\alpha(t) \leq x(t) \leq \beta(t),$$

for $t \in [a, b]$. Then, using the mean value theorem, it can be shown that there is a $t_0 \in (a, b)$ such that

$$|x'(t_0)| = \left| \frac{x(b) - x(a)}{b - a} \right| \leq \lambda.$$

We claim that $|x'(t)| \leq N$ on $[a, b]$. Assume not; then $|x'(t)| > N$ at some points in $[a, b]$. We will only consider the case where there is a $t_1 \in [a, t_0)$ such that $x'(t_1) < -N$. (Another case is considered in Exercise 7.27.) Choose $t_1 < t_2 < t_3 \leq t_0$ so that

$$x'(t_2) = -N, \quad x'(t_3) = -\lambda,$$

and

$$-N < x'(t) < -\lambda, \quad \text{on} \quad (t_2, t_3).$$

On $[t_2, t_3]$,

$$\begin{aligned} |x''(t)| &= |f(t, x(t), x'(t))| \\ &\leq h(|x'(t)|) \\ &= h(-x'(t)). \end{aligned}$$

It follows that

$$-\frac{x'(t)x''(t)}{h(-x'(t))} \leq -x'(t),$$

for $t \in [t_2, t_3]$. Integrating both sides from t_2 to t_3, we get

$$\int_{t_2}^{t_3} \frac{-x'(t)x''(t)}{h(-x'(t))} dt \leq x(t_2) - x(t_3).$$

This implies that

$$\int_{-x'(t_2)}^{-x'(t_3)} \frac{-s}{h(s)} ds = \int_N^\lambda \frac{-s}{h(s)} ds \leq x(t_2) - x(t_3).$$

It follows that

$$\int_\lambda^N \frac{s}{h(s)} ds \leq x(t_2) - x(t_3) \leq \max_{t \in [a,b]} \beta(t) - \min_{t \in [a,b]} \alpha(t),$$

which is a contradiction. $\qquad \square$

Theorem 7.34 *Assume that $f : [a,b] \times \mathbb{R}^2 \to \mathbb{R}$ is continuous and α, β are lower and upper solutions, respectively, of $x'' = f(t,x,x')$ on $[a,b]$ with $\alpha(t) \leq \beta(t)$ on $[a,b]$. Further assume that f satisfies a Nagumo condition with respect to $\alpha(t), \beta(t)$ on $[a,b]$. Assume A, B are constants satisfying*

$$\alpha(a) \leq A \leq \beta(a), \quad \alpha(b) \leq B \leq \beta(b);$$

then the BVP

$$x'' = f(t,x,x'), \quad x(a) = A, \quad x(b) = B \tag{7.26}$$

has a solution satisfying

$$\alpha(t) \leq x(t) \leq \beta(t),$$

for $t \in [a,b]$.

Proof Pick $N_1 > 0$ sufficiently large so that $|\alpha'(t)| < N_1$ and $|\beta'(t)| < N_1$ on $[a,b]$. Let λ be as in Definition 7.32 and pick $N_2 > \lambda$ sufficiently large so that

$$\int_\lambda^{N_2} \frac{s\,ds}{h(s)} > \max_{t \in [a,b]} \beta(t) - \min_{t \in [a,b]} \alpha(t).$$

Then let $N := \max\{N_1, N_2\}$ and let $F(t,x,x')$ be the modification of $f(t,x,x')$ associated with the triple $\alpha(t)$, $\beta(t)$, N. Since F is continuous and bounded on $[a,b] \times \mathbb{R}^2$, we have by Theorem 7.17 that the BVP

$$x'' = F(t,x,x'),$$

$$x(a) = A, \quad x(b) = B$$

has a solution x on $[a,b]$. We claim that $x(t) \leq \beta(t)$ on $[a,b]$. Assume not; then $w(t) := x(t) - \beta(t) > 0$ at some points in $[a,b]$. Since $w(a) \leq 0$ and $w(b) \leq 0$, we get that $w(t)$ has a positive maximum at some point $\xi \in (a,b)$. Hence

$$w(\xi) > 0, \quad w'(\xi) = 0, \quad w''(\xi) \leq 0.$$

This implies that

$$x(\xi) > \beta(\xi), \quad x'(\xi) = \beta'(\xi), \quad x''(\xi) \leq \beta''(\xi).$$

Note that $|x'(\xi)| = |\beta'(\xi)| < N$. But

$$
\begin{aligned}
x''(\xi) - \beta''(\xi) &\geq F(\xi, x(\xi), x'(\xi)) - f(\xi, \beta(\xi), \beta'(\xi)) \\
&= g(\xi, \beta(\xi), \beta'(\xi)) + \frac{x(\xi) - \beta(\xi)}{1 + |x(\xi)|} - f(\xi, \beta(\xi), \beta'(\xi)) \\
&= \frac{x(\xi) - \beta(\xi)}{1 + |x(\xi)|} \\
&> 0,
\end{aligned}
$$

which is a contradiction. Hence $x(t) \leq \beta(t)$ on $[a,b]$. It is left as an exercise to show that $\alpha(t) \leq x(t)$ on $[a,b]$.

Now if $|x'(t)| \leq N$ on $[a, b]$, then x would be our desired solution of the BVP (7.26). As in the proof of Theorem 7.33, we have by the mean value theorem that there is a $t_0 \in (a, b)$ such that

$$|x'(t_0)| = \left| \frac{x(b) - x(a)}{b - a} \right| \leq \lambda < N,$$

where λ is as in Definition 7.32. We claim $|x'(t)| \leq N$ on $[a, b]$. Assume not, then there is a maximal interval $[t_1, t_2]$ containing t_0 in its interior such that $|x'(t)| \leq N$ on $[t_1, t_2]$, where either $t_1 > a$ and $|x'(t_2)| = N$ or $t_2 < b$ and $|x'(t_2)| = N$. We then proceed as in the proof of Theorem 7.33 to get a contradiction. $\square$

Example 7.35 Consider the BVP

$$x'' = x^2 + (x')^4, \quad x(0) = A, \quad x(1) = B. \tag{7.27}$$

Suppose that A, B are positive, and c is a number bigger than both A and B. Then $\alpha(t) := 0$, $\beta(t) := c$ are lower and upper solutions, respectively, of the differential equation in (7.27) on $[0, 1]$ with $\alpha(t) \leq \beta(t)$ on $[0, 1]$. Note that the λ in Definition 7.32 is given by $\lambda = c$. Define the function h by $h(s) := c^2 + s^4$ and note that

$$
\int_\lambda^\infty \frac{s}{h(s)} ds = \int_c^\infty \frac{s}{c^2 + s^4} ds
$$
$$
= \frac{\pi}{4c} - \frac{1}{2c} \arctan c.
$$

The inequality

$$\int_\lambda^\infty \frac{s}{h(s)} ds > c$$

leads to the inequality

$$\frac{\pi}{2} - \arctan c > 2c^2.$$

Let c_0 be the positive solution of

$$\frac{\pi}{2} - \arctan c = 2c^2.$$

Using our calculator, $c_0 \approx .69424$. It follows that if A and B are strictly between 0 and c_0, the BVP has a solution x satisfying

$$0 < x(t) < c_0, \quad t \in [0, 1]$$

(why can we have strict inequalities here?).

$\triangle$

As a final example, we show how upper and lower solutions can be used to verify approximations of solutions of singular perturbation problems of the type studied in Chapter 4.

Example 7.36 Let ϵ be a small positive parameter and consider the BVP

$$\epsilon x'' + 2x' + x^2 = 0,$$
$$x(0) = 3, \quad x(1) = 1.$$

If we substitute a perturbation series

$$x(t) = x_0(t) + \epsilon x_1(t) + \cdots,$$

we obtain

$$2x_0' + x_0^2 = 0,$$

which has the family of solutions

$$x_0(t) = \frac{2}{t + C}.$$

As in Chapter 4, we observe that it is consistent to have a solution with boundary layer at $t = 0$ since for such a solution, the large positive second derivative term $\epsilon x''$ will balance the negative first derivative term $2x'$ in the differential equation. Consequently, we impose the condition $x_0(1) = 1$, which yields

$$x_0(t) = \frac{2}{t + 1}.$$

We will leave it as an exercise (Exercise 7.32) to show that the boundary layer correction at $t = 0$ is $e^{-2t/\epsilon}$. Consequently, a formal approximation of a solution to the BVP is (for small ϵ)

$$A(t, \epsilon) = \frac{2}{t + 1} + e^{-\frac{2t}{\epsilon}}.$$

As a first step in verifying the formal approximation, we define a trial upper solution of the form

$$\beta(t, \epsilon) = \frac{2}{t + 1} + e^{(-2 + a\epsilon)t/\epsilon} + \epsilon f(t, \epsilon),$$

where a is a positive constant and f is a nonnegative continuous function to be determined. Note that

$$\beta(t, \epsilon) - A(t, \epsilon) = e^{-2t/\epsilon} \left(e^{at} - 1 \right) + \epsilon f \geq 0,$$

for $t \geq 0$ and $\epsilon > 0$, and the maximum of $e^{-2t/\epsilon}(e^{at} - 1)$ occurs when $e^{at} = 2/(2 - a\epsilon)$, so for this value of t

$$e^{-\frac{2t}{\epsilon}} \left(e^{at} - 1 \right) = e^{-\frac{2t}{\epsilon}} \left(\frac{a\epsilon}{2 - a\epsilon} \right) = \mathcal{O}(\epsilon),$$

as $\epsilon \to 0$. It follows that

$$\beta(t, \epsilon) - A(t, \epsilon) = \mathcal{O}(\epsilon) \quad (\epsilon \to 0)$$

uniformly for $0 \leq t \leq 1$.

To check if β is an upper solution for the BVP, we compute

$$\epsilon\beta'' \;+\; 2\beta' + \beta^2 \tag{7.28}$$

$$= \; \epsilon\left[\frac{4}{(t+1)^3} + \frac{(a\epsilon-2)^2}{\epsilon^2}e^{(a\epsilon-2)t/\epsilon} + \epsilon f''\right]$$

$$+2\left[-\frac{2}{(t+1)^2} + \frac{a\epsilon-2}{\epsilon}e^{(a\epsilon-2)t/\epsilon} + \epsilon f'\right]$$

$$+\left[\frac{2}{t+1} + e^{(a\epsilon-2)t/\epsilon} + \epsilon f\right]^2.$$

After some simplification, (7.28) reduces to

$$e^{(a\epsilon-2)t/\epsilon}\left[a(a\epsilon-2) + \frac{4}{t+1} + e^{(a\epsilon-2)t/\epsilon} + 2\epsilon f\right] + \frac{4\epsilon}{(t+1)^3}$$

$$+\epsilon\left[\epsilon f'' + 2f' + \frac{4}{t+1}f\right] + \epsilon^2 f^2.$$

Choose f to be a nonnegative solution of

$$\epsilon f'' + 2f' + 4f = -5,$$

specifically,

$$f = \frac{5}{4}\left[e^{-\lambda(1-t)} - 1\right] \quad (0 \le t \le 1),$$

where λ solves $\epsilon\lambda^2 + 2\lambda + 4 = 0$, namely,

$$\lambda = \frac{\sqrt{1-4\epsilon}-1}{\epsilon} = -2 + \mathcal{O}(\epsilon) \quad (\epsilon \to 0).$$

Returning to our calculation, we have that (7.28) is less than or equal to

$$e^{(a\epsilon-2)t/\epsilon}\left[a(a\epsilon-2) + \frac{4}{t+1} + e^{(a\epsilon-2)t/\epsilon} + 2\epsilon f\right]$$

$$+\epsilon\left[\frac{4}{(t+1)^3} + \epsilon f^2 - 5\right].$$

Choose $a = 6$ and ϵ small enough that $\epsilon \le 1/6$, $2\epsilon f \le 1$, and $\epsilon f^2 \le 1$. Then

$$\epsilon\beta'' + 2\beta' + \beta^2 \le 0, \quad (0 \le t \le 1),$$

so β is an upper solution.

We leave it to the reader to check that A itself is a lower solution. It follows from Theorem 7.34 that for positive values of ϵ that are small enough to satisfy the preceding inequalities, the BVP has a solution $x(t,\epsilon)$ such that

$$A(t,\epsilon) \le x(t,\epsilon) \le \beta(t,\epsilon), \quad (0 \le t \le 1).$$

Since $\beta(t,\epsilon) - A(t,\epsilon) = \mathcal{O}(\epsilon)$, we also have

$$0 \le x(t,\epsilon) - A(t,\epsilon) = \mathcal{O}(\epsilon),$$

as $\epsilon \to 0$, uniformly for $0 \le t \le 1$. $\qquad\qquad\qquad\qquad\qquad\triangle$

For a general discussion of the analysis of singular perturbation problems by the method of upper and lower solutions, see Chang and Howes [**7**] and Kelley [**30**].

7.6 Exercises

7.1 Show that if $\lim_{n\to\infty} x_n = x$, in a NLS $\mathbb{X}$, then the sequence $\{x_n\}$ is a Cauchy sequence.

7.2 Show that if $\mathbb{X}$ is the set of all continuous functions on $[a, \infty)$ satisfying $\lim_{t\to\infty} x(t) = 0$, where $\|\cdot\|$ is defined by $\|x\| = \max\{|x(t)| : a \le t < \infty\}$, then $\mathbb{X}$ is a Banach space.

7.3 Assume that $\mathbb{X}$ is the set of all continuous functions on $[a, b]$ and assume ρ is a positive continuous function on $[a, b]$, where $\|\cdot\|$ is defined by $\|x\| = \max\left\{ \frac{|x(t)|}{\rho(t)} : a \le t \le b \right\}$. Show that $\mathbb{X}$ is a Banach space.

7.4 Show that if X is the Banach space of continuous functions on $[0, 1]$ with the max (sup) norm defined by $\|x\| := \max\{|x(t)| : t \in [0, 1]\}$ and the operator $T : X \to X$ is defined by

$$Tx(t) = t + \int_0^t sx(s)ds, \quad t \in [0, 1],$$

then T is a contraction mapping on X.

7.5 Show that if in the contraction mapping theorem (Theorem 7.5), $\alpha \ge 1$, then the theorem is false.

7.6 Find a Banach space $\mathbb{X}$ and a mapping T on $\mathbb{X}$ such that $\|Tx - Ty\| < \|x - y\|$ for all $x, y \in \mathbb{X}$, $x \neq y$, such that T has no fixed point in $\mathbb{X}$.

7.7 Show that if in the contraction mapping thoerem (Theorem 7.5) we replace "T is a contraction mapping" by "there is an integer $m \ge 1$ such that T^m is a contraction mapping," then the theorem holds with the appropriate change in the inequality (7.1).

7.8 For what values of n and m can we apply Theorem 7.6 to conclude that

$$[t^n x']' + t^m x = \frac{1}{t^2 - 1}$$

has a solution that goes to zero as $t \to \infty$?

7.9 Show that the hypotheses of Theorem 7.7 concerning the BVP

$$x'' = \cos x, \quad x(0) = 0, \quad x(1) = 0$$

hold. Assume that $x_0(t) \equiv 0$ and let $\{x_n\}$ be the corresponding sequence of Picard iterates as defined in Theorem 7.5. Find the first Picard iterate x_1 and use (7.1) to find out how good an approximation it is to the actual solution of the given BVP.

7.10 Show that the hypotheses of Theorem 7.7 concerning the BVP

$$x'' = \frac{2}{1+t^2}x, \quad x(0) = 1, \quad x(1) = 2$$

hold. Assume that your initial guess is $x_0(t) = 1 + t^2$ and let $\{x_n\}$ be the corresponding sequence of Picard iterates as defined in Theorem 7.5. Find the first four Picard iterates. Explain your answers. Next find the first Picard iterate x_1 for the BVP

$$x'' = \frac{1}{1+t^2}x, \quad x(0) = 1, \quad x(1) = 2,$$

where again your initial guess is $x_0(t) = 1 + t^2$ and use (7.1) to find out how good an approximation it is to the actual solution of the given BVP.

7.11 Show that the hypotheses of Theorem 7.7 concerning the BVP

$$x'' = -2\sin x, \quad x(0) = 0, \quad x(1) = 1$$

hold. Assume that $x_0(t) = t$ and let $\{x_n\}$ be the corresponding sequence of Picard iterates as defined in Theorem 7.5. Find the first Picard iterate x_1 and use (7.1) to find out how good an approximation it is to the actual solution of the given BVP.

7.12 Use Theorem 7.9 to find L so that if $b - a < L$, then the BVP

$$x'' = -\sqrt{3x^2 + 1}, \quad x(a) = A, \quad x(b) = B$$

has for each $A, B \in \mathbb{R}$ a unique solution.

7.13 Use Theorem 7.9 to find L so that if $b - a < L$, then the BVP

$$x'' = -\frac{9}{1+t^2}x, \quad x(a) = A, \quad x(b) = B,$$

for each $A, B \in \mathbb{R}$, has a unique solution.

7.14 Show that if u is a nontrivial soultion of (7.21) with $u(0) = 0$, then if $u'(t) \neq 0$ on $[0, \infty)$, then $\rho(K, L) := \infty$. Otherwise $\rho(K, L)$ is the first point to the right of 0 where $u'(t)$ is zero.

7.15 Prove the remaining cases in Theorem 7.14.

7.16 (Forced Pendulum Problem) Find a number B so that if $0 < b < B$, then the BVP

$$\theta'' + \frac{g}{l}\sin\theta = F(t), \quad \theta(0) = A, \quad \theta'(b) = m,$$

where g is the acceleration due to gravity, l is the length of the pendulum, and A and m are constants has a unique solution.

7.17 Use Theorem 7.17 and its proof to show that

$$x'' = \frac{x}{x^2 + 1}, \quad x(0) = 0, \quad x(1) = 2$$

has a solution x and to find upper and lower bounds on $x'(t)$.

7.18 Show, as claimed in the proof of Theorem 7.20, that $\alpha(t) \leq x(t)$ on $[a, b]$.

7.19 By adapting the method of Example 7.21, find bounds on a solution of the BVP

$$x'' = \sin x - 2, \quad x(0) = x(1) = 0.$$

7.20 Show that the BVP

$$x'' = -x(2 - x),$$

$$x(0) = 0, \quad x(\pi) = 0$$

has a solution x satisfying $\sin t \leq x(t) \leq 2$ on $[0, \pi]$.

7.21 Show that the BVP

$$x'' = x + x^3,$$

$$x(a) = A, \quad x(b) = B$$

has a solution x satisfying $|x(t)| \leq C$ on $[a, b]$, where $C := \max\{|A|, |B|\}$.

7.22 Consider the differential equation in Example 7.24

$$x'' = f(t, x, x') := |x'|^p,$$

where $0 < p < 1$. Show that there is a solution of this differential equation of the form $x(t) = \kappa |t|^\alpha$ for some $\kappa > 0$ and some $\alpha > 2$.

7.23 What, if anything, can you say about uniqueness or existence of solutions of the given BVPs (give reasons for your answers)?

(i)

$$x'' = \cos(t^2 x^2 (x')^3),$$

$$x(a) = A, \quad x(b) = B$$

(ii)

$$x'' = t^2 + x^3 - (x')^2,$$

$$x(a) = A, \quad x(b) = B$$

(iii)

$$x'' = -e^{4t} + t^2 + \frac{2}{1 + (x')^2},$$

$$x(a) = A, \quad x(b) = B$$

7.24 Show that $x(t) := 4 - \sqrt{4 - t}$ is a solution of the differential equation $x'' = 2(x')^3$ that is bounded on $[0, 4)$ but whose derivative is unbounded on $[0, 4)$.

7.25 For what numbers $a < b$, A, B, does the BVP

$$x'' = \arctan x, \quad x(a) = A, \quad x(b) = B$$

have a unique solution?

7.26 Show that the equation

$$c = \sqrt{2}\cosh\left(\frac{c}{4}\right)$$

has two solutions c_1, c_2 satisfying $0 < c_1 < 2 < c_2 < 14$. Use your calculator to approximate c_1 and c_2. Show that if c is a solution of $c = \sqrt{2}\cosh\left(\frac{c}{4}\right)$, then

$$x(t) := \log\left(\frac{\cosh^2\left(\frac{c}{4}\right)}{\cosh^2\left[\frac{c}{2}(t - \frac{1}{2})\right]}\right)$$

is a solution of the BVP

$$x'' = -e^x, \quad x(0) = 0, \quad x(1) = 0.$$

7.27 In the proof of Theorem 7.33, verify the case where there is a $t_1 \in (t_0, b]$ such that $x'(t_1) < -N$.

7.28 Show that the BVP

$$x'' = -x + x^3 + (x')^2, \quad x(a) = A, \quad x(b) = B,$$

where $0 \le A, B \le 1$, has a solution x satisfying $0 \le x(t) \le 1$ for $t \in [a, b]$.

7.29 Assume that $\alpha, \beta \in C[a, b]$ with $\alpha(t) \le \beta(t)$ on $[a, b]$. Show that if

$$f(t, x, x') := \frac{1}{1 + t^2} + (\sin x)(x')^2,$$

then f satisfies a Nagumo condition with respect to α, β on $[a, b]$.

7.30 Show that if $0 \le A, B \le \frac{\pi}{5}$, then the BVP

$$x'' = x + (x')^4, \quad x(a) = A, \quad x(b) = B$$

has a solution for all sufficiently large b.

7.31 Show that the BVP

$$x'' = \frac{2}{9}(\sin t)x + (x')^4, \quad x(0) = A, \quad x(2) = B,$$

where $-.5 \le A \le .5$, $-.5 \le B \le .5$, has a solution x satisfying $-.5 \le x(t) \le .5$ for $t \in [0, 2]$.

7.32 Verify that the boundary layer correction in Example 7.36 is $e^{-2t/\epsilon}$.

7.33 In Example 7.36, show that the approximation A is itself a lower solution.

7.34 Consider the BVP

$$\epsilon^2 x'' = x + x^3, \quad x(0) = x(1) = 1,$$

where ϵ is a small positive parameter. Show that

$$\beta(t, \epsilon) = e^{-t/\epsilon} + e^{(t-1)/\epsilon}$$

is an upper solution and find a lower solution. Sketch the solution and describe any boundary layers.

7.35 For the singular perturbation problem

$$\epsilon x'' + x' + x^2 = 0, \quad x(0) = 3, \quad x(1) = .5,$$

compute a formal approximation and verify it by the method of upper and lower solutions.

Chapter 8

Existence and Uniqueness Theorems

8.1 Basic Results

In this chapter we are concerned with the first-order vector differential equation

$$x' = f(t, x). \tag{8.1}$$

We assume throughout this chapter that $f : D \to \mathbb{R}^n$ is continuous, where D is an open subset of $\mathbb{R} \times \mathbb{R}^n$.

Definition 8.1 We say x is a *solution* of (8.1) on an interval I provided $x : I \to \mathbb{R}^n$ is differentiable, $(t, x(t)) \in D$, for $t \in I$ and $x'(t) = f(t, x(t))$ for $t \in I$.

Note that if x is a solution of (8.1) on an interval I, then it follows from $x'(t) = f(t, x(t))$, for $t \in I$, that x is continuously differentiable on I.

In the next example we show that a certain nth-order scalar differential equation is equivalent to a vector equation of the form (8.1).

Example 8.2 Assume that D is an open subset of $\mathbb{R} \times \mathbb{R}^n$ and $F : D \to \mathbb{R}$ is continuous. We are concerned with the nth-order scalar differential equation

$$u^{(n)} = F(t, u, u', \cdots, u^{(n-1)}). \tag{8.2}$$

In this equation $t, u, u', \cdots, u^{(n)}$ denote variables. We say a scalar function $u : I \to \mathbb{R}$ is a solution of the nth-order scalar equation (8.2) on an interval I provided u is n times differentiable on I, $(t, u(t), u'(t), \cdots, u^{(n-1)}(t)) \in D$, for $t \in I$, and

$$u^{(n)}(t) = F(t, u(t), u'(t), \cdots, u^{(n-1)}(t)),$$

for $t \in I$. Note that if u is a solution of (8.2) on an interval I, then it follows that u is n times continuously differentiable on I. Now assume that u is a solution of the nth-order scalar equation (8.2) on an interval I. For

W.G. Kelley and A.C. Peterson, *The Theory of Differential Equations:*
Classical and Qualitative, Universitext 278, DOI 10.1007/978-1-4419-5783-2_8,
© Springer Science+Business Media, LLC 2010

$t \in I$, let

$$x(t) = \begin{pmatrix} x_1(t) \\ x_2(t) \\ \vdots \\ x_n(t) \end{pmatrix} := \begin{pmatrix} u(t) \\ u'(t) \\ \vdots \\ u^{(n-1)}(t) \end{pmatrix}.$$

Then

$$x'(t) = \begin{pmatrix} x_1'(t) \\ x_2'(t) \\ \vdots \\ x_n'(t) \end{pmatrix} = \begin{pmatrix} u'(t) \\ u''(t) \\ \vdots \\ u^{(n)}(t) \end{pmatrix}$$

$$= \begin{pmatrix} u'(t) \\ u''(t) \\ \vdots \\ F(t, u(t), u'(t), \cdots, u^{(n-1)}(t)) \end{pmatrix}$$

$$= f(t, x(t)),$$

if we define

$$f(t, x) = f(t, x_1, x_2, \cdots, x_n) = \begin{pmatrix} x_2 \\ x_3 \\ \vdots \\ F(t, x_1, x_2, \cdots, x_n) \end{pmatrix}, \qquad (8.3)$$

for $(t, x) \in D$. Note that $f : D \to \mathbb{R}^n$ is continuous. Hence if u is a solution of the nth-order scalar equation (8.2) on an interval I, then

$$x(t) = \begin{pmatrix} x_1(t) \\ x_2(t) \\ \vdots \\ x_n(t) \end{pmatrix} := \begin{pmatrix} u(t) \\ u'(t) \\ \vdots \\ u^{(n-1)}(t) \end{pmatrix},$$

$t \in I$, is a solution of a vector equation of the form (8.1) with $f(t, x)$ given by (8.3). Conversely, it can be shown that if x defined by

$$x(t) = \begin{pmatrix} x_1(t) \\ x_2(t) \\ \vdots \\ x_n(t) \end{pmatrix},$$

for $t \in I$ is a solution of a vector equation of the form (8.1) on an interval I with $f(t, x)$ given by (8.3); then $u(t) := x_1(t)$ defines a solution of (8.2) on the interval I. Because of this we say that the nth-order scalar equation (8.2) is equivalent to the vector equation (8.1) with f defined by (8.3). $\triangle$

Definition 8.3 Let $(t_0, x_0) \in D$. We say that x is a *solution of the IVP*

$$x' = f(t, x), \quad x(t_0) = x_0, \tag{8.4}$$

on an interval I provided $t_0 \in I$, x is a solution of (8.1) on I, and $x(t_0) = x_0$.

Example 8.4 Note that if $D = \mathbb{R} \times \mathbb{R}^2$, then x defined by

$$x(t) = \begin{pmatrix} \sin t \\ \cos t \end{pmatrix}$$

for $t \in \mathbb{R}$ is a solution of the IVP

$$x' = \begin{pmatrix} x_2 \\ -x_1 \end{pmatrix}, \quad x(0) = \begin{pmatrix} 0 \\ 1 \end{pmatrix},$$

on $I := \mathbb{R}$.   △

Closely related to the IVP (8.4) is the integral equation

$$x(t) = x_0 + \int_{t_0}^{t} f(s, x(s)) \, ds. \tag{8.5}$$

Definition 8.5 We say that $x : I \to \mathbb{R}^n$ is a *solution of the vector integral equation* (8.5) on an interval I provided $t_0 \in I$, x is continuous on I, $(t, x(t)) \in D$, for $t \in I$, and (8.5) is satisfied for $t \in I$.

The relationship between the IVP (8.4) and the integral equation (8.5) is given by the following lemma. Because of this result we say the IVP (8.4) and the integral equation (8.5) are equivalent.

Lemma 8.6 *Assume D is an open subset of $\mathbb{R} \times \mathbb{R}^n$, $f : D \to \mathbb{R}^n$ is continuous, and $(t_0, x_0) \in D$; then x is a solution of the IVP (8.4) on an interval I iff x is a solution of the integral equation (8.5) on an interval I.*

Proof Assume that x is a solution of the IVP (8.4) on an interval I. Then $t_0 \in I$, x is differentiable on I (hence is continuous on I), $(t, x(t)) \in D$, for $t \in I$, $x(t_0) = x_0$, and

$$x'(t) = f(t, x(t)),$$

for $t \in I$. Integrating this last equation and using $x(t_0) = x_0$, we get

$$x(t) = x_0 + \int_{t_0}^{t} f(s, x(s)) \, ds,$$

for $t \in I$. Thus we have shown that x is a solution of the integral equation (8.5) on the interval I.

Conversely assume x is a solution of the integral equation (8.5) on an interval I. Then $t_0 \in I$, x is continuous on I, $(t, x(t)) \in D$, for $t \in I$, and (8.5) is satisfied for $t \in I$. Since

$$x(t) = x_0 + \int_{t_0}^{t} f(s, x(s)) \, ds,$$

for $t \in I$, $x(t)$ is differentiable on I, $x(t_0) = x_0$, and

$$x'(t) = f(t, x(t)),$$

for all $t \in I$. Hence we have shown that x is a solution of the IVP (8.4) on the interval I. □

8.2 Lipschitz Condition and Picard-Lindelof Theorem

In this section we first defined what is meant by a vector function $f : D \to \mathbb{R}^n$ satisfies a *uniform Lipschitz condition with respect to x* on the open set $D \subset \mathbb{R} \times \mathbb{R}^n$. We then state and prove the important Picard-Lindelof theorem (Theorem 8.13), which is one of the main uniqueness-existence theorems for solutions of IVPs.

Definition 8.7 A vector function $f : D \to \mathbb{R}^n$ is said to satisfy a *Lipschitz condition with respect to x* on the open set $D \subset \mathbb{R} \times \mathbb{R}^n$ provided for each rectangle

$$Q := \{(t, x) : t_0 \le t \le t_0 + a, \|x - x_0\| \le b\} \subset D$$

there is a constant K_Q that may depend on the rectangle Q (and on the norm $\| \cdot \|$) such that

$$\|f(t, x) - f(t, y)\| \le K_Q \|x - y\|,$$

for all $(t, x), (t, y) \in Q$.

Definition 8.8 A vector function $f : D \to \mathbb{R}^n$ is said to satisfy a *uniform Lipschitz condition with respect to x* on D provided there is a constant K such that

$$\|f(t, x) - f(t, y)\| \le K \|x - y\|,$$

for all $(t, x), (t, y) \in D$. The constant K is called a *Lipschitz constant for $f(t, x)$ with respect to x on D*.

Definition 8.9 Assume the vector function $f : D \to \mathbb{R}^n$, where $D \subset \mathbb{R} \times \mathbb{R}^n$, is differentiable with respect to components of x. Then the *Jacobian matrix $D_x f(t, x)$ of $f(t, x)$ with respect to x* at (t, x) is defined by

$$D_x f(t, x) = \begin{pmatrix} \frac{\partial}{\partial x_1} f_1(t, x) & \cdots & \frac{\partial}{\partial x_n} f_1(t, x) \\ \frac{\partial}{\partial x_1} f_2(t, x) & \cdots & \frac{\partial}{\partial x_n} f_2(t, x) \\ \cdots & \cdots & \cdots \\ \frac{\partial}{\partial x_1} f_n(t, x) & \cdots & \frac{\partial}{\partial x_n} f_n(t, x) \end{pmatrix}.$$

Example 8.10 If

$$f(t, x) = f(t, x_1, x_2) = \begin{pmatrix} t^2 x_1^3 x_2^4 + t^3 \\ 3t + x_1^2 + x_2^3 \end{pmatrix},$$

then

$$D_x f(t, x) = \begin{pmatrix} 3t^2 x_1^2 x_2^4 & 4t^2 x_1^3 x_2^3 \\ 2x_1 & 3x_2^2 \end{pmatrix}.$$

△

Lemma 8.11 *Assume $D \subset \mathbb{R} \times \mathbb{R}^n$ such that for each fixed t, $D_t := \{x : (t, x) \in D\}$ is a convex set and $f : D \to \mathbb{R}^n$ is continuous. If the Jacobian matrix, $D_x f(t, x)$, of $f(t, x)$ with respect to x is continuous on D, then*

$$f(t, x) - f(t, y) = \int_0^1 D_x f(t, sx + (1 - s)y) \, ds \, [x - y], \qquad (8.6)$$

for all $(t, x), (t, y) \in D$.

Proof Let $(t, x), (t, y) \in D$; then D_t is convex implies that $(t, sx + (1 - s)y) \in D$, for $0 \le s \le 1$. Now for $(t, x), (t, y) \in D$, and $0 \le s \le 1$, we can consider

$$\frac{d}{ds} f(t, sx + (1 - s)y)$$

$$= \frac{d}{ds} \begin{pmatrix} f_1(t, sx_1 + (1 - s)y_1, \cdots, sx_n + (1 - s)y_n) \\ f_2(t, sx_1 + (1 - s)y_1, \cdots, sx_n + (1 - s)y_n) \\ \cdots \\ f_n(t, sx_1 + (1 - s)y_n)y_1, \cdots, sx_n + (1 - s)y_n) \end{pmatrix}$$

$$= \begin{pmatrix} \frac{\partial}{\partial x_1} f_1(\cdots)(x_1 - y_1) + \cdots + \frac{\partial}{\partial x_n} f_1(\cdots)(x_n - y_n) \\ \frac{\partial}{\partial x_1} f_2(\cdots)(x_1 - y_1) + \cdots + \frac{\partial}{\partial x_n} f_2(\cdots)(x_n - y_n) \\ \cdots \\ \frac{\partial}{\partial x_1} f_n(\cdots)(x_1 - y_1) + \cdots + \frac{\partial}{\partial x_n} f_n(\cdots)(x_n - y_n) \end{pmatrix}$$

$$= D_x f(t, sx + (1 - s)y)[x - y],$$

where the functions in the entries in the preceding matrix are evaluated at $(t, sx + (1 - s)y)$. Integrating both sides with respect to s from $s = 0$ to $s = 1$ gives us the desired result (8.6). $\qquad \square$

Theorem 8.12 *Assume $D \subset \mathbb{R} \times \mathbb{R}^n$, $f : D \to \mathbb{R}^n$, and the Jacobian matrix function $D_x f(t, x)$ is continuous on D. If for each fixed t, $D_t := \{x : (t, x) \in D\}$ is convex, then $f(t, x)$ satisfies a Lipschitz condition with respect to x on D.*

Proof Let $\| \cdot \|_1$ be the traffic norm (l_1 norm) defined in Example 2.47 and let $\| \cdot \|$ denote the corresponding matrix norm (see Definition 2.53). Assume that the rectangle

$$Q := \{(t, x) : |t - t_0| \le a, \|x - x_0\|_1 \le b\} \subset D.$$

Let

$$K := \max\{\|D_x f(t, x)\| : (t, x) \in Q\};$$

then using Lemma 8.11 and Theorem 2.54,

$$\|f(t, x) - f(t, y)\|_1 = \|\int_0^1 D_x f(t, sx + (1 - s)y) \, ds \, [x - y]\|_1$$

$$\le \int_0^1 \|D_x f(t, sx + (1 - s)y)\| \, ds \cdot \|x - y\|_1$$

$$\le K\|x - y\|_1,$$

for $(t,x), (t,y) \in Q$. Therefore, $f(t,x)$ satisfies a Lipschitz condition with respect to x on D.   □

Theorem 8.13 (Picard-Lindelof Theorem) *Assume that f is a continuous n-dimensional vector function on the rectangle*

$$Q := \{(t,x) : t_0 \le t \le t_0 + a, \|x - x_0\| \le b\}$$

and assume that $f(t,x)$ satisfies a uniform Lipschitz condition with respect to x on Q. Let

$$M := \max\{\|f(t,x)\| : (t,x) \in Q\}$$

and

$$\alpha := \min\left\{a, \frac{b}{M}\right\}.$$

Then the initial value problem (8.4) has a unique solution x on $[t_0, t_0 + \alpha]$. Furthermore,

$$\|x(t) - x_0\| \le b,$$

for $t \in [t_0, t_0 + \alpha]$.

Proof To prove the existence of a solution of the IVP (8.4) on $[t_0, t_0 + \alpha]$, it follows from Lemma 8.6 that it suffices to show that the integral equation

$$x(t) = x_0 + \int_{t_0}^t f(s, x(s))\, ds \qquad (8.7)$$

has a solution on $[t_0, t_0 + \alpha]$. We define the sequence of *Picard iterates* $\{x_k\}$ of the IVP (8.4) on $[t_0, t_0 + \alpha]$ as follows: Set

$$x_0(t) = x_0, \quad t \in [t_0, t_0 + \alpha],$$

and then let

$$x_{k+1}(t) = x_0 + \int_{t_0}^t f(s, x_k(s))\, ds, \quad t \in [t_0, t_0 + \alpha], \qquad (8.8)$$

for $k = 0, 1, 2, \cdots$. We show by induction that each Picard iterate x_k is well defined on $[t_0, t_0 + \alpha]$, is continuous on $[t_0, t_0 + \alpha]$, and its graph is in Q. Obviously, $x_0(t)$ satisfies these conditions. Assume that x_k is well defined, x_k is continuous on $[t_0, t_0 + \alpha]$, and

$$\|x_k(t) - x_0\| \le b, \quad \text{on} \quad [t_0, t_0 + \alpha].$$

It follows that

$$x_{k+1}(t) = x_0 + \int_{t_0}^t f(s, x_k(s))\, ds, \quad t \in [t_0, t_0 + \alpha]$$

is well defined and continuous on $[t_0, t_0 + \alpha]$. Also,

$$
\begin{aligned}
\|x_{k+1}(t) - x_0\| &\leq \int_{t_0}^{t} |f(s, x_k(s))| \, ds \\
&\leq M(t - t_0) \\
&\leq M\alpha \\
&\leq b,
\end{aligned}
$$

for $t \in [t_0, t_0 + \alpha]$ and the induction is complete.

Let K be a Lipschitz constant for $f(t, x)$ with respect to x on Q. We now prove by induction that

$$
\|x_{k+1}(t) - x_k(t)\| \leq \frac{MK^k(t - t_0)^{k+1}}{(k+1)!}, \quad t \in [t_0, t_0 + \alpha], \tag{8.9}
$$

for $k = 0, 1, 2, \cdots$. We proved (8.9) when $k = 0$. Fix $k \geq 1$ and assume that (8.9) is true when k is replaced by $k - 1$. Using the Lipschitz condition and the induction assumption, we get

$$
\begin{aligned}
\|x_{k+1}(t) - x_k(t)\| &= \left\| \int_{t_0}^{t} [f(s, x_k(s)) - f(s, x_{k-1}(s))] \, ds \right\| \\
&\leq \int_{t_0}^{t} \|f(s, x_k(s)) - f(s, x_{k-1}(s))\| \, ds \\
&\leq K \int_{t_0}^{t} \|x_k(s) - x_{k-1}(s)\| \, ds \\
&\leq MK^k \int_{t_0}^{t} \frac{(s - t_0)^k}{k!} \, ds \\
&= \frac{MK^k(t - t_0)^{k+1}}{(k+1)!},
\end{aligned}
$$

for $t \in [t_0, t_0 + \alpha]$. Hence the proof of (8.9) is complete.

The sequence of partial sums for the infinite series

$$
x_0(t) + \sum_{m=0}^{\infty} [x_{m+1}(t) - x_m(t)] \tag{8.10}
$$

is

$$
\left\{ x_0(t) + \sum_{m=0}^{k-1} [x_{m+1}(t) - x_m(t)] \right\} = \{x_k(t)\}.
$$

Hence we can show that the sequence of Picard iterates $\{x_k(t)\}$ converges uniformly on $[t_0, t_0 + \alpha]$ by showing that the infinite series (8.10) converges uniformly on $[t_0, t_0 + \alpha]$. Note that

$$
\|x_{m+1}(t) - x_m(t)\| \leq \frac{M}{K} \frac{(K\alpha)^{m+1}}{(m+1)!},
$$

for $t \in [t_0, t_0 + \alpha]$ and

$$\sum_{m=0}^{\infty} \frac{M}{K} \frac{(K\alpha)^{m+1}}{(m+1)!} \quad \text{converges.}$$

Hence from the Weierstrass M-test we get that the infinite series (8.10) converges uniformly on $[t_0, t_0+\alpha]$. Therefore, the sequence of Picard iterates $\{x_k(t)\}$ converges uniformly on $[t_0, t_0 + \alpha]$. Let

$$x(t) = \lim_{k \to \infty} x_k(t),$$

for $t \in [t_0, t_0 + \alpha]$. It follows that

$$\|x(t) - x_0\| \le b,$$

for $t \in [t_0, t_0 + \alpha]$. Since

$$\|f(t, x_k(t)) - f(t, x(t))\| \le K \|x_k(t) - x(t)\|$$

on $[t_0, t_0 + \alpha]$,

$$\lim_{k \to \infty} f(t, x_k(t)) = f(t, x(t))$$

uniformly on $[t_0, t_0 + \alpha]$. Taking the limit of both sides of (8.8), we get

$$x(t) = x_0 + \int_{t_0}^{t} f(s, x(s)) \, ds,$$

for $t \in [t_0, t_0 + \alpha]$. It follows that x is a solution of the IVP (8.4).

To complete the proof it remains to prove the uniqueness of solutions of the IVP (8.4). To this end let y be a solution of the IVP (8.4) on $[t_0, t_0+\beta]$, where $0 < \beta \le \alpha$. It remains to show that $y = x$. Since y is a solution of the IVP (8.4) on $[t_0, t_0 + \beta]$, it follows from Lemma 8.6 that y is a solution of the integral equation

$$y(t) = x_0 + \int_{t_0}^{t} f(s, y(s)) \, ds$$

on $[t_0, t_0 + \beta]$. Similarly we can prove by mathematical induction that

$$\|y(t) - x_k(t)\| \le \frac{MK^k(t - t_0)^{k+1}}{(k+1)!}, \tag{8.11}$$

for $t \in [t_0, t_0 + \beta]$, $k = 0, 1, 2, \cdots$. It follows that

$$y(t) = \lim_{k \to \infty} x_k(t) = x(t),$$

for $t \in [t_0, t_0 + \beta]$. $\qquad \square$

Corollary 8.14 *Assume the assumptions in the Picard-Lindelof theorem are satisfied and $\{x_k(t)\}$ is the sequence of Picard iterates defined in the proof of the Picard-Lindelof theorem. If x is the solution of the IVP (8.4), then*

$$\|x(t) - x_k(t)\| \le \frac{MK^k(t - t_0)^{k+1}}{(k+1)!}, \tag{8.12}$$

for $t \in [t_0, t_0 + \alpha]$, *where* K *is a Lipschitz constant for* $f(t,x)$ *with respect to* x *on* Q.

Example 8.15 In this example we maximize the α in the Picard-Lindelof theorem by choosing the appropriate rectangle Q for the initial value problem

$$x' = x^2, \quad x(0) = 1. \tag{8.13}$$

If

$$Q = \{(t,x) : 0 \le t \le a, |x - 1| \le b\},$$

then

$$M = \max\{|f(t,x)| = x^2 : (t,x) \in Q\} = (1 + b)^2.$$

Hence

$$\alpha = \min\left\{a, \frac{b}{M}\right\} = \min\left\{a, \frac{b}{(1+b)^2}\right\}.$$

Since we can choose a as large as we want, we desire to pick $b > 0$ so that $\frac{b}{(1+b)^2}$ is a maximum. Using calculus, we get $\alpha = \frac{1}{4}$. Hence by the Picard-Lindelof theorem we know that the solution of the IVP (8.13) exists on the interval $[0, \frac{1}{4}]$. The IVP (8.13) is so simple that we can solve this IVP to obtain $x(t) = \frac{1}{1-t}$. Hence the solution of the IVP (8.13) exists on $[0, 1)$ [actually on $(-\infty, 1)$]. Note that $\alpha = \frac{1}{4}$ is not a very good estimate. $\triangle$

Example 8.16 Approximate the solution of the IVP

$$x' = \cos x, \quad x(0) = 0 \tag{8.14}$$

by finding the second Picard iterate $x_2(t)$ and use (8.12) to find how good an approximation you get.

First we find the first Picard iterate $x_1(t)$. From equation (8.8) with $k = 0$ we get

$$
\begin{aligned}
x_1(t) &= x_0 + \int_{t_0}^{t} \cos(x_0(s))\, ds \\
&= \int_0^t 1\, ds \\
&= t.
\end{aligned}
$$

From equation (8.8) with $k = 1$ we get that the second Picard iterate $x_2(t)$ is given by

$$
\begin{aligned}
x_2(t) &= x_0 + \int_{t_0}^{t} \cos(x_1(s))\, ds \\
&= \int_0^t \cos s \, ds \\
&= \sin t.
\end{aligned}
$$

To see how good an approximation $x_2(t) = \sin t$ is for the solution $x(t)$ of the IVP (8.14), we get, applying (8.12), that

$$|x(t) - \sin t| \leq \frac{1}{6}t^3.$$

$\triangle$

Corollary 8.17 *Assume D is an open subset of $\mathbb{R} \times \mathbb{R}^n$, $f : D \to \mathbb{R}^n$ is continuous, and the Jacobian matrix $D_x f(t, x)$ is also continuous on D. Then for any $(t_0, x_0) \in D$ the IVP (8.4) has a unique solution on an interval containing t_0 in its interior.*

Proof Let $(t_0, x_0) \in D$; then there are positive numbers a and b such that the rectangle

$$R := \{(t, x) : |t - t_0| \leq a, \ \|x - x_0\|_1 \leq b\} \subset D.$$

In the proof of Theorem 8.12 we proved that $f(t, x)$ satisfies a uniform Lipschitz condition with respect to x on R with Lipschitz constant

$$K := \max\{\|D_x f(t, x)\| : (t, x) \in R\},$$

where $\|\cdot\|$ is matrix norm corresponding to the traffic norm $\|\cdot\|_1$ (see Definition 2.53). Let $M = \max\{\|f(t, x)\| : (t, x) \in R\}$ and let $\alpha := \min\{a, \frac{b}{M}\}$. Then by the Picard-Lindelof theorem (Theorem 8.13) in the case where we use the rectangle R instead of Q and we use the l_1 norm (traffic norm), the IVP (8.4) has a unique solution on $[t_0 - \alpha, t_0 + \alpha]$. $\quad\square$

Corollary 8.18 *Assume A is a continuous $n \times n$ matrix function and h is a continuous $n \times 1$ vector function on an interval I. If $(t_0, x_0) \in I \times \mathbb{R}$, then the IVP*

$$x' = A(t)x + h(t), \quad x(t_0) = x_0$$

has a unique solution.

Proof Let

$$f(t, x) = A(t)x + h(t);$$

then

$$D_x f(t, x) = A(t),$$

and this result follows from Theorem 8.17. $\quad\square$

In Theorem 8.65, we will show that under the hypotheses of Corollary 8.18 all solutions of $x' = A(t)x + h(t)$ exist on the whole interval I. Also, in Theorem 8.65 a bound on solutions will be given.

Corollary 8.19 *Assume D is an open subset of $\mathbb{R} \times \mathbb{R}^n$, the scalar function $F : D \to \mathbb{R}$ is continuous, and $F(t, x_1, x_2, \cdots, x_n)$ has continuous partial* *derivatives with respect to the variables* $x_1, x_2, \cdots, x_n$ *on* D. *Then for any* $(t_0, u_0, u_1, \cdots, u_{n-1}) \in D$ *the IVP*

$$u^{(n)} = F(t, u, u', \cdots, u^{(n-1)}), \tag{8.15}$$

$$u(t_0) = u_0, \quad u'(t_0) = u_1, \quad \cdots, \quad u^{(n-1)}(t_0) = u_{n-1} \tag{8.16}$$

has a unique solution on an interval containing t_0 *in its interior.*

Proof In Example 8.2 we proved that the differential equation (8.15) is equivalent to the vector equation $x' = f(t, x)$, where f is given by (8.3). Note that $f : D \to \mathbb{R} \times \mathbb{R}^n$ and the Jacobian matrix

$$D_x f(t, x) = \begin{pmatrix} 0 & 1 & 0 & \cdots & 0 \\ 0 & 0 & 1 & \cdots & 0 \\ \vdots & \vdots & \ddots & \ddots & \vdots \\ 0 & \cdots & \cdots & 0 & 1 \\ F_{x_1}(t, x_1, \cdots, x_n) & & \cdots & & F_{x_n}(t, x_1, \cdots, x_n) \end{pmatrix}$$

are continuous on D. The initial condition $x(t_0) = x_0$ corresponds to

$$\begin{pmatrix} x_1(t_0) \\ x_2(t_0) \\ \vdots \\ x_n(t_0) \end{pmatrix} = \begin{pmatrix} u(t_0) \\ u'(t_0) \\ \vdots \\ u^{(n-1)}(t_0) \end{pmatrix} = \begin{pmatrix} u_0 \\ u_1 \\ \vdots \\ u_{n-1} \end{pmatrix}$$

and the result follows from Corollary 8.17. $\qquad\qquad\square$

Example 8.20 In this example we apply Corollary 8.19 to the second-order scalar equation

$$u'' = (\sin t)e^u + u^2 + (u')^2.$$

This equation is of the form $u'' = F(t, u, u')$, where

$$F(t, x_1, x_2) = (\sin t)e^{x_1} + x_1^2 + (x_2)^2.$$

Let $D := \mathbb{R}^3$; then D is an open set and $F(t, x_1, x_2)$ is continuous on D. Also, $F_{x_1}(t, x_1, x_2) = (\sin t)e^{x_1} + 2x_1$ and $F_{x_2}(t, x_1, x_2) = 2x_2$ are continuous on D. Hence by Corollary 8.19 we have that every IVP

$$u'' = (\sin t)e^u + u^2 + (u')^2,$$

$$u(t_0) = u_0, \quad u'(t_0) = u_1$$

has a unique solution on an open interval containing t_0. $\qquad\qquad\triangle$

Example 8.21 In this example we apply Corollary 8.19 to the second-order scalar equation

$$u'' = u^{\frac{1}{3}} + 3u' + e^{2t}.$$

This equation is of the form $u'' = F(t, u, u')$, where

$$F(t, x_1, x_2) = x_1^{\frac{1}{3}} + 3x_2 + e^{2t}.$$

It follows that $F_{x_1}(t, x_1, x_2) = \frac{1}{3x_1^{\frac{2}{3}}}$ and $F_{x_2}(t, x_1, x_2) = 3$. If we let D be either the open set $\{(t, x_1, x_2) : t \in \mathbb{R}, \ x_1 \in (0, \infty), \ x_2 \in \mathbb{R}\}$ or the open set $\{(t, x_1, x_2) : t \in \mathbb{R}, \ x_1 \in (-\infty, 0), \ x_2 \in \mathbb{R}\}$, then by Corollary 8.19 we have that for any $(t_0, u_0, u_1) \in D$ the IVP

$$u'' = u^{\frac{1}{3}} + 3u' + e^{2t}, \quad u(t_0) = u_0, \quad u'(t_0) = u_1 \tag{8.17}$$

has a unique solution on an open interval containing t_0. Note that if $u_0 = 0$, then Corollary 8.19 does not apply to the IVP (8.17). $\triangle$

8.3 Equicontinuity and the Ascoli-Arzela Theorem

In this section we define what is meant by an equicontinuous family of functions and state and prove the very important Ascoli-Arzela theorem (Theorem 8.26). First we give some preliminary definitions.

Definition 8.22 We say that the sequence of vector functions $\{x_m(t)\}_{m=1}^{\infty}$ is *uniformly bounded on an interval* I provided there is a constant M such that

$$\|x_m(t)\| \leq M,$$

for $m = 1, 2, 3, \cdots$, and for all $t \in I$, where $\|\cdot\|$ is any norm on $\mathbb{R}^n$.

Example 8.23 The sequence of vector functions

$$x_m(t) = \begin{pmatrix} 2t^m \\ \sin(mt) \end{pmatrix},$$

$m = 1, 2, 3, \cdots$ is uniformly bounded on the interval $I := [0, 1]$, since

$$\|x_m(t)\|_1 = 2|t^m| + |\sin(mt)| \leq M := 3,$$

for all $t \in I$, and for all $m = 1, 2, 3, \cdots$. $\triangle$

Definition 8.24 We say that the family of vector functions $\{x_\alpha(t)\}$, for α in some index set A, is *equicontinuous on an interval* I provided given any $\epsilon > 0$ there is a $\delta > 0$ such that

$$\|x_\alpha(t) - x_\alpha(\tau)\| < \epsilon,$$

for all $\alpha \in A$ and for all $t, \tau \in I$ with $|t - \tau| < \delta$.

We will use the following lemma in the proof of the Ascoli-Arzela theorem (Theorem 8.26).

Lemma 8.25 (Cantor Selection Theorem) *Let $\{f_k\}$ be a uniformly bounded sequence of vector functions on $E \subset \mathbb{R}^n$. Then if D is a countable subset of E, there is a subsequence $\{f_{k_j}\}$ of $\{f_k\}$ that converges pointwise on D.*

Proof If D is finite the proof is easy. Assume D is countably infinite; then D can be written in the form

$$D = \{x_1, x_2, x_3, \cdots\}.$$

Since $\{f_k(x_1)\}$ is a bounded sequence of vectors, there is a convergent subsequence $\{f_{1k}(x_1)\}$. Next consider the sequence $\{f_{1k}(x_2)\}$. Since this is a bounded sequence of vectors, there is a convergent subsequence $\{f_{2k}(x_2)\}$. Continuing in this fashion, we get further subsequences such that

$$f_{11}(x), \ f_{12}(x), \ f_{13}(x), \ \cdots \quad \text{converges at } x_1$$
$$f_{21}(x), \ f_{22}(x), \ f_{23}(x), \ \cdots \quad \text{converges at } x_1, \, x_2$$
$$f_{31}(x), \ f_{32}(x), \ f_{33}(x), \ \cdots \quad \text{converges at } x_1, \, x_2, \, x_3$$
$$\cdots \qquad \cdots \qquad \cdots$$

It follows that the diagonal sequence $\{f_{kk}\}$ is a subsequence of $\{f_k\}$ that converges pointwise on D. $\qquad\qquad\qquad\qquad\qquad\qquad\qquad\qquad\qquad\quad\square$

Theorem 8.26 (Ascoli-Arzela Theorem) *Let E be a compact subset of $\mathbb{R}^m$ and $\{f_k\}$ be a sequence of n-dimensional vector functions that is uniformly bounded and equicontinuous on E. Then there is a subsequence $\{f_{k_j}\}$ that converges uniformly on E.*

Proof In this proof we will use the same notation $\|\cdot\|$ for a norm on $\mathbb{R}^m$ and $\mathbb{R}^n$. If E is finite the result is obvious. Assume E is infinite and let

$$D = \{x_1, x_2, x_3, \cdots\}$$

be a countable dense subset of E. By the Cantor selection theorem (Theorem 8.25) there is a subsequence $\{f_{k_j}\}$ that converges pointwise on D. We claim that $\{f_{k_j}\}$ converges uniformly on E. To see this, let $\epsilon > 0$ be given. By the equicontinuity of the sequence $\{f_k\}$ on E there is a $\delta > 0$ such that

$$\|f_k(x) - f_k(y)\| < \frac{\epsilon}{3}, \tag{8.18}$$

when $\|x - y\| < \delta$, $x, y \in E$, $k \geq 1$. Define the ball about x_i with radius δ by

$$B(x_i) := \{x \in E : \|x - x_i\| < \delta\},$$

for $i = 1, 2, 3, \cdots$. Then $\{B(x_i)\}$ is an open covering of E. Since E is compact there is an integer J such that

$$\{B(x_i)\}_{i=1}^J$$

covers E. Since $\{f_{k_j}(x)\}$ converges pointwise on the finite set

$$\{x_1, x_2, \cdots, x_J\},$$

there is an integer K such that

$$\|f_{k_j}(x_i) - f_{k_m}(x_i)\| < \frac{\epsilon}{3} \tag{8.19}$$

when $k_j, k_m \geq K$, $1 \leq i \leq J$. Now assume $x \in E$; then

$$x \in B(x_{i_0}),$$

for some $1 \leq i_0 \leq J$. Using (8.18) and (8.19), we get

$$
\begin{aligned}
\|f_{k_j}(x) - f_{k_m}(x)\| &\leq \|f_{k_j}(x) - f_{k_j}(x_{i_0})\| \\
&+ \|f_{k_j}(x_{i_0}) - f_{k_m}(x_{i_0})\| + \|f_{k_m}(x_{i_0}) - f_{k_m}(x)\| \\
&< \frac{\epsilon}{3} + \frac{\epsilon}{3} + \frac{\epsilon}{3} \\
&= \epsilon,
\end{aligned}
$$

for $k_j, k_m \geq K$. $\square$

8.4 Cauchy-Peano Theorem

In this section we use the Ascoli-Arzela theorem to prove the Cauchy-Peano theorem (8.27), which is a very important existence theorem.

Theorem 8.27 (Cauchy-Peano Theorem) *Assume $t_0 \in \mathbb{R}$, $x_0 \in \mathbb{R}^n$, and f is a continuous n-dimensional vector function on the rectangle*

$$Q := \{(t, x) : |t - t_0| \leq a, \|x - x_0\| \leq b\}.$$

Then the initial value problem (8.4) has a solution x on $[t_0 - \alpha, t_0 + \alpha]$ with $\|x(t) - x_0\| \leq b$, for $t \in [t_0 - \alpha, t_0 + \alpha]$, where

$$\alpha := \min\left\{a, \frac{b}{M}\right\}$$

and

$$M := \max\{\|f(t, x)\| : (t, x) \in Q\}.$$

Proof For m a positive integer, subdivide the interval $[t_0, t_0 + \alpha]$ into 2^m equal parts so that the interval $[t_0, t_0 + \alpha]$ has the partition points

$$t_0 < t_1 < t_2 \cdots < t_{2^m} = t_0 + \alpha.$$

So

$$t_j = t_0 + \frac{\alpha j}{2^m}, \quad 0 \leq j \leq 2^m.$$

For each positive integer m we define the function x_m (see Figure 1 for the scalar case) recursively with respect to the intervals $[t_j, t_{j+1}]$, $0 \leq j \leq 2^m$, as follows:

$$x_m(t) = x_0 + f(t_0, x_0)(t - t_0), \quad t_0 \leq t \leq t_1,$$

$x_1 = x_m(t_1)$, and, for $1 \leq j \leq 2^m - 1$,

$$x_m(t) = x_j + f(t_j, x_j)(t - t_j), \quad t_j \leq t \leq t_{j+1},$$

where

$$x_{j+1} = x_m(t_{j+1}).$$

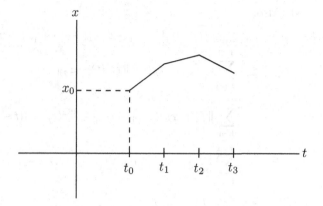

FIGURE 1. Approximate solution $x_m(t)$.

We show by finite mathematical induction with respect to j, $1 \le j \le 2^m$, that $x_m(t)$ is well defined on $[t_0, t_j]$, and

$$\|x(t) - x_0\| \le b, \quad \text{for} \quad t \in [t_0, t_j].$$

First, for $t \in [t_0, t_1]$,

$$x_m(t) = x_0 + f(t_0, x_0)(t - t_0)$$

is well defined and

$$
\begin{aligned}
\|x_m(t) - x_0\| &= \|f(t_0, x_0)\|(t - t_0) \\
&\le \frac{M\alpha}{2^m} \\
&\le b,
\end{aligned}
$$

for $t \in [t_0, t_1]$. Hence the graph of x_m on the first subinterval $[t_0, t_1]$ is in Q. In particular, $x_1 = x_m(t_1)$ is well defined with $(t_1, x_1) \in Q$.

Now assume $1 \le j \le 2^m - 1$ and that $x_m(t)$ is well defined on $[t_0, t_j]$ with

$$\|x_m(t) - x_0\| \le b, \quad \text{on} \quad [t_0, t_j].$$

Since $(t_j, x_j) = (t_j, x_m(t_j)) \in Q$,

$$x_m(t) = x_j + f(t_j, x_j)(t - t_j), \quad t_j \le t \le t_{j+1}$$

is well defined. Also,

$$
\begin{aligned}
\|x_m(t) - x_0\| &= \|[x_m(t) - x_j] + [x_j - x_{j-1}] + \cdots + [x_1 - x_0]\| \\
&\leq \sum_{k=0}^{j-1} \|x_{k+1} - x_k\| + \|x_m(t) - x_j\| \\
&\leq \sum_{k=0}^{j-1} \|f(t_k, x_k)(t_{k+1} - t_k)\| + \|f(t_j, x_j)(t - t_j)\| \\
&\leq \sum_{k=0}^{j-1} \frac{\alpha}{2^m} \|f(t_k, x_k)\| + \|f(t_j, x_j)\|(t - t_j) \\
&\leq \frac{\alpha j}{2^m} M + \frac{\alpha}{2^m} M \\
&\leq \alpha M \\
&\leq b,
\end{aligned}
$$

for $t \in [t_j, t_{j+1}]$. Hence we have shown that $x_m(t)$ is well defined on $[t_0, t_0 + \alpha]$ and

$$\|x_m(t) - x_0\| \leq b$$

on $[t_0, t_0 + \alpha]$. We will show that the sequence $\{x_m\}$ has a subsequence that converges uniformly on $[t_0, t_0 + \alpha]$ to a vector function z and z is a solution of the IVP (8.4) on $[t_0, t_0 + \alpha]$ whose graph on $[t_0, t_0 + \alpha]$ is in Q.

We now claim that for all $t, \tau \in [t_0, t_0 + \alpha]$

$$\|x_m(t) - x_m(\tau)\| \leq M|t - \tau|. \tag{8.20}$$

We will prove this only for the case $t_{k-1} < \tau \leq t_k < t_l < t \leq t_{l+1}$ as the other cases are similar. For this case

$$
\begin{aligned}
\|x_m(t) &- x_m(\tau)\| \\
&= \|[x_m(t) - x_m(t_l)] + \sum_{j=k}^{l-1}[x_m(t_{j+1}) - x_m(t_j)] + [x_m(t_k) - x_m(\tau)]\| \\
&\leq \|x_m(t) - x_m(t_l)\| + \sum_{j=k}^{l-1} \|x_m(t_{j+1}) - x_m(t_j)\| + \|x_m(t_k) - x_m(\tau)\| \\
&\leq \|f(t_l, x_l)\|(t - t_l) + \sum_{j=k}^{l-1} \|f(t_j, x_j)\|(t_{j+1} - t_j) \\
&\quad + \|f(t_{k-1}, x_{k-1})\|(t_k - \tau) \\
&\leq M(t - \tau).
\end{aligned}
$$

Hence (8.20) holds for all $t, \tau \in [t_0, t_0 + \alpha]$.

Since f is continuous on the compact set Q, f is uniformly continuous on Q. Hence given any $\epsilon > 0$ there is a $\delta > 0$ such that

$$\|f(t, x) - f(\tau, y)\| < \epsilon, \tag{8.21}$$

for all $(t, x), (\tau, y) \in Q$ with $|t - \tau| < \delta$, $\|x - y\| < \delta$. Pick m sufficiently large so that

$$\max\left\{ \frac{\alpha}{2^m}, \frac{M\alpha}{2^m} \right\} < \delta.$$

Then, for $t_j \leq t \leq t_{j+1}$, $0 \leq j \leq 2^m - 1$, since

$$\|x_m(t) - x_j\| \leq \frac{M\alpha}{2^m} < \delta$$

and

$$|t_j - t| \leq \frac{\alpha}{2^m} < \delta,$$

we get from (8.21)

$$\|f(t, x_m(t)) - f(t_j, x_j)\| < \epsilon,$$

for $t_j \leq t \leq t_{j+1}$, $0 \leq j \leq 2^m - 1$. Since

$$x'_m(t) = f(t_j, x_j),$$

for $t_j < t < t_{j+1}$, $0 \leq j \leq 2^m - 1$, we get that

$$\|f(t, x_m(t)) - x'_m(t)\| < \epsilon,$$

for $t_j < t < t_{j+1}$, $0 \leq j \leq 2^m - 1$. Hence if

$$g_m(t) := x'_m(t) - f(t, x_m(t)),$$

for $t \in [t_0, t_0 + \alpha]$, where $x'_m(t)$ exists, then we have shown that

$$\lim_{m \to \infty} g_m(t) = 0$$

uniformly on $[t_0, t_0 + \alpha]$, except for a countable number of points.

Fix $t \in [t_0, t_0 + \alpha]$; then there is a j such that $t_j \leq t \leq t_{j+1}$. Then

$$\begin{aligned}
x_m(t) - x_0 &= x_m(t) - x_m(t_0) \\
&= [x_m(t) - x_m(t_j)] + \sum_{k=1}^{j} [x_m(t_k) - x_m(t_{k-1})] \\
&= \int_{t_j}^{t} x'_m(s) \, ds + \sum_{k=1}^{j} \int_{t_{k-1}}^{t_k} x'_m(s) \, ds \\
&= \int_{t_0}^{t} x'_m(s) \, ds
\end{aligned}$$

for $t \in [t_0, t_0 + \alpha]$. Hence

$$x_m(t) = x_0 + \int_{t_0}^{t} [f(s, x_m(s)) + g_m(s)] \, ds, \qquad (8.22)$$

for $t \in [t_0, t_0 + \alpha]$. Since

$$\|x_m(t)\| \leq \|x_m(t) - x_0\| + \|x_0\| \leq \|x_0\| + b,$$

the sequence $\{x_m(t)\}$ is uniformly bounded on $[t_0, t_0 + \alpha]$. Since the sequence $\{x_m(t)\}$ is uniformly bounded and by (8.20) equicontinuous on $[t_0, t_0 + \alpha]$, we get from the Ascoli-Arzela theorem that the sequence $\{x_m(t)\}$ has a uniformly convergent subsequence $\{x_{m_k}(t)\}$ on $[t_0, t_0 + \alpha]$. Let

$$z(t) := \lim_{k \to \infty} x_{m_k}(t),$$

for $t \in [t_0, t_0 + \alpha]$. It follows that

$$\lim_{k \to \infty} f(t, x_{m_k}(t)) = f(t, z(t))$$

uniformly on $[t_0, t_0 + \alpha]$. Replacing m in equation (8.22) by $\{m_k\}$ and letting $k \to \infty$, we get that $z(t)$ is a solution of the integral equation

$$z(t) = x_0 + \int_{t_0}^t f(s, z(s)) \, ds$$

on $[t_0, t_0 + \alpha]$. It follows that $z(t)$ is a solution of the IVP (8.4) with $\|z(t) - x_0\| \le b$, for $t \in [t_0, t_0 + \alpha]$. Similarly, we can show that the IVP (8.4) has a solution $v(t)$ on $[t_0 - \alpha, t_0]$ with $\|v(t) - x_0\| \le b$, for $t \in [t_0 - \alpha, t_0]$. It follows that

$$x(t) := \begin{cases} v(t), & t \in [t_0 - \alpha, t_0], \\ z(t), & t \in [t_0, t_0 + \alpha] \end{cases}$$

is a solution of the IVP (8.4) with

$$\|x(t) - x_0\| \le b$$

on $[t_0 - \alpha, t_0 + \alpha]$.

$\square$

Under the hypotheses of the Cauchy-Peano theorem we get that IVPs have solutions, but they need not be unique. To see how bad things can be, we remark that in Hartman [19], pages 18–23, an example is given of a scalar equation $x' = f(t, x)$, where $f : \mathbb{R} \times \mathbb{R} \to \mathbb{R}$, is continuous, where for every IVP (8.4) there is more than one solution on $[t_0, t_0 + \epsilon]$ and $[t_0 - \epsilon, t_0]$ for arbitrary $\epsilon > 0$.

Theorem 8.28 *Assume D is an open subset of $\mathbb{R} \times \mathbb{R}^n$, $f : D \to \mathbb{R}^n$ is continuous, and K is a compact subset of D. Then there is an $\alpha > 0$ such that for all $(t_0, x_0) \in K$ the IVP (8.4) has a solution on $[t_0 - \alpha, t_0 + \alpha]$.*

Proof For $(t, x), (t, y) \in \mathbb{R} \times \mathbb{R}^n$, define the distance from (t, x) to (t, y) by

$$d[(t, x), (\tau, y)] = \max\{|t - \tau|, \|x - y\|\}.$$

If the boundary of D, $\partial D \ne \emptyset$, set $\rho = d(K, \partial D) > 0$. In this case define

$$K_\rho = \{(t, x) : d[(t, x), K] \le \frac{\rho}{2}\}.$$

If $\partial D = \emptyset$, then let

$$K_\rho = \{(t, x) : d[(t, x), K] \le 1\}.$$

Then $K_\rho \subset D$ and K_ρ is compact. Let

$$M := \max\{\|f(t, x)\| : (t, x) \in K_\rho\}.$$

Let $(t_0, x_0) \in K$; then if $\delta := \min\{1, \frac{\rho}{2}\}$,

$$Q := \{(t, x) : |t - t_0| \le \delta, \|x - x_0\| \le \delta\} \subset K_\rho.$$

Note that

$$\|f(t, x)\| \le M, \quad (t, x) \in Q.$$

Hence by the Cauchy-Peano theorem (Theorem 8.27) the IVP (8.4) has a solution on $[t_0 - \alpha, t_0 + \alpha]$, where $\alpha := \min\{\frac{\rho}{2}, \frac{\rho}{2M}\}$ if $\partial D \ne \emptyset$ and $\alpha := \min\{1, \frac{1}{M}\}$ if $\partial D = \emptyset$. □

8.5 Extendability of Solutions

In this section we will be concerned with proving that each solution of $x' = f(t, x)$ can be extended to a maximal interval of existence. First we define what we mean by the extension of a solution.

Definition 8.29 Assume x is a solution of $x' = f(t, x)$ on an interval I. We say that a solution y on an interval J is an *extension* of x provided $J \supset I$ and $y(t) = x(t)$, for $t \in I$.

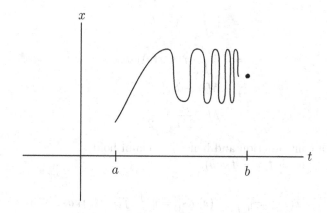

FIGURE 2. Impossible solution of scalar equation $x' = f(t, x)$ on $[a, b)$.

The next theorem gives conditions where a solution of $x' = f(t, x)$ on a half open interval $[a, b)$ can be extended to a solution on the closed interval $[a, b]$. This result implies that there is no solution to the scalar equation $x' = f(t, x)$ of the type shown in Figure 2.

Theorem 8.30 *Assume that f is continuous on $D \subset \mathbb{R} \times \mathbb{R}^n$ and that x is a solution of $x' = f(t, x)$ on the half-open interval $[a, b)$. Assume there is an increasing sequence $\{t_k\}$ with limit b and $\lim_{k \to \infty} x(t_k) = x_0$. Further assume there are constants $M > 0$, $\alpha > 0$, $\beta > 0$ such that $\|f(t, x)\| \le M$ on $D \cap \{(t, x) : 0 < b - t \le \alpha, \|x - x_0\| \le \beta\}$. Furthermore, if $f(b, x_0)$ can be defined so that f is continuous on $D \cup \{(b, x_0)\}$, then x can be extended to be a solution on $[a, b]$.*

Proof Pick an integer N sufficiently large so that

$$0 < b - t_k \leq \alpha, \quad \|x(t_k) - x_0\| < \frac{\beta}{2}, \quad \text{and} \quad 0 < b - t_k \leq \frac{\beta}{2M},$$

for all $k \geq N$. We claim

$$\|x(t) - x_0\| < \beta, \quad \text{for} \quad t_N < t < b.$$

Assume not; then there is a first point $\tau > t_N$ such that

$$\|x(\tau) - x_0\| = \beta.$$

But then

$$
\begin{aligned}
\frac{\beta}{2} &= \beta - \frac{\beta}{2} \\
&< \|x(\tau) - x_0\| - \|x(t_N) - x_0\| \\
&\leq \|x(\tau) - x(t_N)\| \\
&= \left\| \int_{t_N}^{\tau} x'(s) \, ds \right\| \\
&= \left\| \int_{t_N}^{\tau} f(s, x(s)) \, ds \right\| \\
&\leq M|\tau - t_N| \\
&\leq M \cdot \frac{\beta}{2M} = \frac{\beta}{2},
\end{aligned}
$$

which is a contradiction and hence our claim holds.

Note that for $t, \tau \in [t_N, b)$,

$$\|x(t) - x(\tau)\| = \left\| \int_{\tau}^{t} x'(s) \, ds \right\| = \left\| \int_{\tau}^{t} f(s, x(s)) \, ds \right\| \leq M|t - \tau|.$$

It follows that the Cauchy criterion for $\lim_{t \to b-} x(t)$ is satisfied and hence

$$\lim_{t \to b-} x(t) = x_0$$

exists. Define

$$x(b) = x_0 = \lim_{t \to b-} x(t).$$

Now assume that $f(b, x_0)$ is defined so that f is continuous on $D \cup \{(b, x_0)\}$, then

$$
\begin{aligned}
\lim_{t \to b-} x'(t) &= \lim_{t \to b-} f(t, x(t)) \\
&= f(b, x_0).
\end{aligned}
$$

Hence, using the mean value theorem,

$$
\begin{aligned}
x'(b) &= \lim_{t \to b-} \frac{x(b) - x(t)}{b - t} \\
&= \lim_{t \to b-} x'(\xi_t), \quad t < \xi_t < b, \\
&= f(b, x_0) \\
&= f(b, x(b)).
\end{aligned}
$$

Therefore, x is a solution on $[a, b]$. $\square$

Definition 8.31 Assume D is an open subset of $\mathbb{R} \times \mathbb{R}^n$, $f : D \to \mathbb{R}^n$ is continuous, and x is a solution of $x' = f(t, x)$ on (a, b). Then we say (a, b) is a *right maximal interval of existence* for x provided there does not exist a $b_1 > b$ and a solution y such that y is a solution on (a, b_1) and $y(t) = x(t)$ for $t \in (a, b)$. *Left maximal interval of existence* for x is defined in the obvious way. Finally, we say (a, b) is a *maximal interval of existence* for x provided it is both a right and left maximal interval of existence for x.

Definition 8.32 Assume D is an open subset of $\mathbb{R} \times \mathbb{R}^n$, $f : D \to \mathbb{R}^n$ is continuous, and x is a solution of $x' = f(t, x)$ on (a, b). We say $x(t)$ *approaches the boundary* of D, denoted ∂D, as $t \to b-$, write $x(t) \to \partial D$ as $t \to b-$, in case either

(i) $b = \infty$

   or

(ii) $b < \infty$ and for each compact subset $K \subset D$ there is a $t_K \in (a, b)$ such that $(t, x(t)) \notin K$ for $t_K < t < b$.

Similarly (see Exercise 8.20), we can define $x(t)$ approaches the boundary of D as $t \to a+$, write $x(t) \to \partial D$ as $t \to a+$.

Theorem 8.33 (Extension Theorem) *Assume D is an open subset of $\mathbb{R} \times \mathbb{R}^n$, $f : D \to \mathbb{R}^n$ is continuous, and x is a solution of $x' = f(t, x)$ on (a, b), $-\infty \le a < b \le \infty$. Then x can be extended to a maximal interval of existence (α, ω), $-\infty \le \alpha < \omega \le \infty$. Furthermore, $x(t) \to \partial D$ as $t \to \omega-$ and $x(t) \to \partial D$ as $t \to \alpha +$.*

Proof We will just show that x can be extended to a right maximal interval of existence (a, ω) and $x(t) \to \partial D$ as $t \to \omega-$.

   Let $\{K_k\}$ be a sequence of open sets such that the closure of K_k, $\bar{K}_k$, is compact, $\bar{K}_k \subset K_{k+1}$, and $\cup_{k=1}^{\infty} K_k = D$ (see Exercise 8.21).

   If $b = \infty$, we are done, so assume $b < \infty$. We consider two cases:

   Case 1. Assume for all $k \ge 1$ there is a τ_k such that for $t \in (\tau_k, b)$ we have that

$$
(t, x(t)) \notin \bar{K}_k.
$$

Assume there is a $b_1 > b$ and a solution y that is an extension of x to the interval (a, b_1). Fix $t_0 \in (a, b)$; then

$$A := \{(t, y(t)) : t_0 \le t \le b\}$$

is a compact subset of D. Pick an interger k_0 sufficiently large so that

$$\{(t, x(t)) : t_0 \le t < b\} \subset A \subset K_{k_0}.$$

This is a contradiction and hence (a, b) is a right maximal interval of existence for x. Also, it is easy to see that $x(t) \to \partial D$ as $t \to b-$.

Case 2. There is an integer m_0 and an increasing sequence $\{t_k\}$ with limit b such that $(t_k, x(t_k)) \in \bar{K}_{m_0}$, for all $k \ge 1$. Since $\bar{K}_{m_0}$ is compact, there is a subsequence $(t_{k_j}, x(t_{k_j}))$ such that

$$\lim_{j \to \infty} (t_{k_j}, x(t_{k_j})) = (b, x_0)$$

exists and $(b, x_0) \in \bar{K}_{m_0} \subset D$. By Theorem 8.30 we get that we can extend the solution x to $(a, b]$ by defining $x(b) = x_0$. By Theorem 8.28 for each $k \ge 1$ there is a $\delta_k > 0$ such that for all $(t_1, x_1) \in \overline{K}_k$ the IVP

$$x' = f(t, x), \quad x(t_1) = x_1$$

has a solution on $[t_1 - \delta_k, t_1 + \delta_k]$. Hence the IVP $x' = f(t, x)$, $x(b) = x_0$, has a solution y on $[b, b + \delta_{m_0}]$ and so if we extend the definition of x by

$$x(t) = \begin{cases} x(t), & \text{on } (a, b], \\ y(t), & \text{on } (b, b + \delta_{m_0}], \end{cases}$$

then x is a solution on $(a, b+\delta_{m_0}]$. If the point $(b+\delta_{m_0}, x(b+\delta_{m_0})) \in \overline{K}_{m_0}$, then we repeat the process using a solution of the IVP

$$x' = f(t, x), \quad x(b + \delta_{m_0}) = x(b + \delta_{m_0}),$$

to get an extension of the solution x to $(a, b + 2\delta_{m_0}]$, which we also denote by x. Since $\overline{K}_{m_0}$ is compact, there is a first integer $j(m_0)$ such that $(b_1, x(b_1)) \notin \overline{K}_{m_0}$, where

$$b_1 = b + j(m_0)\delta_{m_0}.$$

However, $(b_1, x(b_1)) \in D$ and hence there is an $m_1 > m_0$ such that

$$(b_1, x(b_1)) \in K_{m_1}.$$

Therefore, the extension procedure can be repeated using $\overline{K}_{m_1}$ and the associated δ_{m_1}. Then there is a first integer $j(m_2)$ such that $(b_2, x(b_2)) \notin \overline{K}_{m_1}$, where

$$b_2 = b_1 + j(m_1)\delta_{m_1}.$$

Continuing in this fashion, we get an infinite sequence $\{b_k\}$. We then define

$$\omega = \lim_{k \to \infty} b_k.$$

We claim we have that the solution x has been extended to be a solution of $x' = f(t, x)$ on the interval (a, ω). To see this, let $\tau \in (a, \omega)$; then there is a b_{k_0} such that $\tau < b_{k_0} < \omega$ and x is a solution on $(a, b_{k_0}]$. Since $\tau \in (a, \omega)$

is arbritary, x is a solution on the interval (a, ω). If $\omega = \infty$, we are done. Assume that $\omega < \infty$. Note that the interval (a, ω) is right maximal, since $(b_k, x(b_k)) \notin \overline{K}_{m_{k-1}}$, for each $k \geq 1$ (Why?).

We claim that

$$x(t) \to \partial D \quad \text{as} \quad t \to \omega - .$$

To see this, assume not; then there is a compact set $H \subset D$ and a strictly increasing sequence $\{\tau_k\}$ with limit ω such that

$$(\tau_k, x(\tau_k)) \in H, \quad \text{for} \quad k \geq 1.$$

But by the definition of the sets $\{K_m\}$, there is a $\overline{m}$ such that $H \subset K_{\overline{m}}$ which leads to a contradiction. □

Theorem 8.34 (Extended Cauchy-Peano Theorem) *Assume that D is an open subset of $\mathbb{R} \times \mathbb{R}^n$ and $f : D \to \mathbb{R}^n$ is continuous. Let*

$$Q := \{(t, x) : |t - t_0| \leq a, \ \|x - x_0\| \leq b\} \subset D.$$

Let

$$\alpha = \min\left\{a, \frac{b}{M}\right\},$$

where

$$M = \max\{\|f(t, x)\| : (t, x) \in Q\}.$$

Then **every** *solution of the IVP (8.4) exists on $[t_0 - \alpha, t_0 + \alpha]$.*

Proof Assume x is a solution of the IVP (8.4) on $[t_0, t_0 + \epsilon]$, where $0 < \epsilon < \alpha$. Let $[t_0, \omega)$ be a right maximal interval of existence for x. Since $x(t) \to \partial D$ as $t \to \omega-$ and $Q \subset D$ is compact, there is a $\beta > t_0$ such that

$$x(t) \notin Q,$$

for $\beta \leq t < \omega$. Let t_1 be the first value of $t > t_0$ such that $(t_1, x(t_1)) \in \partial Q$. If $t_1 = t_0 + a$, we are done. So assume

$$\|x(t_1) - x_0\| = b.$$

Note that

$$\begin{aligned}
b &= \|x(t_1) - x_0\| \\
&= \left\| \int_{t_0}^{t_1} x'(s) \, ds \right\| \\
&= \left\| \int_{t_0}^{t_1} f(s, x(s)) \, ds \right\| \\
&\leq \int_{t_0}^{t_1} \|f(s, x(s))\| \, ds \\
&\leq M(t_1 - t_0).
\end{aligned}$$

Solving this inequality for t_1, we get

$$t_1 \geq t_0 + \frac{b}{M} \geq t_0 + \alpha.$$

The other cases of the proof are left to the reader. $\qquad\qquad$ $\square$

Corollary 8.35 *Assume that* $f : \mathbb{R} \times \mathbb{R}^n \to \mathbb{R}^n$ *is continuous and bounded; then every solution of* $x' = f(t, x)$ *has the maximal interval of existence* $(-\infty, \infty)$.

Proof Since f is bounded on $\mathbb{R} \times \mathbb{R}^n$, there is a $M > 0$ such that

$$\|f(t, x)\| \leq M,$$

for $(t, x) \in \mathbb{R} \times \mathbb{R}^n$. Assume x_1 is a solution of $x' = f(t, x)$ with maximal interval of existence (α_1, ω_1), $-\infty \leq \alpha_1 < \omega_1 \leq \infty$. Let $t_0 \in (\alpha_1, \omega_1)$ and let

$$x_0 := x_1(t_0).$$

Then x_1 is a solution of the IVP (8.4). For arbitrary $a > 0$, $b > 0$ we have by Theorem 8.34 that x_1 is a solution on $[t_0 - \alpha, t_0 + \alpha]$, where

$$\alpha = \min\left\{a, \frac{b}{M}\right\}.$$

Since M is fixed, we can make α as large as we want by taking a and b sufficiently large, and the proof is complete. $\qquad\qquad$ $\square$

Theorem 8.36 (Uniqueness Theorem) *Assume* $t_0 \in \mathbb{R}$, $x_0 \in \mathbb{R}^n$ *and the* n-*dimensional vector function* f *is continuous on the rectangle*

$$Q := \{(t, x) : t_0 \leq t \leq t_0 + a, \|x - x_0\| \leq b\}.$$

If the dot product

$$[f(t, x_1) - f(t, x_2)] \cdot [x_1 - x_2] \leq 0,$$

for all (t, x_1), $(t, x_2) \in Q$, *then the IVP* (8.4) *has a unique solution in* Q.

Proof By the Cauchy-Peano theorem (Theorem 8.27), the IVP (8.4) has a solution. It remains to prove the uniqueness. Assume that $x_1(t)$ and $x_2(t)$ satisfy the IVP (8.4) on the interval $[t_0, t_0 + \epsilon]$ for some $\epsilon > 0$ and their graphs are in Q for $t \in [t_0, t_0 + \epsilon]$. Let

$$h(t) := \|x_1(t) - x_2(t)\|^2, \quad t \in [t_0, t_0 + \epsilon],$$

where $\|\cdot\|$ is the Euclidean norm. Note that $h(t) \geq 0$ on $[t_0, t_0 + \epsilon]$ and $h(t_0) = 0$. Also, since $h(t)$ is given by the dot product,

$$h(t) = [x_1(t) - x_2(t)] \cdot [x_1(t) - x_2(t)],$$

we get

$$\begin{aligned}
h'(t) &= 2[x_1'(t) - x_2'(t)] \cdot [x_1(t) - x_2(t)] \\
&= 2[f(t, x_1(t)) - f(t, x_2(t))] \cdot [x_1(t) - x_2(t)] \\
&\leq 0
\end{aligned}$$

on $[t_0, t_0 + \epsilon]$. Thus $h(t)$ is nonincreasing on $[t_0, t_0 + \epsilon]$. Since $h(t_0) = 0$ and $h(t) \geq 0$ we get that $h(t) = 0$ for $t \in [t_0, t_0 + \epsilon]$. Hence

$$x_1(t) \equiv x_2(t)$$

on $[t_0, t_0 + \epsilon]$. Hence the IVP (8.4) has only one solution. $\qquad \square$

The following corollary follows from Theorem 8.36, and its proof is Exercise 8.22.

Corollary 8.37 *Assume $t_0 \in \mathbb{R}$, $x_0 \in \mathbb{R}$ and the scalar function f is continuous on the planar rectangle*

$$Q := \{(t, x) : t_0 \leq t \leq t_0 + a, |x - x_0| \leq b\}.$$

If for each fixed $t \in [t_0, t_0 + a]$, $f(t, x)$ is nonincreasing in x, then the IVP (8.4) has a unique solution in Q.

In the next example we give an application of Corollary 8.37, where the Picard-Lindelof theorem (Theorem 8.13) does not apply.

Example 8.38 Consider the IVP

$$x' = f(t, x), \quad x(0) = 0, \tag{8.23}$$

where

$$f(t, x) := \begin{cases} 0, & t = 0, \ -\infty < x < \infty, \\ 2t, & 0 < t \leq 1, \ -\infty < x < 0, \\ 2t - \frac{4x}{t}, & 0 < t \leq 1, \ 0 \leq x \leq t^2, \\ -2t, & 0 < t \leq 1, \ t^2 < x < \infty. \end{cases}$$

It is easy to see that f is continuous on $[0, 1] \times \mathbb{R}$ and for each fixed $t \in [0, 1]$ $f(t, x)$ is nonincreasing with respect to x. Hence by Corollary 8.37 the IVP (8.23) has a unique solution. See Exercise 8.23 for more results concerning this example.

$\triangle$

8.6 Basic Convergence Theorem

In this section we are concerned with proving the basic convergence theorem (see Hartman's book [19]). We will see that many results depend on this basic convergence theorem; for example, the continuous dependence of solutions on initial conditions, initial points, and parameters.

Theorem 8.39 (Basic Convergence Theorem) *Assume that $\{f_k\}$ is a sequence of continuous n-dimensional vector functions on an open set $D \subset \mathbb{R} \times \mathbb{R}^n$ and assume that*

$$\lim_{k \to \infty} f_k(t, x) = f(t, x)$$

uniformly on each compact subset of D. For each integer $k \geq 1$ let x_k be a solution of the IVP $x' = f_k(t, x)$, $x(t_k) = x_{0k}$, with $(t_k, x_{0k}) \in D$, $k \geq 1$, and $\lim_{k \to \infty} (t_k, x_{0k}) = (t_0, x_0) \in D$ and let the solution x_k have maximal interval of existence (α_k, ω_k), $k \geq 1$. Then there is a solution x of the *limit IVP (8.4) with maximal interval of existence (α, ω) and a subsequence $\{x_{k_j}(t)\}$ of $\{x_k(t)\}$ such that given any compact interval $[\tau_1, \tau_2] \subset (\alpha, \omega)$*

$$\lim_{j \to \infty} x_{k_j}(t) = x(t)$$

uniformly on $[\tau_1, \tau_2]$ in the sense that there is an integer $J = J(\tau_1, \tau_2)$ such that for all $j \geq J$, $[\tau_1, \tau_2] \subset (\alpha_{k_j}, \omega_{k_j})$ and

$$\lim_{j \to \infty, j \geq J} x_{k_j}(t) = x(t)$$

uniformly on $[\tau_1, \tau_2]$. Furthermore,

$$\limsup_{j \to \infty} \alpha_{k_j} \leq \alpha < \omega \leq \liminf_{j \to \infty} \omega_{k_j}.$$

Proof We will only prove that there is a solution of the limit IVP (8.4) with right maximal interval of existence $[t_0, \omega)$ and a subsequence $\{x_{k_j}(t)\}$ of $\{x_k(t)\}$ such that for each $\tau \in (t_0, \omega)$ there is an integer $J = J(\tau)$ such that $[t_0, \tau] \subset (\alpha_{k_j}, \omega_{k_j})$, for $j \geq J$ and

$$\lim_{j \to \infty, j \geq J} x_{k_j}(t) = x(t)$$

uniformly on $[t_0, \tau]$.

Let $\{K_k\}_{k=1}^{\infty}$ be a sequence of open subsets of D such that $\overline{K}_k$ is compact, $\overline{K}_k \subset K_{k+1}$, and $D = \cup_{k=1}^{\infty} K_k$. For each $k \geq 1$, if $\partial D \neq \emptyset$ let

$$H_k := \left\{ (t, x) \in D : d((t, x), \overline{K}_k) \leq \frac{\rho_k}{2} \right\},$$

where $\rho_k := d(\partial D, \overline{K}_k)$, and if $\partial D = \emptyset$ let

$$H_k := \left\{ (t, x) \in D : d((t, x), \overline{K}_k) \leq 1 \right\}.$$

Note that H_k is compact and $\overline{K}_k \subset H_k \subset D$, for each $k \geq 1$. Let $f_0(t, x) := f(t, x)$ in the remainder of this proof. Since

$$\lim_{k \to \infty} f_k(t, x) = f(t, x)$$

uniformly on each compact subset of D, $f_0(t, x) = f(t, x)$ is continuous on D and for each $k \geq 1$ there is an $M_k > 0$ such that

$$\|f_m(t, x)\| \leq M_k \quad \text{on} \quad H_k,$$

for all $m \geq 0$. For each $k \geq 1$ there is a $\delta_k > 0$ such that for all $m \geq 0$ and for all $(\tau, y) \in \overline{K}_k$ every solution z of the IVP

$$x' = f_m(t, x), \quad x(\tau) = y$$

exists on $[\tau - \delta_k, \tau + \delta_k]$ and satisfies $(t, z(t)) \in H_k$, for $t \in [\tau - \delta_k, \tau + \delta_k]$. Since $(t_0, x_0) \in D$, there is an integer $m_1 \geq 1$ such that $(t_0, x_0) \in K_{m_1}$. Let

$$\epsilon_k := \frac{\delta_k}{3}, \quad \text{for} \quad k \geq 1.$$

Since

$$\lim_{k \to \infty} (t_k, x_{0k}) = (t_0, x_0),$$

there is an integer N such that

$$(t_k, x_{0k}) \in K_{m_1} \quad \text{and} \quad |t_k - t_0| < \epsilon_{m_1},$$

for all $k \geq N$. Then for $k \geq N$,

$$[t_0, t_0 + \epsilon_{m_1}] \subset (\alpha_k, \omega_k)$$

and $(t, x_k(t)) \in H_{m_1}$, for $k \geq N$. This implies that the sequence of functions $\{x_k\}_{k=N}^{\infty}$ is uniformly bounded on $[t_0, t_0 + \epsilon_{m_1}]$ and since

$$
\begin{aligned}
\|x_k(\tau_2) - x_k(\tau_1)\| &= \left\| \int_{\tau_1}^{\tau_2} x_k'(s)\, ds \right\| \\
&= \left\| \int_{\tau_1}^{\tau_2} f_k(s, x_k(s))\, ds \right\| \\
&\leq M_{m_1} |\tau_2 - \tau_1|,
\end{aligned}
$$

for all $\tau_1, \tau_2 \in [t_0, t_0 + \epsilon_{m_1}]$, the sequence of functions $\{x_k\}_{k=N}^{\infty}$ is equicontinuous on $[t_0, t_0 + \epsilon_{m_1}]$. By the Ascoli-Arzela theorem (Theorem 8.26) there is a subsequence $\{k_1(j)\}_{j=1}^{\infty}$ of the sequence $\{k\}_{k=N}^{\infty}$ such that

$$\lim_{j \to \infty} x_{k_1(j)}(t) = x(t)$$

uniformly on $[t_0, t_0 + \epsilon_{m_1}]$. This implies that x is a solution of the limit IVP (8.4) on $[t_0, t_0 + \epsilon_{m_1}]$.

Note that

$$\lim_{j \to \infty} (t_0 + \epsilon_{m_1}, x_{k_1(j)}(t_0 + \epsilon_{m_1})) = (t_0 + \epsilon_{m_1}, x(t_0 + \epsilon_{m_1})) \in H_{m_1} \subset D.$$

If $(t_0 + \epsilon_{m_1}, x(t_0 + \epsilon_{m_1})) \in K_{m_1}$, then repeat the process and obtain a subsequence $\{k_2(j)\}_{j=1}^{\infty}$ of $\{k_1(j)\}_{j=1}^{\infty}$ such that

$$\lim_{j \to \infty} x_{k_2(j)}(t)$$

exists uniformly on $[t_1 + \epsilon_{m_1}, t_1 + 2\epsilon_{m_1}]$ and call the limit function x as before. Then x is a solution of the IVP

$$y' = f(t, y), \quad y(t_0 + \epsilon_{m_1}) = x(t_0 + \epsilon_{m_1}).$$

It follows that

$$\lim_{j \to \infty} x_{k_2(j)}(t) = x(t)$$

uniformly on $[t_0, t_0 + 2\epsilon_{m_1}]$. Continuing in this manner, there is a first integer $j(m_1)$ such that an appropriate subsequence converges uniformly to an extended x on $[t_0, t_0 + j(m_1)\epsilon_{m_1}]$ and

$$(t_0 + j(m_1)\epsilon_{m_1}, x(t_0 + j(m_1)\epsilon_{m_1})) \notin K_{m_1}.$$

Pick $m_2 > m_1$ so that

$$(t_0 + j(m_1)\epsilon_{m_1}, x(t_0 + j(m_1)\epsilon_{m_1})) \in K_{m_2}.$$

We then continue this process in the obvious manner to get the desired result. $\qquad\square$

8.7 Continuity of Solutions with Respect to ICs

In this section we are concerned with the smoothness of solutions with respect to initial conditions, initial points, and parameters. Two very important scalar equations that contain a parameter λ are Legendre's equation

$$(1 - t^2)u'' - 2tu' + \lambda(\lambda + 1)u = 0$$

and Bessel's equation

$$t^2 u'' + tu' + (t^2 - \lambda^2)u = 0.$$

Theorem 8.40 (Continuity of Solutions with Respect to Initial Conditions and Parameters) *Assume that D is an open subset of $\mathbb{R} \times \mathbb{R}^n$ and Λ is an open subset of $\mathbb{R}^m$ and f is continuous on $D \times \Lambda$ with the property that for each $(t_0, x_0, \lambda_0) \in D \times \Lambda$, the IVP*

$$x' = f(t, x, \lambda_0), \quad x(t_0) = x_0 \tag{8.24}$$

has a unique solution denoted by $x(t; t_0, x_0, \lambda_0)$. Then x is a continuous function on the set $\alpha < t < \omega$, $(t_0, x_0, \lambda_0) \in D \times \Lambda$.

Proof Assume

$$\lim_{k \to \infty} (t_{0k}, x_{0k}, \lambda_k) = (t_0, x_0, \lambda_0) \in D \times \Lambda.$$

Define

$$f_k(t, x) = f(t, x, \lambda_k),$$

for $(t, x) \in D$, $k \geq 1$. Then

$$\lim_{k \to \infty} f_k(t, x) = f(t, x, \lambda_0)$$

uniformly on compact subsets of D. Let $x_k(t) = x(t; t_{0k}, x_{0k}, \lambda_k)$; then x_k is the solution of the IVP

$$x' = f_k(t, x), \quad x(t_{0k}) = x_{0k}.$$

Let (α_k, ω_k) be the maximal interval of existence for x_k, for $k \geq 1$ and let $x(t; t_0, x_0, \lambda_0)$ be the solution of the limit IVP (8.24) with maximal interval of existence (α, ω). Then by the basic convergence theorem (Theorem 8.39),

$$\lim_{k \to \infty} x_k(t) = \lim_{k \to \infty} x(t; t_{0k}, x_{0k}, \lambda_k))$$
$$= x(t; t_0, x_0, \lambda_0)$$

uniformly on compact subintervals of (α, β) (see Exercise 8.28). The continuity of x with respect to its four arguments follows from this. $\square$

Theorem 8.41 (Integral Means) *Assume D is an open subset of $\mathbb{R} \times \mathbb{R}^n$ and $f : D \to \mathbb{R}^n$ is continuous. Then there is a sequence of vector functions $\{g_k(t, x)\}$, called* integral means, *such that $g_k : \mathbb{R} \times \mathbb{R}^n \to \mathbb{R}^n$, along with its first-order partial derivatives with respect to components of x,* *are continuous and g_k satisfies a uniform Lipschitz condition with respect to x on $\mathbb{R} \times \mathbb{R}^n$, for each $k = 1, 2, 3, \cdots$. Furthermore,*

$$\lim_{k \to \infty} g_k(t, x) = f(t, x)$$

uniformly on each compact subset of D.

Proof Let $\{K_k\}_{k=1}^{\infty}$ be a sequence of open subsets of D such that $\overline{K}_k$ is compact, $\overline{K}_k \subset K_{k+1}$, and $D = \cup_{k=1}^{\infty} K_k$. By the Tietze-Urysohn extension theorem [44], for each $k \geq 1$ there is a continuous vector function h_k on all of $\mathbb{R} \times \mathbb{R}^n$ such that

$$h_k \mid_{\overline{K}_k} = f,$$

and

$$\max\{\|h_k(t, x)\| : (t, x) \in \mathbb{R} \times \mathbb{R}^n\} = \max\{\|f(t, x)\| : (t, x) \in \overline{K}_k\} =: M_k.$$

Let $\{\delta_k\}$ be a strictly decreasing sequence of positive numbers with limit 0. Then for each $k \geq 1$, define the integral mean g_k by

$$g_k(t, x) := \frac{1}{(2\delta_k)^n} \int_{x_1 - \delta_k}^{x_1 + \delta_k} \cdots \int_{x_n - \delta_k}^{x_n + \delta_k} h_k(t, y_1, y_2, \cdots, y_n) \, dy_n \cdots dy_1,$$

where $x = (x_1, x_2, \cdots, x_n)$, for $(t, x) \in \mathbb{R} \times \mathbb{R}^n$. We claim that the sequence $\{g_k(t, x)\}$ satisfies the following:

(i) g_k is continuous on $\mathbb{R} \times \mathbb{R}^n$, for each $k \geq 1$,
(ii) $\|g_k(t, x)\| \leq M_k$ on $\mathbb{R} \times \mathbb{R}^n$, for each $k \geq 1$,
(iii) g_k has continuous first-order partial derivatives with respect to components of x on $\mathbb{R} \times \mathbb{R}^n$, for each $k \geq 1$,
(iv) $\|\frac{\partial g_k}{\partial x_i}\| \leq \frac{M_k}{\delta_k}$ on $\mathbb{R} \times \mathbb{R}^n$, for $1 \leq i \leq n$ and for each $k \geq 1$,
(v) $\lim_{k \to \infty} g_k(t, x) = f(t, x)$ uniformly on each compact subset of D.

We will only complete the proof for the scalar case ($n = 1$). In this case

$$g_k(t, x) = \frac{1}{2\delta_k} \int_{x - \delta_k}^{x + \delta_k} h_k(t, y) \, dy,$$

for $(t, x) \in \mathbb{R} \times \mathbb{R}$. We claim that (i) holds for $n = 1$. To see this, fix $(t_0, x_0) \in \mathbb{R} \times \mathbb{R}$. We will show that $g_k(t, x)$ is continuous at (t_0, x_0). Let $\epsilon > 0$ be given. Since $h_k(t, x)$ is continuous on the compact set

$$Q := \{(t, x) : |t - t_0| \leq \delta_k, |x - x_0| \leq 2\delta_k\},$$

$h_k(t, x)$ is uniformly continuous on Q. Hence there is a $\delta \in (0, \delta_k)$ such that

$$|h_k(t_2, x_2) - h_k(t_1, x_1)| < \frac{\epsilon}{2}, \tag{8.25}$$

for all $(t_1, x_1), (t_2, x_2) \in Q$ with $|t_1 - t_2| \leq \delta$, $|x_1 - x_2| \leq \delta$.

For $|t - t_0| < \delta$, $|x - x_0| < \delta$, consider

$$|g_k(t, x) - g_k(t_0, x_0)|$$

$$\leq \ |g_k(t, x) - g_k(t_0, x)| + |g_k(t_0, x) - g_k(t_0, x_0)|$$

$$\leq \ \frac{1}{2\delta_k} \int_{x-\delta_k}^{x+\delta_k} |h_k(t, y) - h_k(t_0, y)| \, dy$$

$$+ \ \frac{1}{2\delta_k} \left| \int_{x-\delta_k}^{x+\delta_k} h_k(t_0, y) \, dy - \int_{x_0-\delta_k}^{x_0+\delta_k} h_k(t_0, y) \, dy \right|$$

$$\leq \ \frac{\epsilon}{2} + \frac{1}{2\delta_k} \left| \int_{x_0+\delta_k}^{x+\delta_k} h_k(t_0, y) \, dy - \int_{x_0-\delta_k}^{x-\delta_k} h_k(t_0, y) \, dy \right|$$

$$\leq \ \frac{\epsilon}{2} + \frac{1}{2\delta_k} \left| \int_{x_0+\delta_k}^{x+\delta_k} |h_k(t_0, y)| \, dy \right| + \frac{1}{2\delta_k} \left| \int_{x_0-\delta_k}^{x-\delta_k} |h_k(t_0, y)| \, dy \right|$$

$$\leq \ \frac{\epsilon}{2} + \frac{M_k |x - x_0|}{\delta_k}$$

$$< \ \frac{\epsilon}{2} + \frac{\delta M_k}{\delta_k},$$

where we have used (8.25). Hence, if we further assume $\delta < \frac{\epsilon \delta_k}{2M_k}$, then we get that

$$|g_k(t, x) - g_k(t_0, x_0)| < \epsilon,$$

if $|t - t_0| < \delta$, $|x - x_0| < \delta$. Therefore, g_k is continuous at (t_0, x_0). Since $(t_0, x_0) \in \mathbb{R} \times \mathbb{R}$ and $k \geq 1$ are arbitrary, we get that (i) holds for $n = 1$.

To see that (ii) holds for $n = 1$, consider

$$\begin{aligned}
|g_k(t, x)| &= \frac{1}{2\delta_k} \left| \int_{x-\delta_k}^{x+\delta_k} h_k(t, y) \, dy \right| \\
&\leq \frac{1}{2\delta_k} \int_{x-\delta_k}^{x+\delta_k} |h_k(t, y)| \, dy \\
&\leq \frac{1}{2\delta_k} \int_{x-\delta_k}^{x+\delta_k} M_k \, dy \\
&= M_k,
\end{aligned}$$

for all $(t, x) \in \mathbb{R} \times \mathbb{R}$ and for all $k \geq 1$. To see that (iii) and (iv) hold for $n = 1$, note that

$$\frac{\partial g_k}{\partial x}(t, x) = \frac{1}{2\delta_k} [h_k(t, x + \delta_k) - h_k(t, x - \delta_k)]$$

is continuous on $\mathbb{R} \times \mathbb{R}$. Furthermore,

$$\begin{aligned}
\left| \frac{\partial g_k}{\partial x}(t, x) \right| &\leq \frac{1}{2\delta_k} [|h_k(t, x + \delta_k)| + |h_k(t, x - \delta_k)|] \\
&\leq \frac{M_k}{\delta_k},
\end{aligned}$$

for all $(t, x) \in \mathbb{R} \times \mathbb{R}$ and $k \geq 1$. This last inequality implies that $g_k(t, x)$ satisfies a uniform Lipschitz condition with respect to x on $\mathbb{R} \times \mathbb{R}$.

Finally we show that (v) holds for $n = 1$. Let H be a compact subset of D. Then there is an integer $m_0 \geq 1$ such that $H \subset K_{m_0}$. Let $\epsilon > 0$ be given. Fix $(t_0, x_0) \in H$. Let

$$\rho := d(H, \partial K_{m_0}).$$

Since f is uniformly continuous on the compact set $\overline{K}_{m_0}$, there is an $\eta > 0$ such that

$$|f(t_2, x_2) - f(t_1, x_1)| < \epsilon, \tag{8.26}$$

if $(t_1, x_1), (t_2, x_2) \in \overline{K}_{m_0}$ with $|t_1 - t_2| < \eta$, $|x_1 - x_2| < \eta$. Pick an integer $N \geq m_0$ sufficiently large so that $\delta_k \leq \min\{\rho, \eta\}$, for all $k \geq N$. Then for all $k \geq N$,

$$\{(t, x) : |t - t_0| \leq \delta_k, |x - x_0| \leq \delta_k\} \subset \overline{K}_{m_0}.$$

Then for all $k \geq N$,

$$
\begin{aligned}
&|g_k(t_0, x_0) - f(t_0, x_0)| \\
&= \left| \frac{1}{2\delta_k} \int_{x_0 - \delta_k}^{x_0 + \delta_k} h_k(t_0, y) \, dy - f(t_0, x_0) \right| \\
&= \left| \frac{1}{2\delta_k} \int_{x_0 - \delta_k}^{x_0 + \delta_k} h_k(t_0, y) \, dy - \frac{1}{2\delta_k} \int_{x_0 - \delta_k}^{x_0 + \delta_k} f(t_0, x_0) \, dy \right| \\
&\leq \frac{1}{2\delta_k} \int_{x_0 - \delta_k}^{x_0 + \delta_k} |f(t_0, y) - f(t_0, x_0)| \, dy \\
&< \epsilon,
\end{aligned}
$$

where we have used (8.26) in the last step. Since $(t_0, x_0) \in H$ is arbitrary, we get

$$|g_k(t, x) - f(t, x)| < \epsilon,$$

for all $(t, x) \in H$, for all $k \geq N$. Since H is an arbitrary compact subset of D, we get

$$\lim_{k \to \infty} g_k(t, x) = f(t, x)$$

uniformly on compact subsets of D. $\qquad\qquad\qquad\qquad\qquad\qquad\qquad\square$

8.8 Kneser's Theorem

In this section we will prove Kneser's theorem.

Theorem 8.42 (Kneser's Theorem) *Assume D is an open subset of $\mathbb{R} \times \mathbb{R}^n$ and $f : D \to \mathbb{R}^n$ is continuous. If the compact interval $[t_0, c]$ is a subset of the maximal interval of existence for all solutions of the IVP(8.4), then the cross-sectional set*

$$S_c := \{y \in \mathbb{R}^n : y = x(c), \text{ where } x \text{ is a solution of the IVP (8.4) on } [t_0, c]\}$$

is a compact, connected subset of $\mathbb{R}^n$.

Proof To show that S_c is compact we will show that S_c is closed and bounded. First we show that S_c is a closed set. Let $\{y_k\} \subset S_c$ with

$$\lim_{k \to \infty} y_k = y_0,$$

where $y_0 \in \mathbb{R}^n$. Then there are solutions x_k of the IVP (8.4) on $[t_0, c]$ with $x_k(c) = y_k$, $k = 1, 2, 3, \cdots$. By the basic convergence theorem (Theorem 8.39) there is a subsequence $\{x_{k_j}\}$ such that

$$\lim_{j \to \infty} x_{k_j}(t) = x(t)$$

uniformly on $[t_0, c]$, where x is a solution of the IVP (8.4) on $[t_0, c]$. But this implies that

$$y_0 = \lim_{j \to \infty} y_{k_j} = \lim_{j \to \infty} x_{k_j}(c) = x(c) \in S_c$$

and hence S_c is closed.

To see that S_c is bounded, assume not; then there is a sequence of points $\{y_k\}$ in S_c such that

$$\lim_{k \to \infty} \|y_k\| = \infty.$$

In this case there is a sequence of solutions $\{z_k\}$ of the IVP (8.4) such that $z_k(c) = y_k$, $k \geq 1$. But, by the basic convergence theorem (Theorem 8.39), there is a subsequence of solutions $\{z_{k_j}\}$ such that

$$\lim_{j \to \infty} z_{k_j}(t) = z(t)$$

uniformly on $[t_0, c]$, where z is a solution of the IVP (8.4) on $[t_0, c]$. But this implies that

$$\lim_{j \to \infty} z_{k_j}(c) = z(c),$$

which contradicts

$$\lim_{j \to \infty} \|z_{k_j}(c)\| = \lim_{j \to \infty} \|y_{k_j}\| = \infty.$$

Hence we have proved that S_c is a compact set.

We now show that S_c is connected. Assume not; then, since S_c is compact, there are disjoint, nonempty, compact sets A_1, A_2 such that

$$A_1 \cup A_2 = S_c.$$

Let

$$\delta := d(A_1, A_2) > 0.$$

Then let

$$h(x) := d(x, A_1) - d(x, A_2),$$

for $x \in \mathbb{R}^n$. Then $h : \mathbb{R}^n \to \mathbb{R}$ is continuous,

$$h(x) \leq -\delta, \quad \text{for} \quad x \in A_1,$$

and

$$h(x) \geq \delta, \quad \text{for} \quad x \in A_2.$$

In particular, we have
$$h(x) \neq 0$$
for all $x \in S_c$. We will contradict this fact at the end of this proof. Since $f : D \to \mathbb{R}^n$ is continuous, we have by Theorem 8.41 that there is a sequence of vector functions $\{g_k\}$ such that $g_k : \mathbb{R} \times \mathbb{R}^n \to \mathbb{R}^n$ is bounded and continuous and g_k satisfies a uniform Lipschitz condition with respect to x on $\mathbb{R} \times \mathbb{R}^n$, for each $k = 1, 2, 3, \cdots$. Furthermore,
$$\lim_{k \to \infty} g_k(t, x) = f(t, x)$$
uniformly on each compact subset of D. Since $A_i \neq \emptyset$ for $i = 1, 2$ there are solutions x_i, $i = 1, 2$ of the IVP (8.4) with
$$x_i(c) \in A_i.$$

Now define for $i = 1, 2$
$$g_{ik}(t, x) := \begin{cases} g_k(t, x) + f(t_0, x_i(t_0)) - g_k(t_0, x_i(t_0)), & t < t_0, \\ g_k(t, x) + f(t, x_i(t)) - g_k(t, x_i(t)), & t_0 \leq t \leq c, \\ g_k(t, x) + f(c, x_i(c)) - g_k(c, x_i(c)), & c < t, \end{cases}$$
for $(t, x) \in \mathbb{R} \times \mathbb{R}^n$, $k = 1, 2, 3, \cdots$. It follows that g_{ik} is continuous and bounded on $\mathbb{R} \times \mathbb{R}^n$ and satisfies a uniform Lipschitz condition with respect to x on $\mathbb{R} \times \mathbb{R}^n$, for each $i = 1, 2$ and $k = 1, 2, 3, \cdots$. Furthermore,
$$\lim_{k \to \infty} g_{ik}(t, x) = f(t, x)$$
uniformly on each compact subset of D, for $i = 1, 2$. From the Picard-Lindelof theorem (Theorem 8.13) we get that each of the IVPs
$$x' = g_{ik}(t, x), \quad x(\tau) = \xi$$
has a unique solution and since each g_{ik} is bounded on $\mathbb{R} \times \mathbb{R}^n$, the maximal interval of existence of each of these solutions is $(-\infty, \infty)$. Note that the unique solution of the IVP
$$x' = g_{ik}(t, x), \quad x(t_0) = x_0$$
on $[t_0, c]$ is x_i for $i = 1, 2$.

Let
$$p_k(t, x, \lambda) := \lambda g_{1k}(t, x) + (1 - \lambda) g_{2k}(t, x),$$
for $(t, x, \lambda) \in \mathbb{R} \times \mathbb{R}^n \times \mathbb{R}$, $k = 1, 2, 3, \cdots$. Note that for each fixed $\lambda \in \mathbb{R}$, p_k is bounded and continuous and satisfies a uniform Lipschitz condition with respect to x. Furthermore, for each fixed λ,
$$\lim_{k \to \infty} p_k(t, x, \lambda) = f(t, x)$$
uniformly on each compact subset of D. Let for each fixed $\lambda \in \mathbb{R}$, $x_k(\cdot, \lambda)$ be the unique solution of the IVP
$$x' = p_k(t, x, \lambda), \quad x(t_0) = x_0.$$
It follows that $q_k : \mathbb{R} \to \mathbb{R}$, defined by
$$q_k(\lambda) = h(x_k(c, \lambda)),$$

is, by Theorem 8.40, continuous. Since

$$q_k(0) = h(x_k(c, 0)) = h(x_2(c)) \geq \delta,$$

and

$$q_k(1) = h(x_k(c, 1)) = h(x_1(c)) \leq -\delta,$$

there is, by the intermediate value theorem, a λ_k with $0 < \lambda_k < 1$ such that

$$q_k(\lambda_k) = h(x_k(c, \lambda_k)) = 0,$$

for $k = 1, 2, 3, \cdots$. Since the sequence $\{\lambda_k\} \subset [0, 1]$, there is a convergent subsequence $\{\lambda_{k_j}\}$. Let

$$\lambda_0 = \lim_{j \to \infty} \lambda_{k_j}.$$

It follows that the sequence $x_{k_j}(t, \lambda_{k_j})$ has a subsequence $x_{k_{j_i}}(t, \lambda_{k_{j_i}})$ that converges uniformly on compact subsets of $(-\infty, \infty)$ to $\overline{x}(t)$, where $\overline{x}$ is a solution of the limit IVP

$$x' = f(t, x), \quad x(t_0) = x_0.$$

It follows that

$$\lim_{i \to \infty} x_{k_{j_i}}(c, \lambda_{k_{j_i}}) = \overline{x}(c) \in S_c.$$

Hence

$$\lim_{i \to \infty} h(x_{k_{j_i}}(c, \lambda_{k_{j_i}})) = h(\overline{x}(c)) = 0,$$

which is a contradiction.

$$\square$$

8.9 Differentiating Solutions with Respect to ICs

In this section we will be concerned with differentiating solutions with respect to initial conditions, initial points, and parameters.

Theorem 8.43 (Differentiation with Respect to Initial Conditions and Initial Points) *Assume f is a continuous n dimensional vector function on an open set $D \subset \mathbb{R} \times \mathbb{R}^n$ and f has continuous partial derivatives with respect to the components of x. Then the IVP (8.4), where $(t_0, x_0) \in D$, has a unique solution, which we denote by $x(t; t_0, x_0)$, with maximal interval of existence (α, ω). Then $x(t; t_0, x_0)$ has continuous partial derivatives with respect to the components x_{0j}, $j = 1, 2, \cdots, n$ of x_0 and with respect to t_0 on (α, ω). Furthermore, $z(t) := \frac{\partial x(t; t_0, x_0)}{\partial x_{0j}}$ defines the unique solution of the IVP*

$$z' = J(t)z, \quad z(t_0) = e_j,$$

where $J(t)$ is the Jacobian matrix of f with respect to x along $x(t; t_0, x_0)$, that is,

$$J(t) = D_x f(t, x(t; t_0, x_0))$$

and

$$e_j = \begin{pmatrix} \delta_{1j} \\ \delta_{2j} \\ \vdots \\ \delta_{nj} \end{pmatrix},$$

where δ_{ij} is the Kronecker delta function, for $t \in (\alpha, \omega)$. Also, $w(t) := \frac{\partial x(t; t_0, x_0)}{\partial t_0}$ defines the unique solution of the IVP

$$w' = J(t)w, \quad w(t_0) = -f(t_0, x_0),$$

for $t \in (\alpha, \omega)$.

Proof Let $\tau \in (\alpha, \omega)$. Then choose an interval $[t_1, t_2] \subset (\alpha, \omega)$ such that $\tau, t_0 \in (t_1, t_2)$. Since

$$\{(t, x(t; t_0, x_0)) : t \in [t_1, t_2]\} \subset D$$

is compact and D is open, there exists a $\delta > 0$ such that

$$Q_\delta := \{(\bar{t}, \bar{x}) : |\bar{t} - t| \le \delta, \|\bar{x} - x(t; t_0, x_0))\| \le \delta, \text{ for } t \in [t_1, t_2]\}$$

is contained in D. Let $\{\delta_k\}_{k=1}^\infty$ be a sequence of nonzero real numbers with $\lim_{k \to \infty} \delta_k = 0$. Let $x(t) := x(t; t_0, x_0)$ and let x_k be the solution of the IVP

$$x' = f(t, x), \quad x(t_0) = x_0 + \delta_k e_j,$$

where $1 \le j \le n$ is fixed. It follows from the basic convergence theorem (Theorem 8.39) that if (α_k, ω_k) is the maximal interval of existence of x_k, then there is an integer k_0 so that for all $k \ge k_0$, $[t_1, t_2] \subset (\alpha_k, \beta_k)$ and

$$\|x(t) - x_k(t)\| < \delta$$

on $[t_1, t_2]$. Without loss of generality we can assume that the preceding properties of $\{x_k\}$ hold for all $k \ge 1$. Since, for $t_1 < t < t_2$, $0 \le s \le 1$,

$$\|[sx_k(t) + (1 - s)x(t)] - x(t)\|$$
$$= s\|x_k(t) - x(t)\| \le s\delta \le \delta,$$

it follows that

$$(t, sx_k(t) + (1 - s)x(t)) \in Q_\delta,$$

for $t_1 < t < t_2$, $0 \le s \le 1$. By Lemma 8.6,

$$x_k'(t) - x'(t) = f(t, x_k(t)) - f(t, x(t))$$
$$= \int_0^1 D_x f(t, sx_k(t) + (1 - s)x(t)) \, ds \, [x_k(t) - x(t)],$$

for $t_1 < t < t_2$. Let

$$z_k(t) := \frac{1}{\delta_k}[x_k(t) - x(t)];$$

then z_k is a solution of the IVP

$$z' = \int_0^1 D_x f(t, sx_k(t) + (1 - s)x(t)) \, ds \, z, \quad x(t_0) = e_j,$$

where e_j is given in the statement of this theorem. Define $h_k : (t_1, t_2) \times \mathbb{R}^n \to \mathbb{R}^n$ by

$$h_k(t, z) := \int_0^1 D_x f(t, s x_k(t) + (1 - s) x(t)) \, ds \, z,$$

for $(t, x) \in (t_1, t_2) \times \mathbb{R}^n$, $k \geq 1$. Then $h_k : (t_1, t_2) \times \mathbb{R}^n \to \mathbb{R}^n$ is continuous and since

$$\lim_{k \to \infty} x_k(t) = x(t)$$

uniformly on $[t_1, t_2]$, we get that

$$\lim_{k \to \infty} h_k(t, z) = D_x f(t, x(t)) z$$

uniformly on each compact subset of $(t_1, t_2) \times \mathbb{R}^n$. By the basic convergence theorem (Theorem 8.39),

$$\lim_{k \to \infty} z_k(t) = z(t)$$

uniformly on each compact subinterval of (t_1, t_2), where z is the solution of the limit IVP

$$z' = D_x f(t, x(t)) z = D_x f(t, x(t; t_0, x_0)) z, \quad z(t_0) = e_j. \tag{8.27}$$

In particular,

$$\lim_{k \to \infty} \frac{1}{\delta_k} [x_k(\tau) - x(\tau)] = z(\tau),$$

where z is the solution of the IVP (8.27). Since this is true for any sequence $\{\delta_k\}$ of nonzero numbers with $\lim_{k \to \infty} \delta_k = 0$, we get that

$$\frac{\partial x}{\partial x_{0j}}(\tau, t_0, x_0)$$

exists and equals $z(\tau)$. Since $\tau \in (\alpha, \omega)$ was arbitrary, we get that for each $t \in (\alpha, \omega)$,

$$\frac{\partial x}{\partial x_{0j}}(t, t_0, x_0)$$

exists and equals $z(t)$. The rest of this proof is left to the reader. $\qquad \square$

Definition 8.44 The equation $z' = J(t)z$ in Theorem 8.43 is called the *variational equation of $x' = f(t, x)$ along the solution $x(t; t_0, x_0)$*.

Example 8.45 Find $\frac{\partial x}{\partial x_0}(t; 0, 1)$ and $\frac{\partial x}{\partial t_0}(t; 0, 1)$ for the differential equation

$$x' = x - x^2. \tag{8.28}$$

By inspection,

$$x(t, 0, 1) \equiv 1.$$

Here $f(t, x) = x - x^2$, so the variational equation along $x(t; 0, 1)$ is

$$z' = (1 - 2x(t; 0, 1))z,$$

which simplifies to

$$z' = -z.$$

By Theorem 8.43, $z(t) := \frac{\partial x}{\partial x_0}(t; 0, 1)$ solves the IVP

$$z' = -z, \quad z(0) = 1,$$

and $w(t) := \frac{\partial x}{\partial t_0}(t; 0, 1)$ solves the IVP

$$w' = -w, \quad w(0) = -f(0, 1) = 0.$$

It follows that

$$z(t) = \frac{\partial x}{\partial x_0}(t; 0, 1) = e^{-t}$$

and

$$w(t) = \frac{\partial x}{\partial t_0}(t; 0, 1) = 0.$$

It then follows that

$$x(t; 0, 1 + h) \approx he^{-t} + 1,$$

for h close to zero and

$$x(t; h, 1) \approx 1,$$

for h close to zero. Since the equation (8.28) can be solved (see Exercise 8.32) by separating variables and using partial fractions, it can be shown that

$$x(t; t_0, x_0) = \frac{x_0 e^{t-t_0}}{1 - x_0 + x_0 e^{t-t_0}}.$$

It is easy to see (see Exercise 8.32) from this that the expressions for $\frac{\partial x}{\partial x_0}(t; 0, 1)$ and $\frac{\partial x}{\partial t_0}(t; 0, 1)$ given previously are correct. $\triangle$

Corollary 8.46 *Assume that A is a continuous $n \times n$ matrix function and h is a continuous $n \times 1$ vector function on an interval I. Further assume $t_0 \in I$, $x_0 \in \mathbb{R}^n$, and let $x(\cdot; t_0, x_0)$ denote the unique solution of the IVP*

$$x' = A(t)x + h(t), \quad x(t_0) = x_0.$$

Then $z(t) := \frac{\partial x}{\partial x_{0j}}(t; t_0, x_0)$ defines the unique solution of the IVP

$$z' = A(t)z, \quad z(t_0) = e_j,$$

on I. Also $w(t) := \frac{\partial x}{\partial t_0}(t; t_0, x_0)$ defines the unique solution of the IVP

$$w' = A(t)x, \quad w(t_0) = -A(t_0)x_0 - h(t_0).$$

Proof If $f(t, x) := A(t)x + h(t)$, then

$$D_x f(t, x) = A(t)$$

and the conclusions in this corollary follow from Theorem 8.43. $\square$

Example 8.47 Assume that $f(t, u, u', \cdots, u^{(n-1)})$ is continuous and real valued on $(a, b) \times \mathbb{R}^n$ and has continuous first-order partial derivatives with respect to each of the variables $u, u', \cdots, u^{(n-1)}$ on $(a, b) \times \mathbb{R}^n$. Then the IVP

$$u^{(n)} = f(t, u, u', \cdots, u^{(n-1)}), \quad u(t_0) = y_{01}, \quad \cdots, u^{(n-1)}(t_0) = y_{0n}, \quad (8.29)$$

where $t_0 \in (a, b)$, $y_{0k} \in \mathbb{R}$, $1 \leq k \leq n$, is equivalent to the vector IVP

$$y' = h(t, y), \quad y(t_0) = y_0,$$

where

$$h(t, y) := \begin{pmatrix} y_2 \\ y_3 \\ \vdots \\ y_n \\ f(t, y_1, \cdots, y_n) \end{pmatrix}, \quad y_0 := \begin{pmatrix} y_{01} \\ y_{02} \\ \vdots \\ y_{0n} \end{pmatrix}.$$

Then

$$y(t; t_0, y_0) = \begin{pmatrix} u(t; t_0, y_0) \\ u'(t; t_0, y_0) \\ \vdots \\ u^{(n-1)}(t; t_0, y_0) \end{pmatrix},$$

where $u(\cdot; t_0, y_0)$ is the solution of the IVP (8.29). From Theorem 8.43,

$$z(t) := \frac{\partial y}{\partial y_{0j}}(t; t_0, y_0)$$

is the solution of the IVP

$$z' = J(t)z, \quad z(t_0) = e_j,$$

where J is the matrix function

$$\begin{pmatrix} 0 & 1 & 0 & \cdots & 0 \\ 0 & 0 & 1 & \cdots & 0 \\ \vdots & \vdots & \ddots & \ddots & \vdots \\ f_u & f_{u'} & & \cdots & f_{u^{(n-1)}} \end{pmatrix}$$

evaluated at $(t, y(t; t_0, x_0))$. Since

$$\frac{\partial u}{\partial y_{0j}}(t; t_0, y_0)$$

is the first component of

$$z(t) = \frac{\partial y}{\partial y_{0j}}(t; t_0, y_0),$$

we get that

$$v(t) := \frac{\partial u}{\partial y_{0j}}(t; t_0, y_0)$$

solves the IVP

$$v^{(n)} = \sum_{k=0}^{n-1} f_{u^{(k)}}(t, u(t; t_0, y_0), \cdots, u^{(n-1)}(t; t_0, y_0))v^{(k)},$$

$$v^{(i-1)} = \delta_{ij}, \quad i = 1, 2, \cdots, n,$$

where δ_{ij} is the Kronecker delta function. The nth-order linear differential equation

$$v^{(n)} = \sum_{k=0}^{n-1} f_{u^{(k)}}(t, u(t; t_0, y_0), u'(t; t_0, y_0), \cdots, u^{(n-1)}(t; t_0, y_0))v^{(k)} \quad (8.30)$$

is called the variational equation of $u^{(n)} = f(t, u, u', \cdots, u^{(n-1)})$ along $u(t; t_0, y_0)$. Similarly,

$$w(t) := \frac{\partial u}{\partial t_0}(t; t_0, y_0)$$

is the solution of the variational equation satisfying the initial conditions

$$w(t_0) = -y_{02}, \quad \cdots, \quad w^{(n-2)}(t_0) = -y_{0n}, \quad w^{(n-1)}(t_0) = -f(t_0, y_{01}, \cdots, y_{0n}).$$

$$\triangle$$

Example 8.48 If $u(t; t_0, c_1, c_2)$ denotes the solution of the IVP

$$u'' = u - u^3, \quad u(t_0) = c_1, \ u'(t_0) = c_2,$$

find

$$\frac{\partial u}{\partial c_2}(t; 0, 0, 0)$$

and use your answer to approximate $u(t; 0, 0, h)$ when h is small.

We know that

$$v(t) := \frac{\partial u}{\partial c_2}(t; 0, 0, 0)$$

gives the solution of the IVP

$$v'' = f_u(t, u(t; 0, 0, 0), u'(t; 0, 0, 0))v + f_{u'}(t, u(t; 0, 0, 0), u'(t; 0, 0, 0))v',$$

$$v(0) = 0, \quad v'(0) = 1.$$

Since in this example $f(t, u, u') = u - u^3$, we get that $f_u(t, u, u') = 1 - 3u^2$ and $f_{u'}(t, u, u') = 0$. Since by inspection $u(t; 0, 0, 0) \equiv 0$, we get that

$$f_u(t, u(t; 0, 0, 0), u'(t; 0, 0, 0)) = f_u(t, 0, 0) = 1$$

and

$$f_{u'}(t, u(t; 0, 0, 0), u'(t; 0, 0, 0)) = f_{u'}(t, 0, 0) = 0.$$

Hence v solves the IVP

$$v'' = v, \quad v(0) = 0, \quad v'(0) = 1,$$

which implies that $v(t) = \sinh t$. Therefore,

$$\frac{\partial u}{\partial c_2}(t; 0, 0, 0) = \sinh t.$$

It follows that

$$u(t; 0, 0, h) \approx h \sinh t,$$

for small h. Similarly, it can be shown that

$$\frac{\partial u}{\partial c_1}(t, 0, 0, 0) = \cosh t$$

and

$$\frac{\partial u}{\partial t_0}(t, 0, 0, 0) = 0.$$

It then follows that

$$u(t; 0, h, 0) \approx h \cosh t,$$

and

$$u(t; h, 0, 0) \approx 0,$$

for h close to zero. $\triangle$

Theorem 8.49 (Differentiating Solutions with Respect to Parameters) *Assume $f(t, x, \lambda)$ is continuous and has continuous first-order partial derivatives with respect to components of x and λ on an open subset $D \subset \mathbb{R} \times \mathbb{R}^n \times \mathbb{R}^m$. For each $(t_0, x_0, \lambda_0) \in D$, let $x(t; t_0, x_0, \lambda_0)$ denote the solution of the IVP*

$$x' = f(t, x, \lambda_0), \quad x(t_0) = x_0, \tag{8.31}$$

with maximal interval of existence (α, ω). Then $x(t; t_0, x_0, \lambda_0)$ has continuous first-order partial derivatives with respect to components of $\lambda_0 = (\lambda_{01}, \cdots, \lambda_{0m})$ and $z(t) := \frac{\partial x}{\partial \lambda_{0k}}(t; t_0, x_0, \lambda_0), 1 \le k \le m$, is the solution of the IVP

$$z' = J(t; t_0, x_0, \lambda_0)z + g_k(t), \quad z(t_0) = 0,$$

on (α, ω), where

$$J(t; t_0, x_0, \lambda_0) := D_x f(t, x(t; t_0, x_0, \lambda_0), \lambda_0)$$

and

$$g_k(t) = \begin{pmatrix} \frac{\partial f_1}{\partial \lambda_{0k}}(t, x(t; t_0, x_0, \lambda_0), \lambda_0) \\ \frac{\partial f_2}{\partial \lambda_{0k}}(t, x(t; t_0, x_0, \lambda_0), \lambda_0) \\ \vdots \\ \frac{\partial f_n}{\partial \lambda_{0k}}(t, x(t; t_0, x_0, \lambda_0), \lambda_0) \end{pmatrix},$$

for $t \in (\alpha, \omega)$.

Proof Let

$$y := \begin{pmatrix} x_1 \\ \vdots \\ x_n \\ \lambda_1 \\ \vdots \\ \lambda_m \end{pmatrix}, \quad h(t, y) := \begin{pmatrix} f_1(t, x, \lambda) \\ \vdots \\ f_n(t, x, \lambda) \\ 0 \\ \vdots \\ 0 \end{pmatrix}.$$

Then

$$y(t; t_0, x_0, \lambda_0) := \begin{pmatrix} x_1(t; t_0, x_0, \lambda_0) \\ \vdots \\ x_n(t; t_0, x_0, \lambda_0) \\ \lambda_{01} \\ \vdots \\ \lambda_{0m} \end{pmatrix}$$

is the unique solution of the IVP

$$y' = h(t, y), \quad y(t_0) = \begin{pmatrix} x_{01} \\ \vdots \\ x_{0n} \\ \lambda_{01} \\ \vdots \\ \lambda_{0m} \end{pmatrix},$$

with maximal interval of existence (α, ω). It follows from Theorem 8.43 that

$$\tilde{z}(t) := \frac{\partial y \, (t; t_0, x_0, \lambda_0)}{\partial \lambda_{0k}}$$

exists, is continuous on (α, ω), and is the solution of the IVP

$$\tilde{z}' = J(t, y(t; t_0, x_0, \lambda_0))\tilde{z}, \quad \tilde{z}(t_0) = e_{n+k},$$

where e_{n+k} is the unit vector in $\mathbb{R}^{n+m}$ whose $n + k$ component is 1 and

$$J(t, y(t; t_0, x_0, \lambda_0)) = \begin{pmatrix} \frac{\partial f_1}{\partial x_1} & \cdots & \frac{\partial f_1}{\partial x_n} & \frac{\partial f_1}{\partial \lambda_{01}} & \cdots & \frac{\partial f_1}{\partial \lambda_{0m}} \\ \cdots & \cdots & \cdots & \cdots & \cdots & \cdots \\ \frac{\partial f_n}{\partial x_1} & \cdots & \frac{\partial f_n}{\partial x_n} & \frac{\partial f_n}{\partial \lambda_{01}} & \cdots & \frac{\partial f_n}{\partial \lambda_{0m}} \\ 0 & \cdots & 0 & 0 & \cdots & 0 \\ \cdots & \cdots & \cdots & \cdots & \cdots & \cdots \\ 0 & \cdots & 0 & 0 & \cdots & 0 \end{pmatrix},$$

where the entries of this last matrix are evaluated at $(t, x(t; t_0, x_0, \lambda_0), \lambda_0)$. This implies $\tilde{z}_j'(t) = 0$, $n + 1 \leq j \leq n + m$. This then implies that

$$\tilde{z}_j(t) \equiv 0, \quad n + 1 \leq j \leq n + m, \text{ but } j \neq n + k,$$

and

$$\tilde{z}_{n+k}(t) \equiv 1,$$

for $t \in (\alpha, \omega)$. Hence

$$\tilde{z}(t) = \begin{pmatrix} z(t) \\ 0 \\ \vdots \\ 0 \\ 1 \\ 0 \\ \vdots \\ 0 \end{pmatrix},$$

where the $n + k$ component of $\tilde{z}(t)$ is 1. It follows that $z(t) = \frac{\partial x(t; t_0, x_0, \lambda_0)}{\partial \lambda_{0k}}$ is the solution of the IVP

$$z' = J(t; t_0, x_0, \lambda_0) z + g_k(t), \quad z(t_0) = 0,$$

on (α, ω), where

$$J(t; t_0, x_0, \lambda_0) = D_x f(t, x(t; t_0, x_0, \lambda_0), \lambda_0),$$

and

$$g_k(t) = \begin{pmatrix} \frac{\partial f_1}{\partial \lambda_{0k}}(t, x(t; t_0, x_0, \lambda_0), \lambda_0) \\ \frac{\partial f_2}{\partial \lambda_{0k}}(t, x(t; t_0, x_0, \lambda_0), \lambda_0) \\ \vdots \\ \frac{\partial f_n}{\partial \lambda_{0k}}(t, x(t; t_0, x_0, \lambda_0), \lambda_0) \end{pmatrix},$$

for $t \in (\alpha, \omega)$. $\qquad\qquad\qquad\qquad\qquad\qquad\qquad\qquad\qquad\qquad\square$

Similar to Example 8.47, we get the following result.

Example 8.50 Assume that $f(t, u, u', \cdots, u^{(n-1)}, \lambda_1, \cdots, \lambda_m)$ is continuous on $(a, b) \times \mathbb{R}^n \times \mathbb{R}^m$ and has continuous first-order partial derivatives with respect to each of the variables $u, u', \cdots, u^{(n-1)}, \lambda_1, \cdots, \lambda_m$. Then the unique solution $u(t; t_0, y_{01}, y_{02}, \cdots, y_{0n}, \lambda_{01}, \cdots, \lambda_{0m}) = u(t; t_0, y_0, \lambda_0)$ of the initial value problem

$$u^{(n)} = f(t, u, u', \cdots, u^{(n-1)}, \lambda_{01}, \cdots, \lambda_{0m}),$$

$$u(t_0) = y_{01}, \ u'(t_0) = y_{02}, \ \cdots, \ u^{(n-1)}(t_0) = y_{0n},$$

where $(t_0, y_0, \lambda_0) \in (a, b) \times \mathbb{R}^n \times \mathbb{R}^m$ has continuous first-order partial derivatives with respect to $\lambda_{01}, \lambda_{02} \cdots, \lambda_{0m}$ and

$$v(t) := \frac{\partial u}{\partial \lambda_{0j}}(t; t_0, y_0, \lambda_0),$$

$1 \le j \le m$, is the solution of the IVP

$$v^{(n)} = \sum_{k=0}^{n-1} f_{u^{(k)}}\left(t, u(t; t_0, y_0, \lambda_0), \cdots, u^{(n-1)}(t; t_0, y_0, \lambda_0), \lambda_0\right) v^{(k)}$$

$$+ f_{\lambda_j}\left(t, u(t; t_0, y_0, \cdots, u^{n-1}(t; t_0, y_0, \lambda_0), \lambda_0), \cdots, u^{(n-1)}(t; t_0, y_0, \lambda_0), \lambda_0\right),$$

$$v^{(i-1)}(t_0) = 0, \quad i = 1, \cdots, n.$$

$\triangle$

Example 8.51 Let $u(t; t_0, a, b, \lambda)$, $\lambda > 0$, denote the solution of the IVP

$$u'' = -\lambda u, \quad u(t_0) = a, \quad u'(t_0) = b.$$

Then by Example 8.50 we get that

$$v(t) = \frac{\partial u}{\partial \lambda}(t; 0, 1, 0, \lambda)$$

is the solution of the IVP

$$v'' = -\lambda v - \cos(\sqrt{\lambda} t), \quad v(0) = 0, \quad v'(0) = 0.$$

Solving this IVP, we get

$$v(t) = \frac{\partial u}{\partial \lambda}(t; 0, 1, 0, \lambda) = -\frac{1}{2\sqrt{\lambda}} t \sin(\sqrt{\lambda} t).$$

Since

$$u(t; 0, 1, 0, \lambda) = \cos(\sqrt{\lambda} t),$$

we can check our answer by just differentiating with respect to λ. $\triangle$

8.10 Maximum and Minimum Solutions

Definition 8.52 Assume that ϕ is a continuous real-valued function on an open set $D \subset \mathbb{R} \times \mathbb{R}$ and let $(t_0, u_0) \in D$. Then a solution u_M of the IVP

$$u' = \phi(t, u), \quad u(t_0) = u_0, \tag{8.32}$$

with maximal interval of existence (α_M, ω_M) is called a *maximum solution* of the IVP (8.32) in case for any other solution v of the IVP (8.32) on an interval I,

$$v(t) \le u_M(t), \quad \text{on} \quad I \cap (\alpha_M, \omega_M).$$

In a similar way we can also define a *minimum solution* of the IVP (8.32).

Theorem 8.53 *Assume that ϕ is a continuous real-valued function on an open set $D \subset \mathbb{R} \times \mathbb{R}$. Then the IVP (8.32) has both a maximum and minimum solution and these solutions are unique.*

Proof For each $n \ge 1$, let u_n be a solution of the IVP

$$u' = f(t, u) + \frac{1}{n}, \quad u(t_0) = u_0,$$

with maximal interval of existence (α_n, ω_n). Since

$$\lim_{n \to \infty} \left(f(t, u) + \frac{1}{n} \right) = f(t, u)$$

uniformly on each compact subset of D, we have by Theorem 8.39 that there is a solution u of the IVP (8.32) with maximal interval of existence (α_u, ω_u) and a subsequence $\{u_{n_j}(t)\}$ that converges to $u(t)$ in the sense of Theorem 8.39.

Similarly, for each integer $n \geq 1$, let v_n be a solution of the IVP

$$v' = f(t, v) - \frac{1}{n}, \quad v(t_0) = u_0,$$

with maximal interval of existence (α_n^*, ω_n^*). Since

$$\lim_{n \to \infty} \left(f(t, v) - \frac{1}{n} \right) = f(t, v)$$

uniformly on each compact subset of D, we have by Theorem 8.39 that there is a solution v of the IVP (8.32) with maximal interval of existence (α_v, β_v) and a subsequence $v_{n_j}(t)$ that converges to $v(t)$ in the sense of Theorem 8.39. Define

$$u_M(t) = \begin{cases} u(t), & \text{on } [t_0, \omega_u) \\ v(t), & \text{on } (\alpha_v, t_0), \end{cases}$$

and

$$u_m(t) = \begin{cases} v(t), & \text{on } [t_0, \omega_v) \\ u(t), & \text{on } (\alpha_u, t_0). \end{cases}$$

We claim that u_M and u_m are maximum and minimum solutions of the IVP (8.32), respectively. To see that u_M is a maximum solution on $[t_0, \omega_u)$, assume not, then there is a solution z of the IVP (8.32) on an interval I and there is a $t_1 > t_0$, $t_1 \in I \cap [t_0, \omega_u)$ such that

$$z(t_1) > u_M(t_1) = u(t_1).$$

Since

$$\lim_{j \to \infty} u_{n_j}(t_1) = u(t_1),$$

we can pick J sufficiently large so that

$$u_{n_J}(t_1) < z(t_1).$$

Since $z(t_0) = u_0 = u_{n_J}(t_0)$, we can pick $t_2 \in [t_0, t_1)$ such that

$$z(t_2) = u_{n_J}(t_2)$$

and

$$z(t) > u_{n_J}(t)$$

on $(t_2, t_1]$. This implies that $z'(t_2) \geq u'_{n_J}(t_2)$. But

$$
\begin{aligned}
u'_{n_J}(t_2) &= f(t_2, u_{n_J}(t_2)) + \frac{1}{n_J} \\
&> f(t_2, u_{n_J}(t_2)) \\
&= f(t_2, z(t_2)) \\
&= z'(t_2),
\end{aligned}
$$

which gives us our contradiction. There are three other cases that are similar and so will be omitted (see Exercise 8.42). It is easy to see that the maximal and minimal solutions of the IVP (8.32) are unique. Note that since the maximal solution of the IVP (8.32) is unique, we have that any sequence $\{u_n\}$ of solutions of the IVPs

$$
u' = f(t, u) + \frac{1}{n}, \quad u(t_0) = u_0
$$

converges to u_M on $[t_0, \omega_u)$ in the sense of Theorem 8.39. $\qquad \square$

In the remainder of this section we use the following notation.

Definition 8.54 Assume u is defined in a neighborhood of t_0. Then

$$
D^+ u(t_0) := \limsup_{h \to 0+} \frac{u(t_0 + h) - u(t_0)}{h},
$$

$$
D_+ u(t_0) := \liminf_{h \to 0+} \frac{u(t_0 + h) - u(t_0)}{h},
$$

$$
D^- u(t_0) := \limsup_{h \to 0-} \frac{u(t_0 + h) - u(t_0)}{h},
$$

$$
D_- u(t_0) := \liminf_{h \to 0-} \frac{u(t_0 + h) - u(t_0)}{h}.
$$

Theorem 8.55 *Assume that D is an open subset of $\mathbb{R}^2$ and $\phi : D \to \mathbb{R}$ is continuous. Assume that $(t_0, u_0) \in D$, $v : [t_0, t_0 + a] \to \mathbb{R}$ is continuous, $(t, v(t)) \in D$, for $t_0 \leq t \leq t_0 + a$ with $D^+ v(t) \leq \phi(t, v(t))$, and $v(t_0) \leq u_0$; then*

$$
v(t) \leq u_M(t), \quad t \in [t_0, t_0 + a] \cap (\alpha_M, \omega_M),
$$

where u_M is the maximum solution of the IVP (8.32), with maximal interval of existence (α_M, ω_M).

Proof Let v and u_M be as in the statement of this theorem. We now prove that

$$
v(t) \leq u_M(t), \quad t \in [t_0, t_0 + a] \cap (\alpha_M, \omega_M).
$$

Assume not; then there is a $t_1 > t_0$ in $[t_0, t_0 + a] \cap (\alpha_M, \omega_M)$, such that

$$
v(t_1) > u_M(t_1).
$$

By the proof of Theorem 8.53 we know that if $\{u_n\}$ is a sequence of solutions of the IVPs

$$u' = \phi(t, u) + \frac{1}{n}, \quad u(t_0) = x_0,$$

respectively, then

$$\lim_{n \to \infty} u_n(t) = u_M(t)$$

uniformly on compact subintervals of $[t_0, \omega)$. Hence we can pick a positive integer N such that the maximal interval of existence of u_N contains $[t_0, t_1]$ and

$$u_N(t_1) < v(t_1).$$

Choose $t_2 \in [t_0, t_1)$ such that $u_N(t_2) = v(t_2)$ and

$$v(t) > u_N(t), \quad t \in (t_2, t_1].$$

Let

$$z(t) := v(t) - u_N(t).$$

For $h > 0$, sufficiently small,

$$\frac{z(t_2 + h) - z(t_2)}{h} > 0,$$

and so

$$D^+ z(t_2) \geq 0.$$

But

$$
\begin{aligned}
D^+ z(t_2) &= D^+ v(t_2) - u'_N(t_2) \\
&\leq \phi(t_2, v(t_2)) - \phi(t_2, u_N(t_2)) - \frac{1}{N} \\
&= -\frac{1}{N} < 0,
\end{aligned}
$$

which is a contradiction.   □

Similarly, we can prove the following three theorems:

Theorem 8.56 *Assume that D is an open subset of $\mathbb{R}^2$ and $\phi : D \to \mathbb{R}$ is continuous. Assume that $(t_0, u_0) \in D$, $v : [t_0, t_0 + a] \to \mathbb{R}$ is continuous, $(t, v(t)) \in D$, for $t_0 \leq t \leq t_0 + a$, with $D_+ v(t) \geq \phi(t, v(t))$, and $v(t_0) \geq u_0$; then*

$$v(t) \geq u_m(t), \quad t \in [t_0, t_0 + a] \cap (\alpha_m, \omega_m),$$

where u_m is the minimum solution of the IVP (8.32), with maximal interval of existence (α_m, ω_m).

Theorem 8.57 *Assume that D is an open subset of $\mathbb{R}^2$ and $\phi : D \to \mathbb{R}$ is continuous. Assume that $(t_0, u_0) \in D$, $v : [t_0 - a, t_0] \to \mathbb{R}$ is continuous, $(t, v(t)) \in D$, for $t_0 - a \leq t \leq t_0$, with $D^- v(t) \geq \phi(t, v(t))$ and $v(t_0) \leq u_0$; then*

$$v(t) \leq u_M(t), \quad t \in [t_0 - a, t_0] \cap (\alpha_M, \omega_M),$$

where u_M is the maximum solution of the IVP (8.32), with maximal interval of existence (α_M, ω_M).

Theorem 8.58 *Assume that D is an open subset of $\mathbb{R}^2$ and $\phi : D \to \mathbb{R}$ is continuous. Assume that $(t_0, u_0) \in D$, $v : [t_0 - a, t_0] \to \mathbb{R}$ is continuous, $(t, v(t)) \in D$, for $t_0 - a \le t \le t_0$, with $D_- v(t) \le \phi(t, v(t))$, and $v(t_0) \ge u_0$; then*

$$v(t) \ge u_m(t), \quad t \in [t_0 - a, t_0] \cap (\alpha_m, \omega_m),$$

where u_m is the minimum solution of the IVP (8.32), with maximal interval of existence (α_m, ω_m).

We will leave the proof of the following important comparison theorem as an exercise (Exercise 8.46).

Corollary 8.59 *Let D be an open subset of $\mathbb{R}^2$ and assume that $\Psi : D \to \mathbb{R}$ and $\phi : D \to \mathbb{R}$ are continuous with*

$$\Psi(t, u) \le \phi(t, u), \quad (t, u) \in D.$$

If u_M is the maximum solution of the IVP (8.32), with maximal interval of existence (α_M, ω_M), then if v is a solution of $v' = \Psi(t, v)$ with $v(t_0) \le u_0$, then $v(t) \le u_M(t)$ on $[t_0, t_0 + a] \cap (\alpha_M, \omega_M)$.

We can now use Theorems 8.55 and 8.56 to prove the following corollary.

Corollary 8.60 *Assume that D is an open subset of $\mathbb{R}^2$ and $\phi : D \to \mathbb{R}$ is continuous. Assume that $(t_0, u_0) \in D$, and there is a continuously differentiable vector function $x : [t_0, t_0 + a] \to \mathbb{R}^n$ such that $(t, \|x(t)\|) \in D$ for $t_0 \le t \le t_0 + a$ with $\|x'(t)\| \le \phi(t, \|x(t)\|)$; then*

$$\|x(t)\| \le u_M(t), \quad t \in [t_0, t_0 + a] \cap (\alpha_M, \omega_M),$$

where u_M is the maximum solution of the IVP

$$u' = \phi(t, u), \quad u(t_0) = \|x(t_0)\|,$$

with maximal interval of existence (α_M, ω_M). Similarly, if u_m is the minimum solution of the IVP

$$u' = -\phi(t, u), \quad u(t_0) = \|x(t_0)\|,$$

then

$$\|x(t)\| \ge u_m(t), \quad t \in [t_0, t_0 + a] \cap (\alpha_m, \omega_m).$$

Proof Note that for $h > 0$, sufficiently small,

$$\frac{\|x(t + h)\| - \|x(t)\|}{h} \le \frac{\|x(t + h) - x(t)\|}{h} = \left\| \frac{x(t + h) - x(t)}{h} \right\|,$$

implies that

$$D^+ \|x(t)\| \le \|x'(t)\| \le \phi(t, \|x(t)\|),$$

for $t \in [t_0, t_0 + a]$. Hence by Theorem 8.55, we get that

$$\|x(t)\| \le u_M(t), \quad t \in [t_0, t_0 + a] \cap (\alpha_M, \omega_M).$$

Next, for $h > 0$, sufficiently small,

$$\|x(t+h) - x(t)\| \geq \|x(t)\| - \|x(t+h)\| = -\{\|x(t+h)\| - \|x(t)\|\}.$$

Hence for $h > 0$, sufficiently small,

$$\frac{\|x(t+h)\| - \|x(t)\|}{h} \geq -\frac{\|x(t+h) - x(t)\|}{h} = -\left\|\frac{x(t+h) - x(t)}{h}\right\|,$$

and consequently

$$D_+\|x(t)\| \geq -\|x'(t)\| \geq -\phi(t, \|x(t)\|),$$

for $t \in [t_0, t_0 + a]$. It then follows from Theorem 8.56 that

$$\|x(t)\| \geq u_m(t), \quad t \in [t_0, t_0 + a] \cap (\alpha_m, \omega_m).$$

$\square$

Theorem 8.61 (Generalized Gronwall's Inequality) *Assume* $\phi : [t_0, t_0 + a] \times \mathbb{R} \to \mathbb{R}$ *is continuous and for each fixed* $t \in [t_0, t_0 + a]$, $\phi(t, u)$ *is nondecreasing with respect to* u. *Assume that the maximum solution* u_M *of the IVP (8.32) exists on* $[t_0, t_0 + a]$. *Further assume that* $v : [t_0, t_0 + a] \to \mathbb{R}$ *is continuous and satisfies*

$$v(t) \leq u_0 + \int_{t_0}^t \phi(s, v(s))\, ds$$

on $[t_0, t_0 + a]$. *Then*

$$v(t) \leq u_M(t), \quad t \in [t_0, t_0 + a].$$

Proof Let

$$z(t) := u_0 + \int_{t_0}^t \phi(s, v(s))\, ds,$$

for $t \in [t_0, t_0 + a]$. Then $v(t) \leq z(t)$ on $[t_0, t_0 + a]$ and

$$z'(t) = \phi(t, v(t)) \leq \phi(t, z(t)), \quad t \in [t_0, t_0 + a].$$

It follows from Theorem 8.55 that

$$z(t) \leq u_M(t), \quad t \in [t_0, t_0 + a].$$

Since $v(t) \leq z(t)$ on $[t_0, t_0 + a]$, we get the desired result. $\square$

As a consequence to this last theorem, we get the well-known Gronwall's inequality as a corollary.

Corollary 8.62 (Gronwall's Inequality) *Let* u, v *be nonnegative, continuous functions on* $[a, b]$, $C \geq 0$ *be a constant, and assume that*

$$v(t) \leq C + \int_a^t v(s)u(s)\, ds,$$

for $t \in [a, b]$. *Then*

$$v(t) \leq Ce^{\int_a^t u(s)ds}, \quad t \in [a, b].$$

In particular, if $C = 0$, *then* $v(t) \equiv 0$.

Proof This result follows from Theorem 8.61, where we let $\phi(t, w) := u(t)w$, for $(t, w) \in [a, b] \times \mathbb{R}$ and note that the maximum solution (unique solution) of the IVP

$$w' = \phi(t, w) = u(t)w, \quad w(a) = C$$

is given by

$$w_M(t) = Ce^{\int_a^t u(s)ds}, \quad t \in [a, b].$$

$\square$

Theorem 8.63 (Extendability Theorem) *Assume* $\phi : [t_0, t_0 + a] \times \mathbb{R} \to \mathbb{R}$ *is continuous and* u_M *is the maximum solution of the scalar IVP* (8.32), *where* $u_0 \geq 0$, *and assume that* u_M *exists on* $[t_0, t_0 + a]$. *If the vector function* $f : [t_0, t_0 + a] \times \mathbb{R}^n \to \mathbb{R}^n$ *is continuous and*

$$\|f(t, x)\| \leq \phi(t, \|x\|), \quad (t, x) \in [t_0, t_0 + a] \times \mathbb{R}^n,$$

then any solution x *of the vector IVP*

$$x' = f(t, x), \quad x(t_0) = x_0,$$

where $\|x_0\| \leq u_0$, *exists on* $[t_0, t_0 + a]$ *and* $\|x(t)\| \leq u_M(t)$ *on* $[t_0, t_0 + a]$.

Proof Fix x_0 so that $\|x_0\| \leq u_0$. By the extension theorem (Theorem 8.33) and the extended Cauchy–Peano Theorem (Theorem 8.34) there is an $\alpha > 0$ such that all solutions of this IVP exist on $[t_0, t_0 + \alpha]$. Let x be one of these solutions with right maximal interval of existence $[t_0, \omega_x)$. Then

$$
\begin{aligned}
D^+\|x(t)\| &= \limsup_{h \to 0+} \frac{\|x(t+h)\| - \|x(t)\|}{h} \\
&\leq \limsup_{h \to 0+} \left\| \frac{x(t+h) - x(t)}{h} \right\| \\
&= \|x'(t)\| \leq \phi(t, \|x(t)\|).
\end{aligned}
$$

Hence from Theorem 8.55

$$\|x(t)\| \leq u_M(t), \quad t \in [t_0, \omega_x).$$

If $\omega_x \leq t_0 + a$, then by the extended Cauchy-Peano theorem (Theorem 8.34),

$$\lim_{t \to \omega_x} \|x(t)\| = \infty,$$

which is a contradiction. Hence we must have

$$\|x(t)\| \leq u_M(t), \quad t \in [t_0, t_0 + a],$$

and, in particular, the solution x exists on all of $[t_0, t_0 + a]$. $\square$

Corollary 8.64 *Assume that* $\Psi : [0, \infty) \to (0, \infty)$ *is continuous and there is a* $y_0 \in [0, \infty)$ *such that*

$$\int_{y_0}^{\infty} \frac{dv}{\Psi(v)} = \infty.$$

If $f : [t_0, \infty) \times \mathbb{R}^n \to \mathbb{R}^n$ *is continuous and*

$$\|f(t, x)\| \leq \Psi(\|x\|), \quad (t, x) \in [t_0, \infty) \times \mathbb{R}^n,$$

then for all $x_0 \in \mathbb{R}^n$ *with* $\|x_0\| \leq y_0$, *all solutions of the vector IVP*

$$x' = f(t, x), \quad x(t_0) = x_0$$

exist on $[t_0, \infty)$.

Proof Let u be a solution of the IVP

$$u' = \Psi(u), \quad u(t_0) = y_0,$$

with right maximal interval of existence $[t_0, \omega_u)$. If $\omega_u < \infty$, then

$$\lim_{t \to \omega_u -} |u(t)| = \infty.$$

Since $u'(t) = \Psi(u(t)) > 0$, we must have

$$\lim_{t \to \omega_u -} u(t) = \infty.$$

For $t \geq t_0$, $u'(t) = \Psi(u(t))$ implies that

$$\frac{u'(t)}{\Psi(u(t))} = 1.$$

Integrating from t_0 to t, we obtain

$$\int_{t_0}^{t} \frac{u'(s)}{\Psi(u(s))} ds = t - t_0.$$

Letting $v = u(s)$, we get that

$$\int_{y_0}^{u(t)} \frac{dv}{\Psi(v)} = t - t_0.$$

Letting $t \to \omega_u -$, we get the contradiction

$$\infty = \int_{y_0}^{\infty} \frac{dv}{\Psi(v)} = \omega_u - t_0 < \infty.$$

Hence we must have $\omega_u = \infty$ and so u is a solution on $[t_0, \infty)$. It then follows from Theorem 8.63 that all solutions of the vector IVP

$$x' = f(t, x), \quad x(t_0) = x_0$$

exist on $[t_0, \infty)$. $\square$

Theorem 8.65 *Assume A is a continuous $n \times n$ matrix function and h is a continuous $n \times 1$ vector function on I. Then the IVP*

$$x' = A(t)x + h(t), \quad x(t_0) = x_0, \tag{8.33}$$

where $(t_0, x_0) \in I \times \mathbb{R}^n$, has a unique solution x and this solution exists on all of I. Furthermore,

$$\|x(t)\| \leq \left\{ \|x_0\| + \left| \int_{t_0}^{t} \|h(s)\| \, ds \right| \right\} e^{\left| \int_{t_0}^{t} \|A(s)\| \, ds \right|}, \tag{8.34}$$

for $t \in I$.

Proof By Corollary 8.18 we have already proved the existence and uniqueness of the solution of the IVP (8.33), so it only remains to show that the solution of this IVP exists on all of I and that the inequality (8.34) holds. We will just prove that the solution of this IVP exists on I to the right of t_0 and the inequality (8.34) holds on I to the right of t_0. Assume $t_1 > t_0$ and $t_1 \in I$. Let $f(t, x) := A(t)x + h(t)$, for $(t, x) \in I \times \mathbb{R}^n$ and note that

$$\|A(t)x + h(t)\| \leq \|A(t)\|\|x\| + \|h(t)\|$$
$$\leq M_1\|x\| + M_2,$$

for $(t, x) \in [t_0, t_1] \times \mathbb{R}^n$, where M_1, M_2 are suitably chosen positive constants. Since the IVP

$$u' = M_1 u + M_2, \quad u(t_0) = \|x_0\|,$$

has a unique solution that exists on $[t_0, t_1]$, we get from Theorem 8.63 that the solution of the IVP (8.33) exists on $[t_0, t_1]$. Since this holds, for any $t_1 > t_0$ such that $t_1 \in I$ we get that the solution of the IVP (8.33) exists on I to the right of t_0.

Next we show that the solution x of the IVP (8.33) satisfies (8.34) on I to the right of t_0. Since x solves the IVP (8.33) on I, we have that

$$x(t) = x_0 + \int_{t_0}^{t} [A(s)x(s) + h(s)] \, ds,$$

for $t \in I$. It follows that for $t \in I$, $t \geq t_0$

$$\|x(t)\| \leq \|x_0\| + \int_{t_0}^{t} \|A(s)\|\|x(s)\| \, ds + \int_{t_0}^{t} \|h(s)\| \, ds$$
$$= \left\{ \|x_0\| + \int_{t_0}^{t} \|h(s)\| \, ds \right\} + \int_{t_0}^{t} \|A(s)\|\|x(s)\| \, ds.$$

Let $t_1 \in I$ and assume that $t_1 > t_0$. Then for $t \in [t_0, t_1]$,

$$\|x(t)\| \leq \left\{ \|x_0\| + \int_{t_0}^{t_1} \|h(s)\| \, ds \right\} + \int_{t_0}^{t} \|A(s)\|\|x(s)\| \, ds.$$

Using Gronwall's inequality, we get that

$$\|x(t)\| \leq \left\{ \|x_0\| + \int_{t_0}^{t_1} \|h(s)\| \, ds \right\} e^{\int_{t_0}^{t} \|A(s)\| ds},$$

for $t \in [t_0, t_1]$. Letting $t = t_1$, we get

$$\|x(t_1)\| \leq \left\{ \|x_0\| + \int_{t_0}^{t_1} \|h(s)\| \, ds \right\} e^{\int_{t_0}^{t_1} \|A(s)\| ds}.$$

Since $t_1 \in I$, $t_1 > t_0$ is arbitrary, we get the desired result. The remainder of this proof is left as an exercise (see Exercise 8.49).   □

8.11 Exercises

8.1 Assume I is an open interval, $p_k : I \to \mathbb{R}$ is continuous for $0 \leq k \leq n$, $p_n(t) \neq 0$ for $t \in I$, and $h : I \to \mathbb{R}$ is continuous. Show that the nth-order linear equation

$$p_n(t)u^{(n)} + p_{n-1}(t)u^{(n-1)} + \cdots + p_0(t)u = h(t)$$

is equivalent to a vector equation of the form (8.1) with $D := I \times \mathbb{R}^n$. Give what you think would be the appropriate definition of u is a solution of this nth-order equation on I.

8.2 Find the Jacobian matrix of $f(t, x)$ with respect to x for each of the following:

(a) $f(t, x) = \begin{pmatrix} x_1^2 e^{2x_1 x_2} + t^2 \\ 5x_1 x_2^3 \end{pmatrix}$ (b) $f(t, x) = \begin{pmatrix} \sin(x_1^2 x_2^3) \\ \cos(t^2 x_1 x_2) \end{pmatrix}$

(c) $f(t, x) = \begin{pmatrix} x_1^2 t^2 + x_2^2 + x_3^2 \\ x_1 x_2 x_3 \\ 4x_1^2 x_3^4 \end{pmatrix}$ (d) $f(t, x) = \begin{pmatrix} e^{x_1^2 x_2^2 x_3^2} \\ x_1^3 + x_3^2 \\ 4x_1 x_2 x_3 + t \end{pmatrix}$

8.3 Show that $f(t, x) = \frac{x^2}{1+t^2}$ satisfies a Lipschitz condition with respect to x on $\mathbb{R} \times \mathbb{R}$, but does not satisfy a uniform Lipshitz condition with respect to x on $\mathbb{R} \times \mathbb{R}$.

8.4 Show that $f(t, x) = e^{3t} + 3|x|^p$, where $0 < p < 1$, does not satisfy a uniform Lipschitz condition with respect to x on $\mathbb{R}^2$.

8.5 Maximize the α in the Picard-Lindelof theorem by choosing the appropriate rectangle Q concerning the solution of the IVP

$$x' = x^3, \quad x(0) = 2.$$

Then solve this IVP to get the maximal interval of existence of the solution of this IVP.

8.6 Maximize the α in the Picard-Lindelof theorem by choosing the appropriate rectangle Q concerning the solution of the IVP

$$x' = 5 + x^2, \quad x(1) = 2.$$

8.7 Maximize the α in the Picard-Lindelof theorem by choosing the appropriate rectangle Q concerning the solution of the IVP

$$x' = (x+1)^2, \quad x(1) = 1.$$

Then solve this IVP to get the maximal interval of existence of the solution of this IVP.

8.8 Maximize the α in the Picard-Lindelof theorem by choosing the appropriate rectangle Q concerning the solution of the IVP

$$x' = t + x^2, \quad x(0) = 1.$$

8.9 Approximate the solution of the IVP

$$x' = \sin\left(\frac{\pi}{2}x\right), \quad x(0) = 1$$

by finding the second Picard iterate $x_2(t)$ and use (8.12) to find how good an approximation you get.

8.10 Approximate the solution of the IVP

$$x' = 2x^2 - x^3, \quad x(0) = 1$$

by finding the second Picard iterate $x_2(t)$.

8.11 Approximate the solution of the IVP

$$x' = 2x - x^2, \quad x(0) = 1$$

by finding the second Picard iterate $x_2(t)$.

8.12 Approximate the solution of the IVP

$$x' = \frac{x}{1+x^2}, \quad x(0) = 1$$

by finding the second Picard iterate $x_2(t)$ and use (8.12) to find how good an approximation you get.

8.13 Approximate the solution of the IVP

$$x' = \frac{1}{1+x^2}, \quad x(0) = 0$$

by finding the second Picard iterate $x_2(t)$ and use (8.12) to find how good an approximation you get.

8.14 Using Corollary 8.19, what can you say about solutions of IVPs for each of the following?

  (i) $x'' = \sin(tx') + (x-2)^{\frac{2}{3}}$
  (ii) $x''' = t^2 + x + (x')^2 + (x'')^3$

8.15 Show that the sequence of functions $\{x_n(t) := t^n\}, \quad 0 \le t \le 1$, satisfies all the hypotheses of the Ascoli-Arzela theorem (Theorem 8.26) except the fact that this sequence is equicontinuous. Show that the conclusion of the Ascoli-Arzela theorem (Theorem 8.26) for this sequence does not hold.

8.16 Can the Ascoli-Arzela theorem (Theorem 8.26) be applied to the sequence of functions $\{x_n(t) := \sin(nt)\}_{n=1}^{\infty}$, $0 \le t \le \pi$?

8.17 Verify that the sequence of functions $\{x_n(t) := \frac{1}{n}\sin(nt)\}_{n=1}^{\infty}$, is equicontinuous on $\mathbb{R}$.

8.18 Assume $g : [0, \infty) \to \mathbb{R}$ is continuous. Show that if $g'(3) = 0$, then $\{g_n(t) := g(nt) : n \in \mathbb{N}\}$ is not equicontinuous on $[0, \infty)$.

8.19 Assume that $\{x_n\}$ is an equicontinuous sequence of real-valued functions on $[a, b]$, which converges pointwise to x on $[a, b]$. Further assume there is a constant $p \ge 1$ such that for each $n \in \mathbb{N}$, $\int_a^b |x_n(t)|^p dt$ exists, and

$$\lim_{n \to \infty} \int_a^b |x_n(t) - x(t)|^p dt = 0.$$

Show that the sequence $\{x_n\}$ converges uniformly to x on $[a, b]$.

8.20 Write out the definition of $x(t) \to \partial D$ as $t \to a+$ mentioned in Definition 8.32.

8.21 At the beginning of the proof of Theorem 8.33, show how you can define the sequence of open sets $\{K_k\}$ such that the closure of K_k, $\bar{K}_k$, is compact, $\bar{K}_k \subset K_{k+1}$, and $\cup_{k=1}^{\infty} K_k = D$.

8.22 Use Theorem 8.36 to prove Corollary 8.37.

8.23 Find constants α and β so that $x(t) = \alpha t^\beta$ is a solution of the IVP (8.23) in Example 8.38. Show that the the sequence of Picard iterates $\{x_k(t)\}$ [with $x_0(t) \equiv 0$] for the IVP (8.23) does not even have a subsequence that converges to the solution of this IVP. Show directly by the definition of a Lipschitz condition that the $f(t, x)$ in Example 8.38 does not satisfy a Lipschitz condition with respect to x on $[0, 1] \times \mathbb{R}$.

8.24 Show that $x_1(t) := 0$ and $x_2(t) := \left(\frac{2}{3}t\right)^{\frac{3}{2}}$ define solutions of the IVP

$$x' = x^{\frac{1}{3}}, \quad x(0) = 0.$$

Even though solutions of IVPs are not unique, show that the sequence of Picard iterates $\{x_k(t)\}$ [with $x_0(t) \equiv 0$] converges to a solution of this IVP.

8.25 For each constant $x_0 \neq 0$, find the maximal interval of existence for the solution of the IVP $x' = x^3$, $x(0) = x_0$. Show directly that the conclusions of Theorem 8.33 concerning this solution hold.

8.26 Show that the IVP $x'' = -6x(x')^3$, $x(-1) = -1$, $x'(-1) = \frac{1}{3}$, has a unique solution x and find the maximal interval of existence of x. *Hint:* Look for a solution of the given IVP of the form $x(t) = \alpha t^\beta$, where α and β are constants.

8.27 Show that the IVP

$$x' = -x^{\frac{1}{3}} - t^2 \arctan x, \quad x(0) = 0$$

has a unique solution. What is the unique solution of this IVP? Does the Picard-Lindelof theorem apply?

8.28 Show that if $\{x_k\}_{k=1}^{\infty}$ is a sequence of n-dimensional vector functions on $[a, b]$ satisfying the property that every subsequence has a subsequence that converges uniformly on $[a, b]$ to the same function x, then $\lim_{k \to \infty} x_k(t) = x(t)$ uniformly on $[a, b]$.

8.29 If $f(x) = |x|$ for $x \in \mathbb{R}$ and if $\delta > 0$ is a constant, find a formula for the integral mean $g_\delta : \mathbb{R} \to \mathbb{R}$ defined by

$$g_\delta(x) = \frac{1}{2\delta} \int_{x-\delta}^{x+\delta} f(y) \, dy,$$

for $x \in \mathbb{R}$. Use your answer to show that g_δ is continuously differentiable on $\mathbb{R}$. Then show directly that $\lim_{\delta \to 0+} g_\delta(x) = f(x)$ uniformly $\mathbb{R}$.

8.30 Given that

$$f(x) = \begin{cases} 1, & x \geq 0 \\ -1, & x < 0, \end{cases}$$

and $\delta > 0$, find the integral mean

$$g_\delta(x) = \frac{1}{2\delta} \int_{x-\delta}^{x+\delta} f(y) \, dy,$$

for $x \in \mathbb{R}$.

8.31 Find the cross-sectional set S_6 in Kneser's theorem (Theorem 8.42) for each of the following IVPs:

    (i) $x' = x^{\frac{2}{3}}$, $x(0) = 0$
    (ii) $x' = x^2$, $x(0) = \frac{1}{8}$

8.32 Find a formula for the solution $x(t; t_0, x_0)$ of the initial value problem

$$x' = x - x^2, \quad x(t_0) = x_0.$$

Use your answer to find $z(t) = \frac{\partial x}{\partial x_0}(t; 0, 1)$ and $w(t) = \frac{\partial x}{\partial t_0}(t; 0, 1)$. Compare your answers to the results given in Example 8.45.

8.33 Let $x(t; a, b)$ denote the solution of the IVP

$$x' = 8 - 6x + x^2, \quad x(a) = b.$$

Without solving this IVP, find

$$\frac{\partial x}{\partial b}(t; 0, 2) \quad \text{and} \quad \frac{\partial x}{\partial a}(t; 0, 2).$$

Use your answers to approximate $x(t; h, 2)$, when h is close to zero and $x(t; 0, k)$, when k is close to 2.

8.34 Let $x(t; a, b)$ denote the solution of the IVP

$$x' = 1 + x^2, \quad x(a) = b.$$

Use Theorem 8.43 to find

$$\frac{\partial x}{\partial b}(t;0,0) \quad \text{and} \quad \frac{\partial x}{\partial a}(t;0,0).$$

Then check your answers by solving the preceding IVP and then finding these partial derivatives.

8.35 Let $x(t;a,b)$ denote the solution of the IVP

$$x' = \arctan x, \quad x(a) = b.$$

Use Theorem 8.43 to find

$$\frac{\partial x}{\partial b}(t;0,0) \quad \text{and} \quad \frac{\partial x}{\partial a}(t;0,0).$$

8.36 For the differential equation $u'' = u - u^3$, find

$$\frac{\partial u}{\partial y_{01}}(t;0,1,0), \quad \frac{\partial u}{\partial y_{02}}(t;0,1,0), \quad \text{and} \quad \frac{\partial u}{\partial t_0}(t;0,1,0),$$

and use your answers to approximate

$$u(t;0,1+h,0), \quad u(t;0,1,h), \quad \text{and} \quad u(t;h,1,0),$$

respectively, for h close to zero.

8.37 Let $u(t;t_0,a,b)$ denote the solution of the IVP

$$u'' = 4 - u^2, \quad u(t_0) = a, \quad u'(t_0) = b.$$

Using a result in Example 8.47, find $v(t) = \frac{\partial u}{\partial b}(t;0,2,0)$.

8.38 Let $x_k(t,t_0)$, $0 \le k \le n-1$, be the normalized solutions (see Definition 6.18) of $L_n x = 0$ (see Definition 6.1) at $t = t_0$. Derive formulas for $\frac{\partial x_k}{\partial t_0}(t,t_0)$, $0 \le k \le n-1$, in terms of the coefficients of $L_n x = 0$ and the normalized solutions $x_k(t,t_0)$, $0 \le k \le n-1$.

8.39 For each of the following find the normalized solutions $x_k(t,t_0)$, $0 \le k \le n-1$, of the given differential equation at $t = t_0$, and check, by calculating $\frac{\partial x_k}{\partial t_0}(t,t_0)$, $0 \le k \le n-1$, that the formulas that you got in Exercise 8.38 are satisfied.

(i) $x'' + x = 0$
(ii) $x''' - x'' = 0$

8.40 Let $u(t;t_0,a,b,\lambda)$, $\lambda > 0$, denote the solution of the IVP

$$u'' = \lambda u, \quad u(t_0) = a, \quad u'(t_0) = b.$$

Use Example 8.50 to calculate

$$\frac{\partial u}{\partial \lambda}(t;0,0,1,\lambda).$$

Check your answer by finding $u(t;0,0,1,\lambda)$ and differentiating with respect to λ.

8.41 Let $u(t; t_0, a, b, \lambda)$ denote the solution of the IVP

$$u'' + 4u = \lambda, \quad u(t_0) = a, \quad u'(t_0) = b.$$

Use Example 8.50 to find

$$\frac{\partial}{\partial \lambda} u(t; 0, 0, 2, 0).$$

Check your answer by actually finding $u(t; t_0, a, b, \lambda)$ and then taking the partial derivative with respect to λ.

8.42 Near the end of the proof of Theorem 8.53 one of the four cases was considered. Prove one of the remaining three cases.

8.43 Find the maximum and minimum solutions of the IVP

$$x' = x^{\frac{2}{3}}, \quad x(0) = 0,$$

and give their maximal intervals of existence.

8.44 Use Theorem 8.55 to show that if $v : [a, b] \to \mathbb{R}$ is continuous and if $D^+ v(t) \leq 0$ on $[a, b]$, then $v(t) \leq v(a)$ for $t \in [a, b]$.

8.45 Prove Theorem 8.56.

8.46 Prove Corollary 8.59.

8.47 Show that every solution of the IVP $x' = f(x)$, $x(0) = x_0$, where

$$f(x) = \begin{cases} \frac{x}{\sqrt{\|x\|}}, & x \neq 0 \\ 0, & x = 0, \end{cases} \quad x_0 = \begin{pmatrix} 1 \\ 0 \\ \vdots \\ 0 \end{pmatrix},$$

exists on $[0, \infty)$ and find a bound on all such solutions.

8.48 Use Theorem 8.63 to find a lower bound for the right end point of the right maximal interval of existence for the solution of the IVP

$$\begin{aligned} x_1' &= x_1^2 - 2x_1 x_2, \\ x_2' &= x_1 + x_2^2, \\ x_1(0) &= 1, \quad x_2(0) = 0. \end{aligned}$$

8.49 Do the part of the proof of Theorem 8.65, where $t \in I$ and $t < t_0$.

Solutions to Selected Problems

Chapter 1

1.1: The solution of the given IVP is $x(t) = \frac{2}{1 - 2\sin t}$ with maximal interval of existence $\left(-\frac{7\pi}{6}, \frac{\pi}{6}\right)$.

1.2: If $x_0 \leq 0$, the maximal interval of existence is $(-\infty, \infty)$, and if $x_0 > 0$, the maximal interval of existence is $(-d, d)$, where
$$d = \sqrt{e^{\frac{1}{x_0}} - 1}.$$

1.3: The given IVP has among its solutions
$$x_\lambda(t) := \begin{cases} 0, & \text{if} \quad t \leq \lambda \\ \left[\frac{2}{3}(t - \lambda)\right]^{\frac{3}{2}}, & \text{if} \quad t \geq \lambda, \end{cases}$$

for all $\lambda \geq 0$. Theorem 1.3 does not apply since $f_x(t, x) = \frac{1}{3x^{\frac{2}{3}}}$ is not continuous when $x = 0$.

1.4: The maximal interval of existence is $(-frac12, \infty)$.

1.5:

    (i) $x(t) = 3e^{2t} + te^{2t}$, for $t \in \mathbb{R}$

    (ii) $x(t) = 2te^{3t} + 2e^{3t}$, for $t \in \mathbb{R}$

    (iv) $x(t) = t^2 \log t + 2t^2$, for $t \in (0, \infty)$

1.7: Each of the solutions $x(t) = \alpha e^{-2t}$, where $\alpha > 0$ gives you the orbit $(0, \infty)$.

1.10: $N(t) = \frac{K}{1 + e^{-rt}}$

1.13:

    (i) $x(t) = \frac{1}{ct - t^2}$

    (iii) $x^3(t) = 1 + \frac{c}{t^3}$

1.16:

    (i) $x = 0$ is asymptotically stable, $x = \pm 1$ are unstable

    (iii) $x = -2 - \sqrt{2}$ is asymptotically stable, $x = -2 + \sqrt{2}$ is unstable

1.17: $T = 200$ is asymptotically stable.

1.20:

    (i) $F(x) = -\frac{x^3}{3}$

    (iii) $F(x) = x^4 - 4x^2$

1.22:

    (iii) $x = \pm\sqrt{2}$ are asymptotically stable, $x = 0$ is unstable

1.23: $x = b$ is asymptotically stable

1.26:

(ii) $v(t) = \sqrt{\frac{gm}{k}}\left(\frac{1+Ce^{-2\sqrt{\frac{kg}{m}}t}}{1-Ce^{-2\sqrt{\frac{kg}{m}}t}}\right)$, where $C = \frac{v_0 - \sqrt{\frac{gm}{k}}}{v_0 + \sqrt{\frac{gm}{k}}}$

1.30:

(iii) transcritical bifurcation

1.33:

(i) bifurcation at $\lambda = 4$

(ii) bifurcation at $\lambda = 0$

(iii) bifurcation at $\lambda = \pm\frac{2}{3\sqrt{3}}$

1.37:

(i) bifurcation at $\lambda = 0$ and $\lambda = 1$

(ii) bifurcation at $\lambda = 10$ and $\lambda = 14$

Hysteresis occurs in (ii).

Chapter 2

2.2: Follows from the linearity of integration.

2.3:

(i) linearly dependent

(ii) linearly independent

(iii) linearly dependent

2.4: linearly dependent on $(0, \infty)$

2.5:

(i) linearly dependent on $\mathbb{R}$

(iii) linearly independent on $\mathbb{R}$

(v) linearly dependent on $\mathbb{R}$

2.8: linearly dependent on $\mathbb{R}$

2.9: linearly independent on $\mathbb{R}$

2.10: linearly dependent on $\left(-\frac{\pi}{2}, \frac{\pi}{2}\right)$

2.12: $f(x) = x$, $g(x) = |x|$

2.19:

(i) $\lambda^2 - 4\lambda + 11 = 0$

(iii) $\lambda^2 - 3\lambda - 4 = 0$

2.20:

(iii) $x(t) = c_1 e^{-t}\begin{bmatrix} -1 \\ 2 \\ 0 \end{bmatrix} + c_2 e^{-t}\begin{bmatrix} 0 \\ 2 \\ -1 \end{bmatrix} + c_3 e^{8t}\begin{bmatrix} 2 \\ 1 \\ 2 \end{bmatrix}$

2.23:

(i) $x(t) = c_1\begin{bmatrix} -\cos(9t) \\ \sin(9t) \end{bmatrix} + c_2\begin{bmatrix} \sin(9t) \\ \cos(9t) \end{bmatrix}$

(iii)

$$x(t) = c_1 e^t\begin{bmatrix} \sin(2t) \\ \cos(2t) \\ 0 \end{bmatrix} + c_2 e^t\begin{bmatrix} -\cos(2t) \\ \sin(2t) \\ 0 \end{bmatrix} + c_3 e^{3t}\begin{bmatrix} 1 \\ -1 \\ 4 \end{bmatrix}$$

2.26: $\Phi(t) = \begin{bmatrix} e^{4t} & 0 & 0 \\ 0 & -e^{4t} & e^{6t} \\ 0 & e^{4t} & e^{6t} \end{bmatrix}$

2.28: $x(t) = \begin{bmatrix} 4 - 3e^{-t} \\ -3 + 2e^{t} \end{bmatrix}$

2.31:

(iv) $e^{At} = \begin{bmatrix} e^{4t} & 0 & 0 \\ 0 & \frac{1}{2}e^{4t} + \frac{1}{2}e^{6t} & -\frac{1}{2}e^{4t} + \frac{1}{2}e^{6t} \\ 0 & -\frac{1}{2}e^{4t} + \frac{1}{2}e^{6t} & \frac{1}{2}e^{4t} + \frac{1}{2}e^{6t} \end{bmatrix}$

2.33:

(ii) $e^{At} = e^{2t} \begin{bmatrix} \cos(3t) & \sin(3t) \\ -\sin(3t) & \cos(3t) \end{bmatrix}$

(iii) $e^{At} = e^{t} \begin{bmatrix} \cos t - \sin t & 2\sin t \\ -\sin t & \cos t + \sin t \end{bmatrix}$

2.34:

(i) $x(t) = c_1 e^{2t} \begin{bmatrix} 1 \\ 1 \end{bmatrix} + c_2 \begin{bmatrix} -1 \\ 1 \end{bmatrix}$

(ii) $x(t) = c_1 e^{3t} \begin{bmatrix} 1 - t \\ -t \end{bmatrix} + c_2 e^{3t} \begin{bmatrix} t \\ 1 + t \end{bmatrix}$

(iv) $x(t) = c_1 \begin{bmatrix} e^t \\ 0 \\ 0 \end{bmatrix} + c_2 e^t \begin{bmatrix} t \\ 1 \\ t \end{bmatrix} + c_3 e^t \begin{bmatrix} 0 \\ 0 \\ 1 \end{bmatrix}$

2.40: $e^{At} = \begin{bmatrix} \cos t & \sin t \\ -\sin t & \cos t \end{bmatrix}$

2.42:

(i) $x(t) = \begin{bmatrix} 2e^t - 1 \\ 2e^{2t} - 1 \end{bmatrix}$

(ii) $x(t) = \begin{bmatrix} e^{2t} + te^{2t} \\ 2e^{3t} + te^{3t} \end{bmatrix}$

(iii) $x(t) = e^{2t} \begin{bmatrix} \frac{t^3}{6} + 2t - 1 \\ \frac{t^2}{2} + 1 \end{bmatrix}$

2.47:

(i) The trivial solution is unstable on $[0, \infty)$.

(ii) The trivial solution is globally asymtotically stable on the interval $[0, \infty)$.

(iii) The trivial solution is stable on $[0, \infty)$.

2.54: $\|A\|_{\infty} = 4$, $\|A\|_1 = 3$, $\|A\|_2 = \sqrt{10}$, $\mu_{\infty}(A) = 0$, $\mu_1(A) = 1$, and $\mu(A)_2 = -\frac{3}{2} + \frac{\sqrt{10}}{2}$

2.55: $\|A\|_{\infty} = 6$, $\|A\|_1 = 8$, $\mu_{\infty}(A) = 0$ and $\mu_1(A) = -1$

2.60:

(i) $\begin{bmatrix} 0 & 0 \\ 0 & 1 \end{bmatrix}$

(ii) $\begin{bmatrix} \log 2 & 0 \\ 0 & i\pi \end{bmatrix}$

2.61:

(i) The Floquet multiplier is $\mu_1 = 1$.

(iii) The Floquet multiplier is $\mu_1 = e^{-\frac{\pi}{2}}$.

2.63:

(i) The Floquet multipliers are $\mu_1 = e^{\frac{\pi}{2}}$, $\mu_2 = e^{3\pi}$.

(ii) The Floquet multipliers are $\mu_1 = \mu_2 = e^{-2\pi}$.

(iv) The Floquet multipliers are $\mu_1 = e^{-2\pi}$, $\mu_2 = e^{-6\pi}$.

2.64:

(i) The trivial solution is stable on $[0, \infty)$.

(iii) The trivial solution is globally asymptotically stable on the interval $[0, \infty)$.

2.66: The Floquet multipliers are $\mu_1 = -e^{\pi}$, $\mu_2 = -e^{-\pi}$. The trivial solution is unstable on $[0, \infty)$.

2.68: $\mu_1 \mu_2 = e^{4\pi}$

2.69: The Floquet multipliers are $\mu_1 = -e^{\frac{\pi}{2}}$, $\mu_2 = -e^{-\pi}$. The trivial solution is unstable on $[0, \infty)$. The eigenvalues are $\lambda_{1,2}(t) = -\frac{1}{4} \pm i \frac{\sqrt{7}}{4}$.

Chapter 3

3.1: The separatrix is $2x^2 + 2y^2 - x^4 = 1$.

3.3: The points $(0, y)$, for any $y \in \mathbb{R}$, and the points $(x, 0)$, for any $x \in \mathbb{R}$, are equilibrium points.

3.6: $h(x, y) = 4x^2 + xy + 8y^2 + \alpha$

3.7:

(i) The function h defined by $h(x, y) = \frac{x^2 + y^2 - 1}{y}$, $y \neq 0$ is a Hamiltonian function for this system.

(ii) The equations $h(x, y) = \alpha$ lead to the equations $x^2 + (y - \frac{\alpha}{2})^2 = 1 + \frac{\alpha^2}{4}$, which are equations of circles with center at $(0, \frac{\alpha}{2})$, and it is easy to see that these circles pass through the points $(\pm 1, 0)$.

3.8:

(i) $h(x, y) = x^2 - 2xy + e^y$

(ii) Not a Hamiltonian system

(iv) $h(x, y) = x \sin(xy) - x^2 y$

3.9:

(i) Since $\lambda_1 = -3 < \lambda_2 = -2 < 0$, the origin is a stable node.

(ii) Since $\lambda_1 = -2 < 0 < \lambda_2 = 1$, the origin is a saddle point.

3.10:

(i) Since $\lambda_1 = -1 < 0 < \lambda_2 = 3$, the origin is a saddle point.

(ii) Since $\lambda_1 = -1 < 0 < \lambda_2 = 2$, the origin is a saddle point.

3.11:

(i) Since $\lambda_1 = -3 < 0 < \lambda_2 = 2$, the origin is a saddle point.

(ii) Since $\lambda_1 = -2 < \lambda_2 = -1 < 0$, the origin is a stable node.

3.12:

(i) Since $\lambda_{1,2} = \pm 4i$, the origin is a center.

(ii) Since $\lambda_1 = \lambda_2 = 2 > 0$, the origin is an unstable degenerate norm and we are in the case where there is only one linearly independent eigenvector.

(iii) Since $\lambda_{1,2} = -2 \pm 2i$, the origin is a stable spiral point.

(iv) Since $\lambda_{1,2} = 2 \pm i$, the origin is a unstable spiral point.

3.13: The x-nullclines are the lines $x = 0$ and $y = -2x + 2$. The y-nullclines are the lines $y = 0$ and $y = -\frac{1}{2}x + 1$.

3.16: Using $V(x) = x^2$, we get that 0 is asymptotically stable in both cases.

3.17:

(i) The equilibium points $(0,0)$, $(1,1)$ are unstable, while the equilibrium point $(-1,0)$ is asymptotically stable.

(ii) Both the equilibium points $(0,0)$, $(0,-1)$ are unstable.

(iii) Both the equilibium points $(1,1)$, $(-1,-1)$ are unstable.

3.19:

(i) The equilibium point $(0,0)$ is asymptotically stable.

(ii) Both the equilibium points $(\pm 1, 0)$ are unstable. Theorem 3.26 cannot be applied to determine the stability of the equilibrium points $(0, \pm 2)$.

3.21: $\alpha = 3$, $\beta = 5$

3.22: $\alpha = 2$, $\beta = 1$

3.23: The sets $\{(x,y) : 2x^2 + y^2 \le c\}$, $c \ge 0$ are positively invariant.

3.25: No equilibrium points are asymptotically stable.

3.27:

(i) The origin is globally asymptotically stable (all points are attracted to the origin).

(ii) All points inside the circle $x^2 + y^2 = \sqrt{2}$ are attracted to the origin.

3.28: In all three problems the origin is globally asymptotically stable (all points are attracted to the origin).

3.32: The equilibrium points $(0,0)$, $(0,1)$, and $(1,0)$ are unstable, while the equilibrium point $(\frac{2}{3}, \frac{2}{3})$ is asymptotically stable.

3.33: Let $A = x(t) + y(t) + z(t)$, for all t. If $aA > c$, then the equilibrium point $\left(\frac{c}{a}, \frac{A - \frac{c}{a}}{1 + \frac{c}{b}} \right)$ is in the first quadrant and is asymptotically stable.

3.34: The equilibrium point $\left(a(b - \frac{1}{a-1}), \frac{1}{a-1} \right)$ is asymptotically stable.

3.36:

(i) This is a Hamiltonian system with Hamiltonian $h(x,y) = x^3 - 3xy + y^3$.

3.49: Applying the Poincaré-Bendixon theorem one can show there is a periodic orbit between the circles $r = \frac{1}{2}$ and $r = 2$.

Chapter 4

4.1: $x(t,\epsilon) = 3e^{-2t} + \epsilon\left(\frac{9}{2}(e^{-4t} - e^{-2t})\right) + \cdots$

4.2: $x(t,\epsilon) = \frac{e^{-t}}{1+\epsilon(1-e^{-t})}$

4.3:

(i) $x_0(t) = e^{-t}$, $x_1(t) = \frac{1}{2}e^{-3t} - \frac{1}{2}e^{-t}$

(ii) $x(t,\epsilon) = e^{-t} + \epsilon\left(\frac{1}{2}e^{-3t} - \frac{1}{2}e^{-t}\right) + \epsilon^2\left(\frac{3}{8}e^{-t}(1 - e^{-2t})^2\right) + \cdots$

4.4:

(i) $x_0(t) = e^t$, $x_1(t) = e^t - e^{2t}$

(ii) $x(t,\epsilon) = e^t + \epsilon\left(e^t - e^{2t}\right) + \epsilon^2\left(e^t(1 - e^t)^2\right) + \cdots$

4.14: $u(x,t) = A\cos(x - t) + B\sin(x - t)$

Chapter 5

5.1:

(i) $\left((1 - t^2)x'\right)' + n(n + 1)x = 0$

(iii) $\left(te^{-t}x'\right)' + ae^{-t}x = 0$

(v) $\left((t + 3)^2 x'\right)' + \lambda x = 0$

5.10:

(i) $x(t,s) = e^s e^{2t} - e^{2s}e^t$

(ii) $x(t,s) = te^{5s}e^{5t} - se^{5s}e^{5t}$

(iii) $x(t,s) = s^2 t^3 - s^3 t^2$

(iv) $x(t,s) = \frac{1}{2}e^{2t} - \frac{1}{2}e^{2s}$

(v) $x(t,s) = \sin(t - s)$

5.11:

(i) $x(t) = e^t - e^{2t} + te^{2t}$

(iii) $x(t) = \frac{1}{5}t^2 - \frac{1}{4}t^3 + \frac{1}{20}t^7$

(v) $x(t) = 4 - 4\cos t$

5.12:

(i) $x(t,s) = e^s e^{2t} - e^{2s}e^t$

(ii) $x(t,s) = te^{5s}e^{5t} - se^{5s}e^{5t}$

(iii) $x(t,s) = s^2 t^3 - s^3 t^2$

(iv) $x(t,s) = \frac{1}{2}e^{-2s} - \frac{1}{2}e^{-2t}$

(v) $x(t,s) = \sin(t - s)$

5.13: $x(t,s) = st\sin(\log(\frac{t}{s}))$

5.14:

(i) $x(t) = t^2$

(ii) $x(t) = \frac{1}{9} + \frac{17}{9}\cos(3t)$

5.16:

(i) $\lambda_n = 4n^2$, $x_n(t) = \sin(2nt)$, $n \in \mathbb{N}$

(ii) $\lambda_n = (2n - 1)^2$, $x_n(t) = \cos((2n - 1)t)$, $n \in \mathbb{N}$

(iii) $\lambda_n = n^2\pi^2$, $x_n(t) = \sin(n\pi \ln t)$, $n \in \mathbb{N}$

(v) $\lambda_n = 1 + n^2\pi^2$, $x_n(t) = \frac{1}{t}\sin(n\pi \ln t)$, $n \in \mathbb{N}$

5.17: $\alpha = -\frac{7}{10}$

5.21: $\lambda_0 = 0$, $x_0(t) = 1$, and $\lambda_n = \frac{n^2\pi^2}{9}$, $x_n(t) = c_1\cos\left(\frac{n\pi t}{3}\right) + c_2\sin\left(\frac{n\pi t}{3}\right)$, $n \geq 1$

5.22: You get a **Storm-Louisville** problem.

5.23:

(i) $u(x,t) = e^{-k\lambda t}X(x)$, where X is a solution of $X'' + \lambda X = 0$

(iii) $u(r,\theta) = R(r)\Theta(\theta)$, where R is a solution of $r^2 R'' + rR' - \lambda R = 0$ and Θ is a solution of $\Theta'' + \lambda\Theta = 0$

5.31:

(i) $e^{-3t}\left(e^t\left(e^{-3t}x\right)'\right)' = 0$

(ii) $e^{-t}\left(e^{2t}\left(e^{-t}x\right)'\right)' = 0$

(iii) $\frac{1}{t^4}\left(t^3\left(\frac{1}{t^4}x\right)'\right)' = 0$

5.32:

(i) $e^{-2t}\left(e^{-t}\left(e^{-2t}x\right)'\right)' = 0$

(ii) $e^t\left(e^{-2t}\left(e^t x\right)'\right)' = 0$

(iii) $\frac{1}{t^2}\left(\frac{1}{t}\left(\frac{1}{t^2}x\right)'\right)' = 0$

5.33:

(i) A general solution is given by $x(t) = c_1 e^{-2t} + c_2 t$.

(ii) A general solution is given by $x(t) = c_1 t + c_2 \int_1^t \frac{1+s}{s^2 e^s}\,ds$.

(iii) A general solution is given by $x(t) = c_1(1+t) + c_2(t^2 - t)$.

(iv) A general solution is given by $x(t) = c_1 e^t + c_2(t+1)$.

5.34: A general solution is given by $x(t) = c_1 \frac{\sin t}{\sqrt{t}} + c_2 \frac{\cos t}{\sqrt{t}}$.

5.37:

(i) $\left(\frac{d}{dt} + 3\right)e^{-5t}\left(\frac{d}{dt} - 3\right)x = 0$

(ii) $\left(\frac{d}{dt} + 1\right)\left(\frac{d}{dt} - 1\right)x = 0$

(iii) $\left(\frac{d}{dt} + \frac{2}{t}\right)\frac{1}{t^5}\left(\frac{d}{dt} - \frac{2}{t}\right)x = 0$

5.41:

(i) $z(t) = e^{2t}$ and $z(t) = \frac{Ce^{6t} - 3e^{2t}}{Ce^{4t} + 1}$

(ii) $z(t) = 2e^{-4t}$ and $z(t) = \frac{1 + 2C + 2t}{Ce^{4t} + te^{4t}}$

(iii) $z(t) = 1$ and $z(t) = \frac{Ct^2 - 1}{Ct^2 + 1}$

(iv) $z(t) = 2\cot(2t)$ and $z(t) = \frac{-2\sin(2t) + 2C\cos(2t)}{\cos(2t) + C\sin(2t)}$

(v) $z(t) = e^{-4t}$ and $z(t) = \frac{C + 3e^{2t}}{Ce^{4t} + e^{6t}}$

5.42:

(i) $z_m(t) = -3e^{2t}$

(ii) $z_m(t) = 2e^{-4t}$

(iii) $z_m(t) = -1$

(iv) $z_m(t) = 2\cot(2t)$

(v) $z_m(t) = e^{-4t}$

5.43:

(i) oscillatory by the Fite-Wintner theorem

(ii) oscillatory by Example 5.82

5.45:

(i) $x_0(t) = \cos t$

5.49: $Q[x_1] \approx -.377$

5.50:

   (i) no local minimums

   (ii) no local maximums

   (iii) no local maximums and no local minimums

5.55: The global proper minimum value of Q is

$$Q[t + \frac{2}{t}] = 6 < Q[3] = 9 \log 2 \approx 6.24.$$

5.56: $I[\theta] = \int_a^b \{\frac{1}{2} mL^2 [\theta'(t)]^2 + mgL \cos \theta(t)\} \, dt$

5.58:

   (i) linear homogeneous

   (ii) linear homogeneous

   (iii) linear homogeneous

   (iv) linear nonhomogeneous

   (v) nonlinear

5.59:

$$G(t, s) = \begin{cases} -\int_s^b \frac{1}{p(\tau)} d\tau, & a \leq t \leq s \leq b \\ -\int_t^b \frac{1}{p(\tau)} d\tau, & a \leq s \leq t \leq b \end{cases}$$

5.61:

$$G(t, s) = \begin{cases} \frac{1}{3}(s - 2)(t + 1), & 0 \leq t \leq s \leq 1 \\ \frac{1}{3}(t - 2)(s + 1), & 0 \leq s \leq t \leq 1 \end{cases}$$

5.62:

   (i) $x(t) = \frac{1}{12} t(t^3 - 1)$

   (ii) $x(t) = \frac{1}{3} e^t + \frac{4}{9} e^{-2t} - \frac{7}{9}$

   (iii) $x(t) = \frac{31}{15} e^{2t} - \frac{21}{10} e^{3t} + \frac{1}{30} e^{8t}$

5.64:

   (i) $x(t) = \frac{7}{3} e^t - \frac{11}{6} e^{2t} + \frac{1}{2} e^{3t}$

   (ii) $x(t) = e^t - \frac{1}{\sqrt{2}} e^{\frac{\pi}{8}} \sin(2t)$

   (iii) $x(t) = t^4 + t^5 - 2t^3$

5.68:

   (i) $T = \sqrt[4]{12}$

   (ii) $T = \sqrt{3}$

   (iii) $T \approx 1.15$

5.70:

   (i) $x(t) = 4$

   (ii) $u(t, s) = \frac{1}{2} \cos(t - s) - \frac{1}{2} \sin(t - s)$, $\quad v(t, s) = \frac{1}{2} \cos(t - s) + \frac{1}{2} \sin(t - s)$, $\quad x(t) = 2$

Chapter 6

6.8:

   (i) $x(t, s) = e^{t-s} - t + s - 1$

   (ii) $x(t, s) = \frac{1}{2} e^{t-s} + \frac{1}{2} e^{-(t-s)} - 1$

(iii) $x(t,s) = \frac{t^2(t-s)}{s^2}$

6.9: $x(t,s) = \frac{(t-s)^{n-1}}{(n-1)!}$

6.10:

(i) $y(t) = -\frac{1}{3}t^3 - t^2 - 2t - 2 + 2e^t$

(ii) $y(t) = \frac{1}{3} - \frac{1}{4}e^t + \frac{1}{8}e^{-t} + \frac{1}{24}e^{3t}$

6.19: $x_0(t,0) = \cos t$, $x_1(t,0) = \sin t$

6.20: $x_0(t,0) = 1$, $x_1(t,0) = t$, and $x_2(t,0) = e^t - t - 1$

6.21:

(i) $G(t,s) := \begin{cases} -\frac{1}{6}t^3(1-s)^3, & 0 \le t \le s \le 1 \\ \frac{1}{6}[(t-s)^3 - t^3(1-s)^3], & 0 \le s \le t \le 1 \end{cases}$

(ii) $G(t,s) := \begin{cases} -\sin t \cos s, & 0 \le t \le s \le \frac{\pi}{2} \\ -\cos t \sin s, & 0 \le s \le t \le \frac{\pi}{2} \end{cases}$

(iii) $G(t,s) := \begin{cases} -\frac{1}{6}t^3, & 0 \le t \le s \le 1 \\ -\frac{1}{2}t^2 s + \frac{1}{2}ts^2 - \frac{1}{6}s^3, & 0 \le s \le t \le 1 \end{cases}$

(iv) $G(t,s) := \begin{cases} -\sin t(\sin s + \cos s), & 0 \le t \le s \le \frac{\pi}{4} \\ -\sin s(\sin t + \cos t), & 0 \le s \le t \le \frac{\pi}{4} \end{cases}$

6.22:

(i) $y(t) = t^4 - t^3$

(ii) $y(t) = e^t - \cos t - e^{\frac{\pi}{2}}\sin t$

(iii) $y(t) = t^4 - 4t^3$

(iv) $y(t) = t - \sqrt{2}\sin t$

6.23:

(i) $e^{3t}\left(e^{-t}\left(e^{-t}\left(e^{-t}x\right)'\right)'\right)' = 0$

6.24:

(iii) $\left(\left(\frac{1}{t}(tx)'\right)'\right)' = 0$, $\{u_1(t) = \frac{1}{t}, u_2(t) = t, u_3(t) = t^2\}$

6.26: $p_1 \equiv 0$, $p_0 = \overline{p_0}$.

Chapter 7

7.9: $x_1(t) = \frac{1}{2}t(t-1)$, $|x(t) - x_1(t)| \le \frac{1}{56}$, for $0 \le t \le 1$

7.10: All the Picard iterates for the first BVP are the same $x_n(t) = 1+t^2$, $n = 1, 2, 3, \cdots$. This is because $x(t) = 1+t^2$ is the solution of this BVP. For the second BVP, $x_1(t) = \frac{1}{2}t^2 + \frac{1}{2}t + 1$, and $|x(t) - x_1(t)| \le \frac{1}{56}$, for $t \in [0,1]$.

7.12: $L = \frac{\pi}{\sqrt[4]{3}}$

7.13: $L = \frac{\pi}{3}$

7.16: $B = \frac{\pi}{2}\sqrt{\frac{l}{g}}$

7.17: $-\frac{1}{2}t + \frac{7}{4} \le x'(t) \le \frac{1}{2}t + \frac{9}{4}$

7.20: Apply Theorem 7.20

7.23:

(i) By Theorem 7.17 every such BVP has a solution.

(ii) By Theorem 7.23 every such BVP has at most one solution.

(iii) By Theorem 7.25 every such BVP has at most one solution.

7.31: In the definition of the Nagumo condition, take $h(s) = \frac{1}{9} + s^4$ and apply Theorem 7.34.

Chapter 8

8.2:

(iii) $D_x f(t, x) = \begin{pmatrix} 2t^2 x_1 & 2x_2 & 2x_3 \\ x_2 x_3 & x_1 x_3 & x_1 x_2 \\ 8x_1 x_3^4 & 0 & 16x_1^2 x_3^3 \end{pmatrix}$

8.5: $\alpha = \frac{1}{27}$ The solution of the given IVP is defined by $x(t) = \frac{2}{\sqrt{1-8t}}$, whose maximal interval of existence is $(-\infty, \frac{1}{8})$.

8.6: $\alpha = \frac{1}{10}$

8.7: $\alpha = \frac{1}{8}$ The solution of the given IVP is defined by $x(t) = \frac{2t-1}{3-2t}$, whose maximal interval of existence is $(-\infty, \frac{3}{2})$.

8.8: The number α is the positive solution of $4x^3 + 4x - 1 = 0$.

8.9: $x_2(t) = 1 + \frac{2}{\pi} \sin(\frac{\pi}{2}t)$

8.10: $x_2(t) = 1 + t + \frac{1}{2}t^2 - \frac{1}{3}t^3 - \frac{1}{4}t^4$

8.11: $x_2(t) = -\frac{1}{3}t^3 + t + 1$

8.12: $x_2(t) = 1 + \log(1 + \frac{1}{2}t + \frac{1}{8}t^2)$, $|x(t) - x_2(t)| \le \frac{1}{12}t^3$

8.13: $x_2(t) = \arctan t$, $|x(t) - x_2(t)| \le \frac{9}{128}t^3$

8.23: The solution of the IVP (8.23) is given by $x(t) = \frac{1}{3}t^2$.

8.26: The solution is given by $x(t) = t^{\frac{1}{3}}$ and the maximal interval of existence for this solution is $(-\infty, 0)$.

8.27: The unique solution is given by $x(t) \equiv 0$. The Picard-Lindelof theorem does not apply because $f(t, x)$ does not satisfy a Lipschitz condition in any rectangle with center $(0, 0)$.

8.29:

$$g_\delta(x) = \begin{cases} \frac{x^2 + \delta^2}{2\delta}, & |x| < \delta \\ |x|, & |x| \ge \delta \end{cases}$$

8.30:

$$g_\delta(x) = \begin{cases} 1, & x \ge \delta \\ \frac{1}{\delta}x, & |x| \le \delta \\ -1, & x \le -\delta \end{cases}$$

8.31:

(i) $S_6 = [0, 8]$

(ii) $S_6 = \{\frac{1}{2}\}$

8.32: $x(t; t_0, x_0) = \frac{x_0 e^{t-t_0}}{1 - x_0 + x_0 e^{t-t_0}}$, $z(t) = \frac{\partial x}{\partial x_0}(t; 0, 1) = e^{-t}$, and $w(t) = \frac{\partial x}{\partial t_0}(t; 0, 1) = 0$

8.33: $\frac{\partial x}{\partial b}(t; 0, 2) = e^{-2t}$, $\frac{\partial x}{\partial a}(t; 0, 2) = 0$, $x(t; h, 2) \approx 2$, when h is close to zero, and $x(t; 0, k) \approx 2 + (k-2)e^{-2t}$, when k is close to 2

8.34: $\frac{\partial x}{\partial b}(t, 0, 0) = \sec^2 t$ and $\frac{\partial x}{\partial a}(t, 0, 0) = -\sec^2 t$

8.35: $\frac{\partial x}{\partial b}(t, 0, 0) = e^t$ and $\frac{\partial x}{\partial a}(t, 0, 0) = 0$

8.37: $v(t) = \frac{\partial u}{\partial b}(t; 0, 2, 0) = \frac{1}{2}\sin(2t)$

8.38: $\frac{\partial x_0}{\partial t_0}(t, t_0) = p_0(t_0)x_{n-1}(t, t_0)$ and $\frac{\partial x_k}{\partial t_0}(t, t_0) = -x_{k-1}(t, t_0) + p_k(t_0)x_{n-1}(t, t_0)$ for $1 \le k \le n - 1$

8.39:

(i) $x_0(t, t_0) = \cos(t - t_0)$, $x_1(t, t_0) = \sin(t - t_0)$

(ii) $x_0(t, t_0) = 1$, $x_1(t, t_0) = t - t_0$, $x_2(t, t_0) = e^{t-t_0} - t + t_0 - 1$

8.40:

$$\frac{\partial u}{\partial \lambda}(t; 0, 1, 0, \lambda) = \frac{t}{2\lambda} \cosh(\sqrt{\lambda}t) - \frac{1}{2\lambda\sqrt{\lambda}} \sinh(\sqrt{\lambda}t)$$

8.41: $\frac{\partial u}{\partial \lambda}(t, 0, 0, 2, 0) = -\frac{1}{4}\cos(2t) + \frac{1}{4}$

8.43: $x_M(t) = \begin{cases} \frac{1}{27}t^3, & \text{on } [0, \infty) \\ 0, & \text{on } (-\infty, 0) \end{cases}$

$$x_m(t) = \begin{cases} 0, & \text{on } [0, \infty) \\ \frac{1}{27}t^3, & \text{on } (-\infty, 0) \end{cases}$$

8.47: $\|x(t)\| \le \frac{1}{4}t^2 + t + 1$, $t \in [0, \infty)$

Bibliography

[1] C. Ahlbrandt and J. Hooker, A variational view of nonoscillation theory for linear differential equations, Differential and Integral Equations (Iowa City, Iowa, 1983/Argonne, Ill., 1984), 1–21, Univ. Missouri-Rolla, Rolla, MO, 1985.

[2] C. Ahlbrandt and A. Peterson, Discrete Hamiltonian Systems: Difference Equations, Continued Fractions, and Riccati Equations, Kluwer Academic Publishers, Boston, 1996.

[3] P. Bailey, L. F. Shampine, P. E. Waltman, Nonlinear Two Point Boundary Value Problems, Academic Press, New York, 1968.

[4] M. Bohner and A. Peterson, Dynamic Equations on Time Scales: An Introduction with Applications, Birkhäuser, Boston, 2001.

[5] C. M. Bender and S. A. Orszag, Advanced Mathematical Methods for Scientists and Engineers, McGraw-Hill, New York, 1978.

[6] G. F. Carrier and C. E. Pearson, Ordinary Differential Equations, Ginn/Blaisdell, Waltam, MA., 1968.

[7] K. W. Chang and F. A. Howes, Nonlinear Singular Perturbation Phenomena: Theory and Applications, Springer-Verlag, New York, 1984.

[8] C. Chicone, Ordinary Differential Equations with Applications, Springer, New York, 1999.

[9] R. A. Churchill and J. W. Brown, Fourier Series and Boundary Value Problems, McGraw-Hill, New York, 1987.

[10] E. A. Coddington and N. Levinson, Theory of Ordinary Differential Equations, McGraw-Hill, New York, 1955.

[11] W. A. Coppel, Disconjugacy, Lecture Notes in Mathematics, Springer–Verlag, Berlin, 1971.

[12] R. L. Devaney, An Introduction to Chaotic Dynamical Systems, Addison-Wesley, Redwood City, CA, 1989.

[13] W. Eckhaus, Asymptotic Analysis of Singular Perturbations, North-Holland, Amsterdam, 1979.

[14] L. Edelstein-Keshet, Mathematical Models in Biology, Random House, New York, 1988.

[15] R. A. Fisher, The wave of advance of advantageous genes, *Ann. Eugenics* 7 (1937), 353–369.

[16] R. Grimshaw, Nonlinear Ordinary Differential Equations, CRC Press, Boca Raton, 1993.

[17] J. Guckenheimer and P. Holmes, Nonlinear Oscillations, Dynamical Systems, and Bifurcations of Vector Fields, Springer-Verlag, New York, 1983.

[18] P. Hartman, Principal solutions of disconjugate n-th order linear differential equations, *American Journal of Math.* 77 (1955), 475–483.

[19] P. Hartman, Ordinary Differential Equations, John Wiley, New York, 1964.

[20] E. Hille, Analytic Function Theory, Ginn and Company, New York, 1959.

[21] G.W. Hill, On the part of the motion of the lunar perigee which is a function of the mean motions of the sun and moon, *Acta Math.* 8 (1886), 1–36.

[22] M. Hirsch, Systems of differential equations that are competitive or cooperative IV: structural stability in three dimensional systems, *SIAM J. Math. Anal.* 21 (1990), 1225–1234.

[23] M. Hirsch and S. Smale, Differential Equations, Dynamical Systems, and Linear Algebra, Academic Press, San Diego, 1974.

[24] R. Horn and C. Johnson, Matrix Analysis, Cambridge University Press, New York, 1995.

[25] M. Holmes, Introduction to Perturbation Methods, Springer-Verlag, New York, 1995.

[26] F. Howes, Singular Perturbations and Differential Inequalities, *Memoirs AMS,* No. 168, 1976.

[27] J. H. Hubbard and B. H. West, Differential Equations: a Dynamical Systems Approach, Higher-Dimensional Systems, Springer-Verlag, New York, 1995.

[28] G. Iooss and D. D. Joseph, Elementary Stability and Bifurcation Theory, Undergraduate Texts in Mathematics, Springer-Verlag, New York, 1980.

[29] W. Kelley, Travelling wave solutions of reaction-diffusion equations with convection, *World Scientific Series in Applicable Analysis* 1 (1992), 355–363.

[30] W. Kelley, Perturbation problems with quadratic dependence on the first derivative, *Nonlinear Analysis* 51 (2002), 469–486.

[31] W. G. Kelley and A. C. Peterson, Difference Equations: An Introduction with Applications, 2nd ed., Academic Press, San Diego, 2001.

[32] M. Liapunov, Problème Général de la Stabilté du Movement, Princeton University Press, Princeton, 1947.

[33] J. David Logan, Applied Mathematics, 2nd ed., Wiley Interscience, New York, 1997.

[34] E. Lorenz, Deterministic non-periodic flow, *J. Atmos. Sci.* 20 (1963), 131–141.

[35] W. Magnus and S. Walker, Hill's Equation, Interscience Publishers, New York, 1966.

[36] Markus and Yamabe, Global stability criteria for differential systems, *Osaka Math. J.* 12 (1960), 305–317.

[37] J. D. Murray, Mathematical Biology, Springer-Verlag, New York, 1989.

[38] R. E. O'Malley, Jr., Singular Perturbation Methods for Ordinary Differential Equations, Springer-Verlag, New York, 1991.

[39] H. -O. Peitgen, H. Jürgens, and D. Saupe Chaos and Fractals: New Frontiers in Science, Springer-Verlag, New York, 1992.

[40] E. C. Pielou, An Introduction to Mathematical Ecology, Wiley, New York, 1969.

[41] L. Prandtl, Über Flüssigkeits-bewegung bei kleiner Reibung, Verhandlungen, III. Int. Math. Kongresses, Tuebner, Leipzig, 484–491.

[42] W. T. Reid, Applied Mathematics, 2nd ed., Wiley Interscience, 1997.

[43] W. T. Reid, Oscillation criteria for linear differential systems with complex coefficients, *Pacific J. Math.* 6 (1956), 733–751.

[44] H. L. Royden, Real Analysis, 2nd ed., Academic Press, San Diego, 1968.

[45] H. Sagan, Boundary and Eigenvalue Problems in Mathematical Physics, John Wiley, New York, 1963.

[46] George F. Simmons and Steven G. Krantz Differential Equations: Theory, Technique, and Practice The Walter Rudin Student Series in Advanced Mathematics McGraw Hill, Boston, 2007.

[47] D. R. Smith, Singular-Perturbation Theory, Cambridge University Press, Cambridge, 1985.

[48] H. L. Smith, Monotone Dynamical Systems: An Introduction to the Theory of Competitive and Cooperative Systems, American Mathematics Society, Providence, RI, 1995.

[49] W. Trench, Canonical forms and principle systems for general disconjugate equations, *Trans. Amer. Math. Soc.* 189 (1974), 319–327.

[50] Y. Udea, Randomly transitional phenomena in the system governed by Duffing's equation, *J. Stat. Phys.* 20 (1979), 181–196.

[51] F. Verhulst, Nonlinear Differential Equations and Dynamical Systems, Springer, New York, 1996.

[52] P. F. Verhulst, Notice sur la loi que la population suit dans son accroissement, *Corr. Math. et Phys.* 10 (1838), 113–121.

[53] S. Wiggins, Introduction to Applied Nonlinear Dynamical Systems and Chaos, Springer-Verlag, New York, 1990.

[54] F. A. Williams, Theory of combustion in laminar flows, *Ann. Rev. Fluid Mech.* 3 (1971), 171–188.

Index

α-limit set, 111
ω-limit set, 111, 123

Abel's formula, 195, 306
absolute error, 163
action integral, 249
adjoint
  nth-order linear, 305
  vector equation, 302
admissible
  function, 233
  variation, 240
algebraic boundary layer function, 184
approaches the boundary, 365
Ascoli-Arzela theorem, 357
asymptotically stable, 9, 57
autonomous
  first-order scalar, 5
  system, 87
averaging
  method, 178

basic convergence theorem, 369
Bendixson-Dulac theorem, 129
Bernoulli's equation, 18
Bessel's equation, 372
bifurcation, 14, 21
  diagram, 15
  pitchfork, 16, 138
  saddle node, 14
  transcritical, 15, 116
big oh, 164
boundary condition
  linear homogeneous, 278
  linear nonhomogeneous, 278
  nonlinear, 278
boundary layer, 179
boundary layer correction, 179
brine problem, 20
bubble problem, 116

calculus of variations, 240
Cantor selection theorem, 356
carrying capacity, 8, 12, 187
Cauchy function
  $(p(t)x')' = 0$, 201
  nth-order, 284
  example, 201
  self-adjoint equation, 199
Cauchy sequence, 309
Cauchy-Peano theorem, 358
  extended, 367
Cayley-Hamilton theorem, 43
center, 105
characteristic equation, 31
chemostat
  example, 151
  model, 154
compact interval, 214
companion matrix, 24
comparison theorem
  $(k, n - k)$ conjugate BVP, 295
  for BVP, 264
  self-adjoint equations, 217
competition model, 154, 159
completing the square lemma, 233
continuity
  with respect to initial conditions, 372
  with respect to parameters, 372
contraction mapping
  definition, 309
contraction mapping theorem, 309
coupled vibrations, 25
cycle, 120
  limit, 120

degenerate node, 99, 101
disconjugate
  nth-order, 294
  criterion, 265
  definition, 212

419

divergence
  vector field, 129
dominant solution, 224
double zero, 212
double-well oscillator, 160
Duffing equation, 189

eigenfunction, 205
eigenpair, 205
  definition, 31
eigenvalue
  definition, 31
  simple, 205
  SLP, 205
eigenvector
  definition, 31
epidemic
  example, 151
  model, 154
equicontinuous, 356
equilibrium point, 89
  definition, 6
  isolated, 6
error
  absolute, 163
  relative, 163
Euler-Lagrange equation, 243
existence and uniqueness theorem
  self-adjoint equation, 193
  linear vector equation, 25
  scalar case, 2
existence-uniqueness
  matrix case, 37
existence-uniqueness theorem
  $(k, n - k)$ BVP, 287
  $(k, n - k)$ right-focal BVP, 295
  nth-order linear equations, 281
  autonomous system, 88
exponential boundary layer function, 184
extendability theorem, 393
extension, 363
extension theorem, 365

fast variable, 181
first variation, 241
Fisher's equation, 168
  with convection, 189
Fite-Wintner theorem, 232
fixed point, 309
Floquet
  theorem, 69
  multiplier, 70
  system, 65
  theory, 65

Floquet exponent, 86
Floquet multiplier
  example, 70
forced pendulum problem, 341
formal adjoint
  vector equation, 302
formally self-adjoint
  scalar nth-order, 305
  vector equation, 302
fundamental lemma of the calculus of
  variations, 243
fundamental matrix
  definition, 40
  theorem, 41

gene activation, 21
general solution, 29
generalized Gronwall's inequality, 392
generalized logistic equation, 11
glycolysis, 158
  model, 156
gradient, 109
gradient system, 111, 127, 153
Green's formula, 196
Green's function, 253, 259, 260, 263
  $(k, n - k)$ BVP, 289
  jump condition, 279
Green's function
  General Two Point BVP, 253
  Periodic BVP, 269
Green's theorem, 306
Gronwall's inequality, 392
  generalized, 392

Hamilton's principle, 249
Hamiltonian function, 94
Hamiltonian system, 93
heteroclinic orbit, 93
Hill's equation, 74
homoclinic orbits, 92
Hopf bifurcation, 133
  example, 158
hystereses, 16

index
  of a curve, 128
initial point, 2
initial value, 2
inner product, 195
  complex-valued functions, 306
  weight function, 206
integral equation, 347
integral means, 372

Jacobi equation, 244

Jacobian matrix, 348
Jordan canonical form, 65

Keplerian model, 173
Kneser's theorem, 375, 399

Lagrange bracket, 195
Lagrange identity, 195, 305
Lagrangian, 249
LaSalle Invariance Theorem, 112
LaSalle invariance theorem
    problem, 153, 154
left maximal interval of existence, 365
Legendre's equation, 372
Legendre's necessary condition, 245
    problems, 277
Leibniz rule, 202
Liapunov function, 109
    problem, 152
    strict, 109
Liapunov's Inequality, 265
Liapunov's stability theorem, 109
limit cycle, 120
limit cycles
    example, 156
linear differential equation
    first-order, 4
linear operator
    definition, 26
linear space
    definition, 26
linear system, 23
linear vector equation, 23
linearly dependent
    vector functions, 28
    vectors, 27
Liouville's formula, 282
Liouville's theorem, 37
nth-order scalar, 282
Lipshitz condition
    definition, 348
local section, 122
logistic
    equation, 8
logistic equation, 18
    generalized, 11
logistic growth, 8, 87
    with harvesting, 21
Lorenz system, 159
lower solution, 326
Lozinski measure, 61

Mammana factorization, 228
Mathematica, 145
Mathieu's equation, 74, 175, 190

matrix equation
    first order, 36
matrix exponential
    properties, 49
matrix log
    example, 67
    theorem, 66
matrix norm, 349
max norm, 313
maximal interval of existence, 2, 88, 365
maximum solution, 387
method of averaging, 178
method of multiple scales, 178
minimum solution, 387
    Riccati equation, 236
model
    chemostat, 154
    competition, 154, 159
    epidemic, 154
    glycolysis, 156
multiple scales
    method, 178

Nagumo condition, 334
Newton's law of cooling, 5, 19
Newton's law of motion, 1
nondimensional, 164
nonextendability
    example, 3
nonoscillatory, 213
nonsingular
    matrix, 40
nonuniqueness
    example, 3
norm
    Euclidean, 56
    max, 56, 313
    traffic, 56
    vector case, 56
normalized solutions, 292
normed linear space, 309
nullclines, 95

orbit, 7, 88
order of differential equation, 1
ordinary differential equation, 1
orthogonal
    weight function, 206
oscillatory, 213

pendulum equation, 93, 278
pendulum problem, 92, 189
    friction, 112, 117, 152
per capita growth rate, 8
period doubling, 159

route to chaos, 140
periodic boundary conditions, 275
periodic solution, 122, 155, 172
perturbation
    methods, 161
    series, 161
perturbations
    singular, 178
phase line diagram, 7
phase plane diagram, 90
Picard iterate
    exercise, 397
Picard iterates, 310, 350
Picard-Lindelof theorem, 350
Picone identity
    scalar case, 217
pitchfork bifurcation, 16, 138
plankton example, 328
Poincaré-Bendixson
    example, 123
    theorem, 122
Poincaré-Bendixson theorem
    example, 156
Polya factorization, 219
    nth-order, 298
positive definite
    quadratic functional, 234
positively invariant, 111
potential energy function, 9
predator-prey
    example, 116, 125, 154, 159
principal solutions, 301
proper global minimum, 241
Putzer algorithm
    example, 46
    example, 45, 47
    theorem, 43

quadratic functional, 233

Rössler
    attractor, 160
    system, 159
radioactive decay, 1
recessive solution, 224
reduction of order
    exercise, 276
relative error, 163
renormalization, 174
Riccati equation, 229
Riccati operator, 229
Riccati substitution, 230
right maximal interval of existence, 365
Routh-Hurwitz criterion, 134

saddle point, 103, 117
saddle-node bifurcation, 14
saturation level, 12
second variation, 241
self-adjoint
    scalar nth-order, 305
    vector equation, 302
self-adjoint equation, 192
semigroup property, 108
sensitive dependence on initial conditions, 140
separation of variables, 207
separatrix, 92
separtrix
    problem, 149
shock layer, 184
shooting method, 326
similar matrices, 77
simple zero, 212
simply connected domain
    definition, 129
singular perturbation, 178
singular Sturm-Liouville problem, 275
skew Hermitian, 302
solution, 1
    initial value problem, 2
    integral equation, 347
    vector equation, 345
    vector IVP, 347
species example, 151
spiral point, 106
    stable, 107
    unstable, 108
spring problem
    coupled masses, 25
spurious, 186
stability theorem
    constant matrix case, 57
stable, 9, 57
    asymptotically, 9, 57
    globally asymptotically, 57
    limit cycle, 120
    unstable, 9
stable manifold, 117
stable node, 98
stable spiral point, 106
stablility theorem, 113
strange attractor, 140
strict Liapunov function, 109
Sturm comparison theorem, 217
Sturm separation theorem, 213
Sturm-Liouville, differential equation, 204

Taylor's theorem, 307

terminal velocity, 20, 179
trace of matrix
    definition, 37
trajectory, 88
transcritical bifurcation, 15, 116
transient chaos, 139
traveling wave, 169, 170, 188
Trench factorization, 223
    nth-order, 301
trivial solution
    vector equation, 31

uniform Lipshitz condition
    definition, 348
    exercise, 396
uniformly bounded
    definition, 356
uniqueness theorem, 368
unitary matrix, 308
unstable, 9, 57
unstable manifold, 117

unstable node, 97
unstable spiral point, 106
upper solution, 326

van der Pol, 158
van der Pol equation, 132
van der Pol's equation, 189
variation of constants formula
    example for vector case, 52
    vector case, 51
    first-order scalar, 4
    higher order, 285
    self-adjoint equation, 202
variational equation, 380, 383
Velhurst
    equation, 8

Weierstrass integral formula, 247
Wronskian, 195
Wronskian determinant
    k functions, 282